MACHINE SHOP TRAINING COURSE

MACHINE SHOP TRAINING COURSE

VOLUME II

A Treatise on Machine Shop Practice in Two Volumes, Including Fundamental Principles, Methods of Adjusting and Using Different Types of Machine Tools, with Typical Examples of Work, Measuring Instruments and Gages, Cutting Screw Threads by Different Processes, Thread Grinding, Gear Cutting, Precision Toolmaking Methods, Typical Shop Problems with Solutions, and Miscellaneous Facts Relating to the Art of Machine Construction.

BY

FRANKLIN D. JONES

INDUSTRIAL PRESS INC.

200 MADISON AVENUE, NEW YORK 10016

Fifth Edition
14 16 18 2Ø 19 17 15

PRINTED IN THE UNITED STATES OF AMERICA

CONTENTS OF VOLUME 2

Cutting Threads in Holes by Tapping 1
Cutting External Screw Threads with Dies 15
Forming Screw Threads by Milling and Grinding . . 33
Forming Screw Threads by Rolling Process . . . 51
Planing Process and Its General Application . . . 62
Milling Flat, Curved and Irregular Surfaces . . . 100
Dividing Circumferences into Equal Parts by Indexing . 135
Generating Helical Grooves with Index-Head . . . 174
Operations Requiring Angular Adjustment of Index-Head 191
Milling Irregular Contours by Reproducing Shape of Model 200
Cutting Spur Gears by Milling with Formed Cutters . 212
Milling Bevel, Helical and Worm-Gears 241
Cutting Very Large Spur Gears and Bevel Gears . . 262
Generating Methods of Forming Gear Teeth . . . 278
Grinding Cylindrical and Tapering Parts 314
Surface Grinding and Types of Machines Used . . . 350
Internal Grinding Including Centerless Method . . 357
Grinding Milling Cutters and Reamers 371
Lapping and Other Precision Finishing Processes . . 396
Broaching Internal and External Surfaces 422
Chipping, Filing, Scraping and Hand Grinding . . . 440
Tool Steels and Other Metal-Cutting Materials . . 470
Heat-Treatment of Steels Used for Metal-Cutting Tools . 476
Numerical Control 493
Definitions of Shop Terms in General Use . . . 513
Index 561

The "MACHINE SHOP TRAINING COURSE" consists of two volumes. Contents of Volume 1 will be found on the page following.

CONTENTS OF VOLUME 1

Principles Underlying All Metal-Cutting Operations
Lathes and Their Principal Mechanical Features
Turning Cylindrical Parts in Lathe
Taper Turning in Lathe and Taper Attachments
Chucks and Faceplates; Drilling and Boring in Lathe
Single-Point Tool Forms and Tool Grinding
Principles Governing Speeds and Feeds for Metal Cutting
Cooling and Lubricating Fluids for Metal-Cutting Tools
Screw Thread Standards and Their Application
Cutting Screw Threads in the Lathe
How to Calculate Change Gears for Thread Cutting
Measuring Pitch Diameters of Screw Threads
Turret Lathes and Machines of Automatic Type
Vertical Boring and Turning Machines
General Practice in Drilling and Reaming Holes
Precision Methods of Spacing or Locating Holes
Cylinder Boring and Precision Jig Boring
Controlling Accuracy in Interchangeable Manufacture
Different Classes of Fits for Assembled Machine Parts
Calipers, Micrometers and Other Measuring Instruments
Fixed Gages for Checking the Sizes of Duplicate Parts
Gaging Tapering Parts and Measuring Angles
Precision Gage Blocks and Their Application
Generating Surfaces and Angles by Precision Methods
Engineering Standards Applied in Machine Building
How to Read Blueprints or Mechanical Drawings
Miscellaneous Rules and Formulas
Shop Safety

Cutting Threads in Holes by Tapping

A tap has thread-shaped teeth which cut a thread of similar form and pitch in a drilled hole. Thus a tap may be compared to a screw which has teeth formed on it and is hardened so that it will cut metals such as cast iron, steel, brass, etc. The rotation for tapping may be applied either to the tap or to the work. Before using a tap, a hole is first drilled slightly larger than the root diameter of the thread, by using a tap drill, as indicated at *A*, Fig. 1. The hole is then threaded by screwing a tap into it, as at *B*, after which a stud is inserted as at *C*, or possibly a machine screw or cap-screw.

Taps are employed for most internal thread-cutting operations, especially in the smaller holes, and most of the smaller external threads are cut with dies. The extent to which taps are applied for internal threading gradually diminishes as the diameter of the threaded part increases above a certain size, which also applies to the use of dies in cutting external threads.

What conditions govern the use of taps?

Practically all small holes that are to receive machine screws, studs, cap-screws, etc., are threaded by means of taps, but tools of this class are not suitable for cutting a great many of the larger screw threads. It is impracticable to draw a definite dividing line on the basis of diameter alone, since there are frequently other important conditions that affect the problem. Whether or not taps may be used to advantage often depends upon: (1) Number of parts requiring screw threads or amount of work; (2) diameter and pitch of thread; and (3) the relation between thread-

1

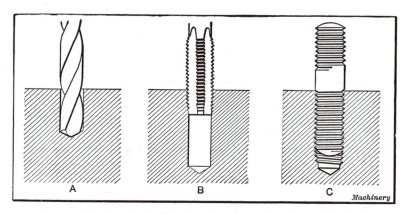

Fig. 1. (A) Tap Drill. (B) Tap for Threading Hole. (C) Stud Inserted in Threaded Hole

cutting and other machining operations. If cutting the thread is the principal or only operation, and especially if the parts are required in large numbers, a tap might be used even for quite large diameters in order to cut the threads quickly. Pipe fittings, which have comparatively fine threads, are an example.

The equipment that is available in a shop is another important point. No experienced manufacturer or shop superintendent would consider making or buying high-priced taps for a certain job unless the number of parts to be tapped were large enough to warrant the expenditure. For cutting a few threads, the engine lathe or turret lathe might be more economical and often preferable regardless of the quantity, either because of the diameter of the thread, pitch, or the relation of the screw thread to other machined surfaces.

What is a tap drill and the general rule for determining tap drill diameters?

The drill used for drilling holes which are to be tapped is known as a *tap drill*. The general rule is to use a tap drill that will leave about three-fourths of the standard thread depth in the tapped hole, as illustrated by the diagram *A*, Fig. 2. In other words, the tap drill is somewhat larger than the root diameter of the screw thread; consequently, a complete thread is not formed because there are practical advantages in reducing the thread depth, as explained later.

Rule.—To find the tap drill diameter in inches, subtract from the outside diameter of the tap an amount equal to 1 divided by the number of threads per inch.

Example 1.—Find the tap drill diameter for a 3/4-inch American Standard screw thread of the coarse-pitch series. This diameter has 10 threads per inch; hence,

$$\text{Tap drill diameter} = \frac{3}{4} - \frac{1}{10} = 0.65 \text{ inch}$$

In the practical application of this rule, the nearest commercial drill size is always used. In this particular case, a 21/32-inch drill is almost the exact equivalent of 0.65 inch.

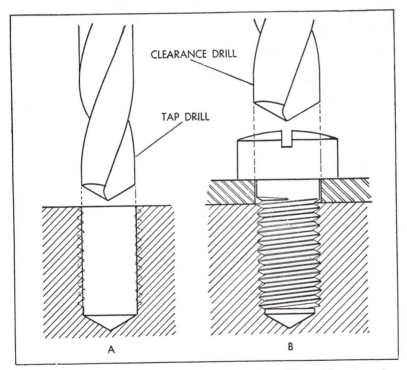

Fig. 2. (A) Tap Drill Diameter is Larger than Minor Diameter of Thread. (B) Clearance Drill is Slightly Larger than Body Diameter of Screw

Tap Drills and Clearance Drills for Machine Screws

Size of Screw		No. of Threads per Inch	Tap Drills		Clearance Hole Drills			
					Close Fit		Free Fit	
No. or Diam.	Decimal Equiv.		Drill Size	Decimal Equiv.	Drill Size	Decimal Equiv.	Drill Size	Decimal Equiv.
0	.060	80	3/64	.0469	52	.0635	50	.0700
1	.073	64	53	.0595	48	.0760	46	.0810
		72	53	.0595				
2	.086	56	50	.0700	43	.0890	41	.0960
		64	50	.0700				
3	.099	48	47	.0785	37	.1040	35	.1100
		56	45	.0820				
4	.112	36*	44	.0860	32	.1160	30	1285
		40	43	.0890				
		48	42	.0935				
5	.125	40	38	.1015	30	.1285	29	.1360
		44	37	.1040				
6	.138	32	36	.1065	27	.1440	25	.1495
		40	33	.1130				
8	.164	32	29	.1360	18	.1695	16	.1770
		36	29	.1360				
10	.190	24	25	.1495	9	.1960	7	.2010
		32	21	.1590				
12	.216	24	16	.1770	2	.2210	1	.2280
		28	14	.1820				
14	.242	20*	10	.1935	D	.2460	F	.2570
		24*	7	.2010				
1/4	.250	20	7	.2010	F	.2570	H	.2660
		28	3	.2130				
5/16	.3125	18	F	.2570	P	.3230	Q	.3320
		24	I	.2720				
3/8	.375	16	5/16	.3125	W	.3860	X	.3970
		24	Q	.3320				
7/16	.4375	14	U	.3680	29/64	.4531	15/32	.4687
		20	25/64	.3906				
1/2	.500	13	27/64	.4219	33/64	.5156	17/32	.5312
		20	29/64	.4531				

*Screws marked with asterisk (*) are not in the American Standard but are from the former A.S.M.E. Standard and are still used—especially the No. 4 size with 36 threads per inch.

Example 2.—Find the tap drill diameter for a No. 10 American Standard machine screw having 24 threads per inch. The outside diameter of the No. 10 size is 0.190 inch.

$$\text{Tap drill diameter} = 0.190 - \frac{1}{24} = 0.148 \text{ inch}$$

In this case, the nearest commercial drill size is the No. 25 which has a diameter of 0.1495 inch (see also the accompanying table containing tap drill sizes for machine screws).

Why is the tap drill diameter important?

Tapping troubles are often caused by using tap drills that are too small in diameter. For ordinary manufacturing not more than 75 or 80 per cent of the standard thread depth is necessary, and for some classes of work not more than 50 per cent is required. Tests have demonstrated that a full depth of thread in a nut is *practically* no stronger than 75 per cent depth of thread, and that 75 per cent depth of thread is only 20 per cent stronger than 50 per cent depth of thread. From 75 per cent depth of thread, the power required to tap rapidly increases as a full depth of thread is approached, and the danger of tap breakage is much greater.

Tap manufacturers often find that users of taps actually punch or drill the holes to be tapped of a smaller diameter than the root diameter of the thread, so that the end of the tap must act as a reamer before the thread can be cut. Frequently the tap will ream clear through the nut.

Depth of tapped hole to obtain strength equal to or greater than stud is as follows: For steel, depth $d=1\frac{1}{4}D$ (D=diameter of stud); for cast iron, $d=1\frac{3}{4}D$; for aluminum, $d=2D$; for magnesium, $d=2\frac{1}{2}D$.

Is the tap drill size affected by the material to be tapped?

Tap drill sizes for machine screws, in particular, should be varied according to the material to be tapped and the depth of the tapped hole. In general, for holes where the screws enter more than one and one-half times the diameter, one-half of the full thread is usually sufficient. Soft, tough materials, such as copper, Norway iron, drawn aluminum, etc., should have a larger hole for the tap than the hard, crystalline materials, such as cast metals. If the tap drill size is not increased, the tops of the threads of the softer materials will be torn off, thus actually decreasing the effective depth of the thread of the tapped hole. On the other hand, if the hole is originally drilled large, the tap will, when cutting tenacious materials, especially after the keen edge has been slightly dulled by use, reduce the size of the hole by drawing the metal at the top of the thread, thereby increasing the depth of the threads.

What is the meaning of the term "clearance drill"?

When a machine screw or cap-screw is used to fasten one machine part to another, the hole in the untapped part is drilled slightly larger in diameter than the outside or body diameter of the screw. This is done in order to provide a slight amount of clearance, as illustrated by diagram *B*, Fig. 2; hence, the drill used is known as a clearance drill. Assume that No. 6 machine screws are to be used for attaching a plate to a casting. The outside diameter of this screw is 0.138 inch, and the size of the clearance drill might be No. 27 or No. 25. (See accompanying table which gives the sizes of both tap drills and clearance drills.) Theoretically, if a No. 25 clearance drill was used, the clearance would equal 0.1495 — 0.1380 = 0.0115 inch. Actually, the clearance would generally be somewhat greater due to the tendency of a drill to cut slightly larger than its size. Some might prefer the No. 27 drill size which would leave a clearance of 0.006 inch. When there are a number of holes in a plate, the larger clearance might be preferable to compensate for slight errors between the holes in the plate and the tapped holes in a casting or other part.

In tapping holes, how is the tap rotated?

When tapping is done on a small scale, the tap is commonly turned with a hand wrench, but, when considerable tapping must be done, the general practice is to rotate the tap by power. There are many appliances for machine or power tapping which differ considerably in their construction, but most of them operate on practically the same principle.

The machines used principally for tap driving include drilling machines, turret lathes, automatic screw machines, and special tapping machines. The type of machine or attachment used for tapping may depend altogether upon the relation of the tapping operation to other operations. The number of holes to be tapped and the advantages or disadvantages of doing both drilling and tapping on one machine should also be considered. Regular drilling machines are extensively used for tapping, because the latter operation naturally follows the drilling of the hole. In many cases, the work is so large and cumbersome that there is a decided advantage in tapping it on the same machine used for drilling. The drill press is also used for tapping a great many small parts, but when large numbers of pieces are required, special tapping machines are frequently used. Tapping operations in turret lathes, automatic screw machines, horizontal and vertical boring mills naturally follow the turning, drilling or boring operations in these machines. A special tapping operation is thus avoided and greater accuracy is secured between the tapped hole and other machined surfaces.

How many taps are generally used in tapping by hand?

The taps used for hand-tapping operations are usually made in sets of three, and are known as *taper, plug,* and *bottoming* taps or Nos. 1, 2 and 3. The point of the taper tap is turned down to the diameter at the bottom of the thread for a length of three or four threads, and then several threads are chamfered or tapered. The rest of the tap is cylindrical. On the plug tap, three threads are chamfered or tapered at the point, whereas, on the bottoming tap, only about one thread is chamfered, so that the end of the latter has practically no taper. When tapping holes which extend through the part being tapped, it is common practice to turn the taper tap clear through the piece, but when the tapped hole is comparatively deep and especially if the material is hard or tough, the taper and plug taps are frequently used alternately; for instance, if the taper tap sticks after being screwed down into the hole, then the plug tap is inserted to relieve the long cut made by the taper end of the first tap. When a hole is "blind" or closed at the bottom, and it is desired to tap as near to the bottom as possible, the taper and plug taps are followed by the bottoming tap, which, because of its short chamfer at the end, will cut a thread practically to the bottom of the hole. If the tapped hole must be square with a finished surface, care must be taken when starting the tap to see that it stands perpendicular to this surface. A double-ended form of tap wrench or one having two handles should be used if there is room, and care should be taken to press evenly on both handles, especially when starting the tap, to prevent it from being started at an angle. After the tap has been screwed in a short distance, its position can be tested by using a small try-square.

When taps are driven by power, are they held rigidly?

The method of holding and driving a tap may vary according to the type of machine used for tapping, the kind of tap, or the class of work. The tap-holder may be designed to hold the tap rigidly or to allow it to slip or stop in case the resistance to tapping becomes excessive. Tap-holders may also be arranged to provide a certain amount of lost motion or floating movement either in a longitudinal or a lateral direction the same as die-holders. The object of the longitudinal movement is to allow the tap to advance according to the lead of its own thread and independently of the advance movement of the driving spindle or slide to which it may be attached. The lateral floating movement is intended to compensate for any lack of alignment between the hole being tapped and the tap-driving spindle or holder, although some taps as well as dies that are supposed to have this self-adjusting feature do not line up properly with a hole that is out of line because of the frictional resistance of the driving pins or lugs, which prevents a free floating movement. Tap-holders for solid or non-collapsing taps may be of the releasing or non-releasing type. Another type of tap-holder is designed especially to permit inserting or removing taps easily and quickly. There are also tap-holding chucks or tapping attachments that are equipped with a mechanism for reversing the rotation of a non-collapsing tap to back it out of a hole. Many tap-holders are designed along the same general lines as die-holders.

Why are some tap-holders arranged to slip when tapping resistance is high?

Frictional chucks, arranged to allow the tap to slip before it is strained to the breaking point, are often used for holding power-driven taps. The need for a frictional drive may arise when the tap strikes the bottom of a "blind" hole; or if the hole before tapping is not large enough, the tap may be subjected to excessive strains. These holders are adjustable to obtain the proper amount of frictional resistance.

How is the tap rotation reversed when the tapping operation is completed?

When drilling machines are used for tap driving, the reverse movement for backing taps out of holes after the thread is cut to the required depth may be obtained either by reversing the spindle of the machine or by using a special chuck or attachment designed to reverse the motion of the tap. Many drilling machines are not equipped with a mechanism for reversing the rotation of the spindle and a special attachment is then required. The Errington automatic - reverse tapping chuck or attachment is shown applied to a drilling machine in Fig. 3.

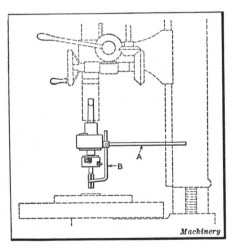

Fig. 3. Diagram Illustrating General Arrangement of Errington Automatic-reverse Tapping Chuck and Application to Drilling Machine

This chuck has a taper shank which is inserted in the spindle of the machine. The body of the chuck and the gage *B*, which may be used to control the depth to which holes are tapped, are both prevented from rotating with the machine spindle by a rod *A*. When tapping a hole, the regular feed-lever or handwheel of the machine is used to lower the spindle as the tap passes down into the hole; when the lower end of stop-rod *B* comes into contact with the face of the work, the direct forward motion drive is released and the tap stops revolving. The machine spindle is then raised; this movement, by the engagement of gearing within the chuck body, causes the tap to back out of the hole rapidly. Raising or lowering the machine table or work would have the same effect.

The tap-holder used in conjunction with this automatic reversing chuck may or may not have an adjustable friction drive for safeguarding taps against breakage. For tapping holes in cast iron, brass, etc., or where blind holes are deep enough to prevent the tap from striking the bottom, the frictional form of holder may not be required. The latter is intended especially for tapping in steel or for bottom tapping.

What are the chief features of tapping mechanisms on drilling machines?

The tapping mechanism of a drilling machine for use with non-collapsing taps may be arranged not only to reverse the rotation of a tap for backing it out of the hole, but also to safeguard the tool against excessive torsional stresses by means of an adjustable friction located somewhere between the tap and the source of power, which slips in case an unusual amount of resistance is encountered. A frictional chuck may be used for this purpose, or an adjustable friction may be introduced somewhere in the spindle driving mechanism of the machine. A tap reversing mechanism of the kind that is incorporated in the design of a drilling machine, permits tapping at a reduced speed, and then backing out the tap at a relatively high speed. The forward and reverse motions are controlled by a lever, which operates a clutch connecting with the reversing gears.

Is a solid or non-collapsing tap always reversed for backing it out of the tapped hole?

The removal of a solid non-collapsing tap from a hole in which a thread has been cut usually is done by backing the tap out of the hole, but if the tap remains stationary while tapping, as on turret lathes and many automatic screw machines, the work-spindle reverses for backing out the tap. Another method of removing a tap that is common in screw machine practice does not require a reversal either of the tap or of the part being tapped. While cutting the thread, the tap revolves in the same direction as the work, but at a slower speed; consequently, the work tends to screw itself onto the tap just as though the tap were stationary and the work were revolving at a speed equal to the difference between the speed of the tap-spindle and the work-spindle. The opposite effect is obtained by revolving the tap faster than the work and in the same direction; this causes the tap to back out of the hole at a rate equal to the difference between the tap-spindle and work-spindle speeds. (This same method is applied to dies.) These changes of speed are controlled automatically.

What are the advantages of ground thread taps?

Taps which have ground threads are often used in preference to those with cut threads both to obtain greater precision and larger production or lower cost per tapped hole. While ground-thread taps are more expensive than cut-thread taps, the former will produce many more tapped holes before sharpening is required. The so-called *commercial ground-thread taps* may be used when a Class 2 fit is required, whereas the *precision ground-thread taps* usually are employed for Class 3 fits. (The different classes of screw thread fits included in the American Standard are listed in section Different Classes of Fits for Assembled Machine Parts, Volume 1.)

When a cut-thread tap is hardened, more or less distortion occurs. Tap manufacturers, for example, may allow a lead error of plus or minus 0.003 inch per inch for cut-thread taps, whereas, for ground-thread taps, a common lead error is only plus or minus 0.0005 inch per inch. High-speed steel ground-thread taps are extensively used in the automotive industry. In all tapping operations, the degree of accuracy obtained and also the cost per tapped hole, depend not only upon the tap itself, but upon the condition and type of tapping machine used, the use of efficient work-holding fixtures, and of the proper cutting fluid.

What are the advantages of collapsing taps?

The advantages of collapsing taps, as compared with the solid or non-collapsing type for machine tapping, are similar to the advantages of self-opening dies in contrast with the non-opening type. When a solid tap is removed by reversing either the tap or the work, the threads cut by the tap may be injured by the chips which wedge in between the tool and the finished thread. The time wasted while the tap is backing out often greatly reduces the rate of production. On the other hand, solid taps cost less and are applicable for tapping numerous studs, screws, and bolt holes that are too small for taps of the collapsible type. The collapsing of taps may be done either automatically by the engagement of a collar, gage-plate, or lever on the tap with the surface of the work or a fixed stop; the collapsing action may also occur soon after the travel of the turret-slide is discontinued as the result of relative motion between parts of the tap similar to that which occurs when a stop is used. Collapsing taps are similar in principle to self-opening dies, except that the action is reversed, the tap chasers moving inward in a radial direction, instead of outward, in order to clear the thread as the tap is withdrawn.

How are collapsing taps reset?

The method of resetting collapsing taps varies according to the construction of the tap and may depend, in some cases, upon the type of machine used for the tapping operations. Collapsing taps of the non-revolving class, which are intended for use in turrets, etc., commonly have a lever projecting from the side which is used for resetting the tap after it has collapsed. This lever may be operated by hand, or the machine, if an automatic type, may be equipped with some form of projecting part arranged to engage the closing lever as the tap and turret-slide are withdrawn after the tapping operation. Some collapsing taps designed to revolve have a sliding flange or collar which engages a suitable collar or stop and serves to reset the tap.

What are the characteristic features of tapping machines?

Tapping machines may be intended either for general tapping operations or for use on one class of work, like the tapping of nuts. The designs vary considerably, including vertical and horizontal designs in single- and multiple-spindle types. The spindles may be adjusted on some machines for varying the center-to-center distance, or they may be fixed in the case of machines used exclusively for tapping duplicate parts. Tapping machines also vary in regard to the mechanism for obtaining the forward and reverse motions of the tap-spindle and the method of controlling these motions. A simple arrangement for obtaining the two motions is by means of a clutch which is interposed between two pulleys revolving in opposite directions and is alternately engaged with these pulleys. The clutch may be controlled by (1) a hand-lever connecting with the clutch; (2) a foot-lever connecting with the clutch; (3) pushing the work and its fixture forward until contact is made with a stop-rod or lever which shifts the clutch for backing out the tap; (4) pushing the work against the tap while tapping and by pulling in the opposite direction for backing out the tap, the clutch being shifted by the direct thrust from the part being tapped and the resulting longitudinal motion of the tap-spindle. The latter method is applied only to machines used for the lighter classes of work. The characteristic features of well-designed tapping machines are convenience of control and, for small tapping operations, a sensitive drive that will transmit enough power for operating the tap under normal conditions, but not enough to break it in case the resistance to rotation becomes excessive.

For what classes of work are special tapping machines used?

In automotive plants or wherever large numbers of duplicate parts must be tapped, a special machine may be designed to tap simultaneously all the holes in cylinder blocks or other parts. For example, thirty holes are tapped simultaneously in cylinder blocks on the multiple-spindle machine illustrated in Fig. 4. There are twenty-two tapping spindles on the upper head, which moves downward automatically after the cylinder block has been pushed into the fixture. At the same time, the rear head, equipped with eight additional tapping spindles, advances. All of the crankshaft, bearing-stud, oil-pan, valve-cover, and oil-pressure plunger holes are tapped in this one operation.

This machine is shown as an example of one that is designed around a particular product. These "single purpose" machines are also applied to drilling and certain other operations when the number of duplicate parts is large enough to warrant using them.

Fig. 4. Multiple-spindle Tapping Machine

Cutting External Screw Threads with Dies

A great many external screw threads, especially in the smaller sizes, are cut by means of dies because tools of this class not only cut threads very rapidly but, when properly made and used, are capable of producing screws that meet most commercial requirements as to accuracy. Dies are efficient as a means of cutting screw threads, because they usually finish the thread complete in one cut. In cutting threads of coarse pitch, considerable metal is removed when a single cut is taken, and although these heavy cuts may be distributed between three or four teeth at the throat of the die, the difficulty of obtaining smooth, accurate screw threads increases for the larger pitches. Dies may be used for pitches up to one-quarter or one-third inch, or even larger, but die threading operations of this kind are not common. When dies are used for cutting screws of coarse pitch and of relatively small diameter, the torsional or twisting strain on the work and the resulting effect on the accuracy of the screw may be so great that the use of a lathe or thread milling machine is preferable, if not necessary.

What classes of work are threaded by means of dies?

Dies are applied in general to the threading of a large percentage of the small and medium-sized screw threads, ranging from small screws, bolts, and studs up to heavy screws four or five inches in diameter, or larger, in some cases. These die-cut screw threads include those of the rougher grades represented by ordinary studs, bolts, etc., and also a great deal of the more accurate work on machine parts generally. Work of the stud and bolt class is almost invariably threaded by means of dies, but the application

15

of dies to the better grades of screw-thread cutting depends upon conditions. For instance, dies ordinarily are used on turret lathes and semi-automatic or automatic screw machines for external threading operations, because they provide an efficient method not only of forming a screw thread, but also of doing it without removing the work from the machine. On the contrary, dies are rarely used on the engine lathe, which has its own screw-cutting mechanism and is designed primarily for a wide range of work rather than for manufacturing duplicate parts in quantity like the turret lathe and automatic screw machine. In some instances, the use of dies on the turret lathe or screw machine is not practicable, as, for example, when there is a shoulder between the surface to be threaded and the turret; then the thread may be formed by rolling or by cutting with an attachment.

While there is no well-defined line showing just where the practical application of dies ends, there are certain limiting factors which should be considered. In general, as the pitch of the thread and its diameter increase beyond certain limits, the use of dies decreases. Whether or not a die could be used to advantage might depend upon the type of machine employed for the other operations, such as turning, etc., and the number of pieces to be threaded or the degree of accuracy required.

What types of dies are used for cutting screw threads?

Dies may be divided into two general classes, namely, those that are removed from the screw thread by being backed off or unscrewed, and those that may be opened so that the cutting edges clear the screw thread, thus permitting the die to be removed merely by traversing it over the work in a lengthwise direction. The non-opening dies are capable in some cases of hand adjustment, but the object of this adjustment is to vary the size of the die, so far as machine threading operations are concerned. The types of non-opening dies in common use may be designated as (1) solid dies, or those that are rigid and incapable of any adjustment for varying the diameter; (2) flexible dies, or those that are split in one or more places and may be ad-

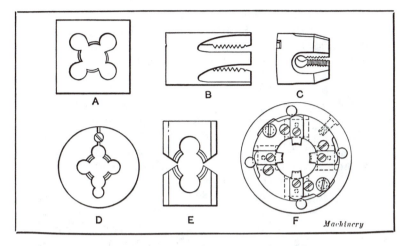

Fig. 1. Different Types of Threading Dies

justed to some extent by compressing or expanding; (3) sectional dies, or those formed of two adjustable sections; (4) rigid adjustable dies having inserted chasers that may be adjusted radially within certain limits either for maintaining a standard size or for varying the size slightly.

One form of solid die is shown at *A*, Fig. 1. This simple form is used for some bolt and pipe threading operations and for the rougher classes of screw cutting. The accuracy of the die is likely to be impaired considerably by hardening, and there is no means of reducing the diameter as it gradually increases on account of wear. Some solid dies are round or of hexagonal shape. The latter are similar to a nut except that they have flutes on the inside to form cutting teeth. Dies of this kind are used principally for repair work, as they are easily turned with a wrench and are convenient for truing or recutting battered threads.

What types of solid dies are adjustable?

One-piece dies, having enough flexibility to permit adjustment, are made in different shapes. Three forms are shown in Fig. 1. The spring screw die is shown at *B*, the "acorn" die at *C*, and the round split die (sometimes known as a "button" die) at *D*. A great many threading dies of these general types are used on automatic screw machines for cutting the smaller sizes of screws. The die at *B* has projections or prongs on which the cutting edges are formed. These prongs are quite flexible and must be held in position when the die is in use by some form of external ring or clamp.

The lands or cutting ends of the "acorn" die *C* are shorter and wider than those of a spring die, which increases their strength against torsional or twisting strains. The part of each land or prong just back of the cutting teeth is made comparatively thin to give the necessary flexibility for radial adjustment. The die is contained within an adjustable cap which is screwed onto the die-holder. The outer end of each die prong is beveled to fit a corresponding conical surface on the inside of the adjusting cap. Radial adjustment of the lands for varying the size of the die is obtained by simply turning the adjustable cap in one direction or the other after loosening the locking nut. The acorn die, like a spring die, may readily be resharpened on the radial faces, the grinding being done in the flute, the same as with a spring die.

Round split or button dies (one form of which is shown at *D*, Fig. 1) are used principally on turret lathes and on hand and automatic screw machines; they are applied to the same general classes of work as spring screw threading dies. The button dies are more rigid than the spring dies, if the latter are adjusted by the usual ring or clamp, but they are not so easily sharpened on the radial cutting faces. The initial cost of the button die, however, is less than that of a spring die and the former is not distorted as much in hardening.

For what purpose are two-part dies commonly used?

The sectional type of die formed of two separate parts (*E*, Fig. 1) is extensively used in die-stocks for hand threading operations on bolts, pipe, etc., and also in some power-driven screw-cutting machines. The outside shape of these dies varies. The form shown has sides with a double bevel to give the two die sections a tight grip in the holder and permit reversing the position of the die for cutting close to a shoulder. Some dies of the two-piece form are round and are used in preference to the round type shown at *D*, which is split on one side only. The advantages claimed for the two-piece round type are that it does not lose its shape the same as a die which springs together on one side only, and it can be ground more easily.

Why are many dies provided with a set of removable thread-cutting chasers?

The chaser type of die is preferable to a solid split or spring die especially for cutting comparatively large screw threads, partly because it is capable of more accurate work. The one-piece die is always subject to slight distortion in hardening, especially when made in large sizes. The use of inserted chasers is also an advantage, because a practically new die may be obtained by inserting a new set of chasers, when, as a result of repeated grinding and wear, this becomes necessary; the body of the die may be used permanently.

The construction of these dies varies considerably. The particular design shown at *F*, Fig. 1, has four chasers, which are backed up at the ends by hardened pins carried by a heavy adjusting ring. A loose or tight fit may be obtained with a set of chasers by means of an adjusting screw which changes the position of the adjusting ring and pins relative to the oval-backed chasers. The die may be adjusted either to compensate for the wear of the chasers and maintain a standard size or to vary the diameter slightly from the standard. Dies of the adjustable chaser type may be used in preference to the automatic or self-opening class (to be described later) for cutting screw threads that are so short that the time saved by the opening dies is too small to offset the difference in cost. The adjustable chaser design, in general, is also more rigid than the self-opening die and better adapted for taking heavy cuts on threads of coarse pitch when the thread is to be finished by a single cut.

What is the tangential-chaser type of die?

The tangential-chaser type of die-head has the chasers set in a tangential position relative to the work as illustrated by the right-hand diagram, Fig. 2, instead of in a radial position as shown by the left-hand diagram. The

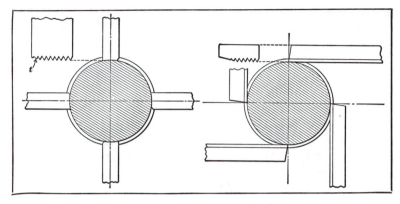

Fig. 2. Radial and Tangential Chasers for Threading Dies

teeth extend throughout the length of the chasers and they
are in the form of straight parallel ridges corresponding
in cross-section to the shape of the screw threads they are
intended to cut. These chasers are sharpened by grinding
on the ends, and as they become shorter as the result of
repeated grinding, they are simply adjusted to the correct
cutting position. The throat of the chaser and the front
teeth form the thread, while the rear teeth extend across
the center line and take a bearing on the work just back of
the cutting end of the tool. This is to make the four chasers
act as a lead nut for controlling the forward movement of
the die in accordance with the lead of the thread being cut.
By using right- and left-hand holders the same chasers may
be employed for cutting right- and left-hand screw threads.
Fig. 3 shows a tangential-chaser type of die threading the
stems of motor exhaust valves. The screw thread diameter
is 1/2 inch; the number of threads per inch, 20, the pitch
diameter tolerance, 0.0015 inch.

What is the "throat" of thread-cutting die chasers?

The leading side of each chaser in a die-head is usually
beveled as shown in Fig. 4. This beveled edge is known as
the "throat" of the chaser and serves to begin the cut gradually when the die is first starting a thread and also when
it advances. At *A* is shown a chaser at the point of start-

ing a cut, and *B* represents the position of the chaser after it has advanced far enough to begin forming a "full" or complete thread. The throat of the chaser not only inclines relative to the axis of the die (or screw being cut), but it is given clearance back of the cutting edge in a circumferential direction. In some cases, the throat angle *a* must be abrupt in order to cut a full thread close to a shoulder. Aside from a requirement of this kind, the throat should preferably be ground so that the work of cutting

Fig. 3. Machine Equipped with Tangential Type of Die

a thread to the full depth is distributed over at least two or three teeth on the leading side of the die. It is common practice to grind the throat at such an angle that the beveled part extends from the root or base of the leading or most advanced tooth in the set of chasers back to the third tooth, which may be slightly beveled. The throats of some dies extend back over four, five, or more teeth, although the shorter chamfer is more common. Each chaser should be ground to the same angle and so that each throat will be the same distance from the axis of the die-head.

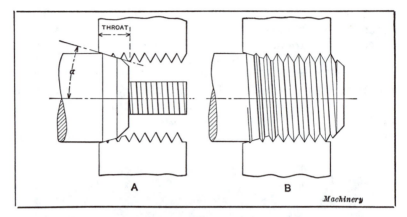

Fig. 4. Diagrams Illustrating How Throat of Die Serves to Start
Thread-cutting Operation Gradually by Distributing
Work Between Several Chaser Teeth

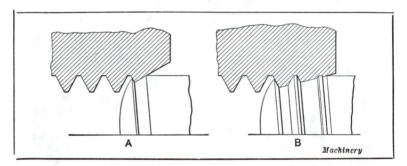

Fig. 5. (A) Throat that Extends Considerably Beyond Roots of Chaser
Teeth so that it is not Effective. (B) Throat that is Effective

Why are the throats of some die chasers ineffective in distributing the work?

A method of chamfering dies or die chasers that sometimes causes trouble is illustrated at *A*, Fig. 5. The die
chaser may appear to have considerable chamfer, but when
the relation between the throat of the die and the work is
considered, it will be seen that there is little *effective* chamfer or throat. This is due to the fact that the beveled edge
extends beyond or outside of the root or bottoms of the
chaser teeth, where it cannot be effective when a screw

thread is being cut; consequently the entire work of cutting the thread groove is imposed upon the leading tooth of each chaser. The chamfer, or throat, to be effective should begin near the root of the leading tooth (as indicated at *B*), so that it assists in the work of cutting the thread groove. If the die is held in a floating holder and the work is likely to be considerably out of line, the edges can be beveled beyond the root of the teeth somewhat, but this extra chamfer should not be considered as a part of the throat.

What causes a die to advance at the rate required for cutting a thread of correct pitch?

Most dies are self-propelling or self-leading and move along as the screw is cut, except when the die is held stationary and the work is revolved, in which case the action is reversed but is similar so far as the practical result is concerned. This motion of the die relative to the screw or vice versa is due to the fact that the action of the die is similar to that of a nut. As one tooth cuts a thread groove, the next successive tooth follows in the same groove, and even though it may cut the groove deeper, it serves in part to steady the die like the section of a nut and force it to advance at a rate equal approximately at least to the pitch of the screw thread.

Is the movement of a die ever controlled by a lead-screw?

While most dies are "self-leading," it is sometimes advisable to control positively the longitudinal motion of the die relative to the work. This control may be utilized merely to start the die, or the arrangement may be such that the longitudinal motion of the die is controlled positively throughout the entire screw-cutting operation. This positive action may be derived from a lead-screw or from a cam, depending upon the type of machine. A lead-screw is sometimes applied to a threading machine of the bolt-cutter type, especially when cutting square threads, or special forms such as the ratchet thread. For screw-cutting operations of this kind, if the die follows its own lead, the accumulated error is often considerable; that is, the lead error between two threads might be small, but the total error in the length of the screw, or a section of it, might be considerable. By using a lead-screw, the die is prevented from increasing or decreasing the lead. If more than one cut is required, as in cutting a screw of coarse pitch, an indicator or thread-chasing dial of the type used on engine lathes for catching the thread is convenient on a die threading machine having a lead-screw.

The general method of cutting threads on the automatic screw machine is to use a cam that starts a die onto the work and then allows the turret-slide to lag behind somewhat so that the die can lead itself on. Lost motion in the die-holder allows the die to follow its own lead or move independently of the turret-slide.

Whenever a lead-screw is used in conjunction with a die, it is very important to have the positive feeding movement of the carriage and work, per revolution of the die-head, equal to the pitch of the chaser teeth in the die. The function of the lead-screw is simply to prevent the die from either gaining or losing in pitch by so controlling the movement of the carriage that its advance per revolution of the die corresponds to the pitch or lead of the screw thread. The rate of the carriage movement may be controlled either by means of change-gearing, through which the lead-screw is driven from the main spindle, or by the use of a lead-screw which corresponds to the pitch of the thread to be cut.

What are the main features of automatic or self-opening dies?

The different designs of automatic or self-opening dies differ in regard to the mechanism for opening the die chasers at the completion of a cut, the method of closing the chasers to the cutting position after removing the die, the method of supporting the chasers against radial thrusts, and other features. Self-opening dies, in general, are formed of two main sections. One section, which includes the shank and inner part of the die body, is attached to the turret, spindle, or other part of the machine. These two main sections have a certain relative motion for opening the die or releasing the chasers from the work and for closing the chasers to the working position. This motion for operating the die may either be parallel to the axis of the die, rotary, or helical. The radial motion of chasers of the radial type, for opening or closing the die, is commonly derived or controlled either from cam surfaces or the conical surface of a sleeve in contact with the chasers.

The three general methods for opening dies of this class automatically are by stopping the travel of the turret at a predetermined point; by the engagement of an outside tripping finger, latch, or lever on the die-head with a fixed stop or plate; and by the engagement of the end of the work with a tripping plate located inside the die. Most self-opening dies are of the non-revolving type, the die remaining stationary while the part to be threaded rotates. Dies of this class commonly have a hand-lever for opening or closing them. Some dies of the automatic class are designed to be revolved.

How is a non-opening die removed after cutting a screw thread?

After a screw thread has been cut with a non-opening die, in most cases it is necessary to back it off, and this may be done in three different ways: (1) the rotation of the work may be reversed after the thread is cut; (2) the die itself may be reversed, thus unscrewing it from the threaded part; (3) the die may be revolved in the same direction as the work, but at a somewhat slower rate of speed while cutting the thread and then at a faster rate so that the die backs off the threaded part while it still continues to revolve in the same direction as the work. This third method is employed on some automatic screw machines.

Are dies practicable for cutting screw threads of coarse pitch?

The maximum pitch of thread that can be cut satisfactorily with a threading die may depend upon several factors, such as the design of the die in regard to rigidity, the accuracy required in the screw thread, the degree of finish or smoothness, the relation between the pitch of the thread and the diameter or strength of the screw to resist the strain of cutting, and the condition of the chasers, especially as to sharpness. As a general rule, when the pitch is coarser than four or five threads per inch, the difficulty of cutting threads with dies increases rapidly, although some dies are used successfully on screw threads having two or three threads per inch or less in extreme cases. Dies of special design could be constructed for practically any pitch if the size of the die, its cost, and the power for driving it were regarded as secondary, and the screw blank were strong enough to resist the cutting strains. If the screw diameter is relatively small in proportion to the pitch, there may be considerable distortion of the screw due to the torsional strains set up when cutting the thread with a die. For this reason, dies for coarse pitches work better when cutting threads on screws which are large enough in diameter to resist the torsional or twisting strains. As a general rule, if the number of threads per inch is only one or two less than the standard number for a given diameter, the screw blank will be strong enough to permit cutting the thread with a die without excessive distortion. When a coarse thread is cut by the milling process, the cutting is done much more gradually and there is not the same difficulty due to a twisting action that is encountered when using a die of coarse pitch on comparatively small diameters.

What types of dies are used for cutting tapering threads?

Tapering threads may be cut by using dies of different types. These include (1) dies with chasers which taper to correspond to the taper on the work and are arranged to move outward radially as the die traverses toward the large end of the taper; (2) dies of the solid or non-opening class which have the same taper as that required on the work, assuming that the length of the threaded part does not exceed the length of the cutting edges on the die or the width of the chasers; (3) dies intended for parallel threads but arranged to open radially for producing a taper thread. The first type of die referred to, which has tapering chasers that move outward in accordance with the taper, is preferable for the most accurate work. The solid or non-opening die, which simply tapers to correspond to the taper screw thread to be cut, is not only limited to comparatively short screw threads but does not produce a very satisfactory thread because ridges are left wherever each cutting edge

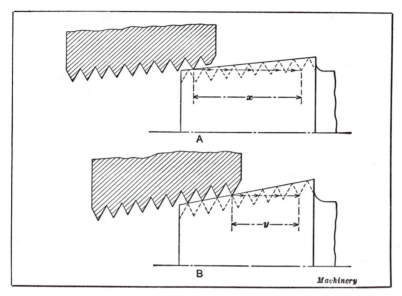

Fig. 6. Views Illustrating How Longitudinal Movement of Solid Die for Taper Threading is Reduced as Taper Increases

stops cutting. The use of a solid tapering die also subjects the work to considerable torsional strain (which might be objectionable) because there are more cutting edges at work at the same time and the power required to turn the die increases as the length and taper increases. The greater the length the more cutting edges will be at work simultaneously, and the more abrupt the taper the less the number of revolutions of the die for completing the thread. The effect of the taper on the number of revolutions for completing the thread is illustrated at A and B, Fig. 6. In one case the die moves inward a distance x before the teeth cut to full depth, but when the taper is more abrupt, as shown at B, the die only needs to move longitudinally a distance y, in order to cut the full depth of thread.

The third method mentioned, by which a die for straight thread cutting is used for tapering work, may be applied to cutting threads having a slight taper, but is not recommended for the usual classes of taper work, which include such parts as wash-out plugs for steam boilers, the ends of faucets or cocks of various kinds, buffing lathe spindles, etc. A tapering die of the non-opening class does not require a throat or chamfered teeth on the leading side unless the taper is slight, because several cutting edges begin work at the same time, as indicated at *B*, so that there is no need for a throat, since the object of the chamfered edges in a straight die is to start the cut gradually between several teeth, instead of having one leading tooth cut the thread groove to the full depth.

How are self-opening dies designed for taper thread cutting?

A self-opening die-head for taper threading is so designed that the chasers move outward at a rate depending upon the angle of taper, until the thread is cut, and then move outward suddenly to clear the work. The radial movement of taper-threading die chasers is ordinarily controlled by means of a tapering former plate, which allows the cam or scroll ring of the die-head to turn slowly as the die advances, the rate of this turning movement depending upon the angle of the former plate relative to the axis of the die. The former plate serves about the same purpose as the adjustable slide or bar of a taper attachment for the engine lathe. Taper-threading dies of the self-opening class may be either of the inside or outside trip type.

Is it practicable to cut square threads with dies?

The cutting of square threads with dies is usually regarded by die manufacturers as a difficult proposition, and unless the die is made very carefully, unsatisfactory results are obtained. While square threads have largely been replaced by the Acme form, some manufacturers prefer the square thread, and occasionally they are cut by means of dies. The chasers for square threading dies that are to be self-leading should have teeth that are slightly relieved on the sides by lapping. Unless there is a little side relief, the chaser teeth bind and are frequently broken. If the die is to be self-leading, however, the side relief or clearance must be very slight, as otherwise the die will not be properly supported and will cut a very inaccurate thread as to lead. A square-thread die should preferably be controlled by a lead-screw and the chaser teeth should be given enough side clearance to prevent binding. As the Acme thread is superior to the square thread and may readily be cut with dies or taps, the use of dies for square threads is not common.

Are die-holders designed to hold dies rigidly?

Die-holders may be arranged to hold the die rigidly or to permit either a longitudinal movement or a combined longitudinal and universal "floating" movement. Many of the die-holders used on machine tools or in connection with power-driven threading operations allow the die a limited amount of motion in the direction of its axis, so that it will be free to follow its own lead and will not be retarded by a backward pull of the tool-slide to which it may be attached. For instance, when cutting a thread with a die in a hand-operated turret lathe, the turret-slide is moved up until the die has started on the work. If the die-holder has longitudinal motion, the turret can be allowed to lag behind somewhat without interfering with the forward motion of the die. When the machine is of the automatic type, the cam operating the turret-slide is generally designed to start the die on the work and then the slide is allowed to travel a little slower than the die which governs its own motion independently.

Some die-holders are arranged to give the die a free floating movement in a direction perpendicular to the axis, so that, if the part to be threaded is not exactly in line with the die, the latter can center itself relative to the axis of the work and cut a concentric thread on it. While every holder having the axial motion also has some lateral play, this is very slight in some holders and is a special feature in others. A lateral or universal float is especially desirable for dies used on parts that are chucked either by hand or from a magazine attachment for the threading operation. This floating motion is found on holders for both non-opening and self-opening dies.

What is a non-releasing type of die-holder?

The die-holders used for solid or non-opening dies may be of the rigid type, the floating non-releasing type, or the

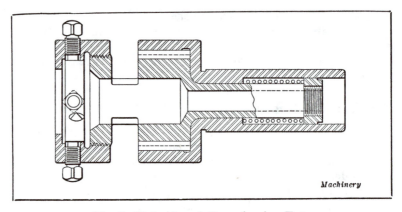

Fig. 7. Die-holder of Non-releasing Type

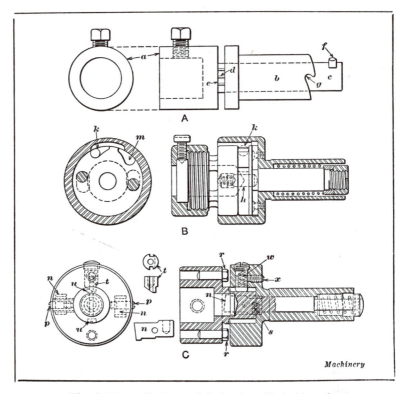

Fig. 8. Three Designs of Releasing Die-holders for
Holding Non-opening Dies

releasing type. For turret lathe and automatic screw machine work, the non-releasing type is used extensively, although the releasing design is preferable under certain conditions.

One design of non-releasing die-holder, which is often used for thread cutting on automatic screw machines with non-opening dies, is shown in Fig. 7. The arrangement is such that the part to which the die is attached can move in a lengthwise direction relative to the shank. The die, however, is not released or permitted to rotate with the work after the thread is cut as with the releasing type of holder.

Why is the releasing type of die-holder used for some thread-cutting operations?

The releasing type of die-holder (which is intended only for non-opening dies) is used when it is necessary to govern closely the length of the thread, as, for example, when cutting a thread close to a shoulder. With this type of die-holder, the die is released after the thread has been cut to the required length. The action is as follows: When the forward motion of the turret-slide discontinues, the rotation of the screw thread draws one section of the die-holder farther forward until the driving connection between the two sections disengages; the die then revolves with the work as long as the latter continues to run forward. When the spindle is reversed, the die starts to rotate backward with it, but this reverse movement is stopped automatically by the die-holder, and the stationary die is then backed off the screw as the spindle continues its reverse rotation.

If the reversal of the machine is controlled by the operator, as in a hand screw machine, a releasing die-holder should be used, because, if the machine is not reversed at the instant a die of the non-opening type reaches the limit of its forward travel, the thread may be stripped or the die broken in attempting to cut close to a shoulder. When the releasing type of holder is applied to the threading spindle of a multiple-spindle automatic screw machine, if the threading operation is completed before the other operations, the releasing device permits the die to revolve loosely until all the operations are completed.

What types of releasing die-holders are used?

There are various designs or types of releasing die-holders. Three designs are illustrated in Fig. 8 merely to show the general principle of operation. A simple form is shown at A. The holder a has a shank c which passes through the sleeve b. This sleeve is held in the turret. When the die is cutting, the holder a is prevented from rotating with the work by the engagement of lugs d and e. When these lugs separate at the end of the cut, the die revolves with the work until the rotation is reversed; then pin f, as it revolves with shank c, engages notch g as the turret-slide is returned, and stops further rotation of the die, which is backed off the threaded section.

The holder illustrated at B operates on the same general principle as the one just described, although the construction is quite different. The driving connection between the two main sections of the holder is through the pins h. When the turret-slide stops and the die-holder is drawn forward, these pins h are disengaged, thus releasing the die. As soon as the work-spindle reverses, a ball k slides out of the deep part of the pocket in which it normally rests, thus locking the two main sections together while the die is being removed. The ball is inserted in pocket m when the die-holder is used for cutting left-hand threads.

The main part of the die-holder shown at C, which is held in the turret, carries two bevel-ended driving lugs n, which are mounted upon pins p, so that they are free to swivel. These driving lugs engage pins r in the die-holder proper; when the latter is drawn forward at the end of the cut or after the turret-slide stops moving, lugs n, acted upon by springs s, swivel until the beveled ends are practically in a plane at right angles to the axis of the holder. This swiveling movement provides clearance between lugs n and pins r so that the parts cannot strike against each other while the die is revolving with the work. When the spindle reverses, the beveled plunger t drops into slot u and holds the die stationary. This plunger is backed up by spring w. To change this die-holder for cutting left-hand threads, the driving lugs n are turned over, thus reversing the position of the straight driving side. The plunger t, which is held in position by the end of screw x, is also reversed.

Forming Screw Threads by Milling and Grinding

When screw threads are cut by milling, the thread groove is formed by a revolving milling cutter shaped to conform to the shape of the thread to be milled. Thread grinding is similar in principle to milling in that a rotating milling cutter is used in one case and a rotating grinding wheel in the other. While we may not think of a grinding wheel as a cutting tool, it does, in fact, consist of innumerable small cutting edges, and grinding is a true cutting process which removes metal by producing minute chips.

How is a screw thread milled by the single-cutter method?

There are two general methods of forming screw threads by milling, which may be designated as the single-cutter and the multiple-cutter methods. The way a single cutter is used is indicated by the diagram, Fig. 1. The profile of the cutter or the shape of its cutting edge conforms to the sectional shape of the thread groove. This cutter should revolve as fast as possible without dulling the cutting edges excessively, in order to mill a smooth thread and prevent unevenness such as would result with a slow-moving cutter, on account of the tooth spaces. As the cutter rotates, the part on which a thread is to be milled is also revolved, but at a very slow rate (a few inches per minute), since this rotation of the work is practically a feeding movement. The cutter is ordinarily set to the full depth of the thread groove and finishes a single thread in one passage, although deep threads of coarse pitch may need two or even three cuts. For work of this kind a roughing cut is sometimes taken with a special cutter which is somewhat narrower than the

33

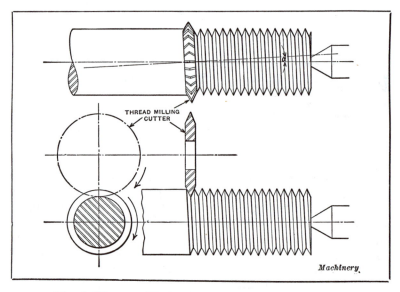

THREAD MILLING
CUTTER

Machinery

**Fig. 1. Diagram Illustrating Method of Milling Screw
Thread with a Single Cutter**

finishing cutter. Fig. 2 shows the milling of a double-thread
worm.

Whenever a single cutter is used, the axis of the cutter is
inclined to some angle a instead of being parallel to the axis
of the screw, in order to locate the cutter in line with the
thread groove at the point where the cutting action takes
place.

Rule.—Tangent of angle a, Fig. 1, = lead of screw
thread ÷ pitch circumference of screw.

The helical thread groove is generated in practically the
same way as when an engine lathe is used. The single cutter
process is especially applicable to the milling of large screw
threads of coarse pitch, and either single or multiple threads.
For fine pitches and short threads, the multiple-cutter
method (described in the next paragraph) is preferable,
because it is more rapid. The milling of taper screw threads
may be done on a single-cutter type of machine by traversing the cutter laterally as it feeds along in a lengthwise
direction, the same as when using a taper attachment on an
engine lathe.

Why are multiple cutters used for some thread milling operations?

A thread milling method which requires the use of a multiple cutter is illustrated by the diagrams *A* and *B*, Fig. 3. This multiple cutter is practically a series of single cutters, although formed of one solid piece of steel, at least so far as the cutter proper is concerned. The annular rows of teeth do not lie in a helical path, like the teeth of a hob or tap, but coincide with planes which are perpendicular to the axis of the cutter. If the cutter had helical teeth the same as a gear hob, it would have to be geared to revolve in a certain fixed ratio with the screw being milled, but a cutter having annular teeth may rotate at any desired speed, while the screw blank is rotated slowly to provide a suitable rate of feed. (The multiple cutters used for thread milling

Fig. 2. Milling a Double-thread Worm of 3½ Inches Diameter

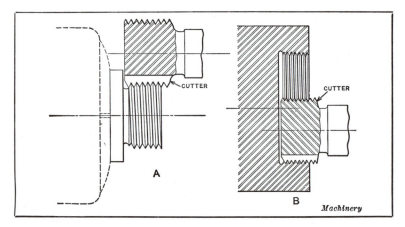

Fig. 3. Examples of External and Internal Thread Milling with
a Multiple Type of Cutter

are frequently called "hobs," but in this section the term hob will be applied only to cutters having helical teeth.)

The object in using a multiple cutter instead of a single cutter is to finish a screw thread complete in approximately one revolution of the work, a slight amount of over-travel being allowed to insure milling the thread to the full depth where the cut joins the starting point. In order to finish the thread complete in about one revolution, it is necessary to use a cutter which is at least one or two threads or pitches wider than the thread to be milled. In using a multiple cutter, it is simply fed in to the full thread depth and then either the cutter or screw blank is moved in a lengthwise direction a distance equal to the lead of the thread. Since there is an annular row of cutting teeth for each thread groove, this movement equal to the lead is sufficient to finish the entire thread in one revolution of the work, plus whatever additional movement there might be due to the over-travel. If an exceptionally smooth thread were required, the work might be revolved two revolutions and the cutter be traversed a distance equal to twice the lead of the thread. During the first revolution the thread would be rough-milled and a light finishing cut would then be taken while the work made a second revolution. Sketch B, Fig. 3, illustrates the application of a multiple cutter to internal thread milling.

It is apparent that the length of the thread that can be milled by the multiple-cutter method is limited. A cutter supported at one end only would be deflected considerably if the length of the cutting end and the "over-hang" were increased to any great extent. Since the cutter is milling along the entire screw thread at the same time, the lateral thrust is relatively large as compared with a single cutter operating on a thread of corresponding pitch. For these reasons, the multiple cutter is used for milling comparatively short threads and usually medium or fine pitches. The sketches in Fig. 3 represent typical examples of external and internal work for which the multiple-cutter type of thread milling machine has proved very efficient, although its usefulness is not confined to shoulder work and "blind" holes. The milling cutter is used frequently in preference to a tap, because it produces a smoother thread, especially if the metal has soft stringy spots. Fig. 4 illustrates the milling of external threads (see samples of work in foreground).

In using a multiple cutter, is it inclined to obtain alignment with thread groove?

In using multiple cutters either for internal or external thread milling, the axis of the cutter is set parallel with the axis of the work, instead of inclining the cutter to suit the helix angle of the thread, as when using a single cutter. Theoretically, this is not the correct position for a cutter, since each cutting edge is revolving in a plane at right angles to the screw's axis while milling a thread groove of helical form. It might be supposed that there would be serious interference between the cutter and the thread, and as a result a decided change in the standard thread form. In practice, however, the defect is very slight and may be disregarded except when milling threads which incline considerably relative to the axis like a thread of multiple form and large helix angle. Threads which have steeper sides than the American Standard or Whitworth forms should ordinarily be milled with a single cutter, assuming that the milling process is preferable to other methods. For instance, in milling an Acme thread which has an included angle between the sides of 29 degrees, there might be considerable interference if a multiple cutter were used, unless the screw thread diameter were large enough in proportion to the pitch to prevent such interference. If an attempt were made to mill a square thread with a multiple cutter, the results would be unsatisfactory owing to the interference. If a multiple cutter, in any case, were inclined to align it with the thread groove, the same as is done when the single form of cutter is employed, the advantage of the multiple type would be lost, and instead of finishing the thread in one revolution of the screw blank, it would be necessary to traverse the cutter along the entire length of the thread.

Interference between the cutter and work is more pronounced when milling internal threads, because the cutter does not clear itself so well. Experiments have shown that

Fig. 4. Milling an External Thread with Multiple Cutter

multiple cutters for internal work should preferably not exceed one-third the diameter of the hole to be threaded. A cutter that is one-quarter the diameter of the thread will do very satisfactory work. It is preferable to use as small a cutter as practicable, either for internal or external work, not only to avoid interference, but to reduce the strain on the driving mechanism.

What are the main features of thread milling machines?

Some thread milling machines are so designed that the cutter-slide or carriage is traversed along a horizontal bed by a lead-screw which is connected to the work-spindle through change-gears, the arrangement being practically the same as on an engine lathe. With this general type of machine, the cutter (which may be single or multiple) moves along one side of the work, while the latter rotates but does not move axially. This order is reversed in another general type of milling machine which is designed to move the part on which a thread is to be milled, in an axial or lengthwise direction, while the cutter-slide remains stationary, except when it is traversed laterally or at right angles to the screw thread for moving the cutter in or out of the working position. A machine of this kind may have a work-table which is traversed by a gear-driven lead-screw, or the traversing motion may be imparted to the work-holding spindle either by the direct action of a lead-screw or by a lead-screw and gearing combined. When a lead-screw is applied directly to the work-spindle, the lead of its thread is the same as the lead of the thread to be milled, and different lead-screws are used for milling threads of different pitches. Some of these "duplicating" machines are designed especially for milling threads on large numbers of duplicate parts, and they are less complicated than a machine equipped with change-gears of whatever ratios are required for a range of pitches. Other machines which derive the traversing motion directly from a lead-screw are so constructed that one lead-screw may easily be replaced with another of different pitch. Such machines are intended for general application, but lead-screws are changed for milling threads of different pitch, instead of change-gears, as in the case of the other general type of machine mentioned. Thread milling machines embodying the general principles of operation outlined differ, of course, in regard to the details of construction.

Under what conditions is thread milling applied in preference to some other method?

Each standard method of cutting threads, whether by milling, by means of taps and dies, or by using a single-point tool in the lathe has its own advantages when applied under favorable conditions. A thread milling machine may be used (1) because the pitch of the thread is too coarse for cutting with a die, (2) because the milling process is more efficient than using a single-point tool in a lathe. When making comparison between thread-cutting processes, it is also essential to consider the relation that may exist between the thread-cutting operation and other operations which may precede it. To illustrate this point, a lathe may be inferior to a thread milling machine for cutting a thread of a certain size and pitch, and yet the lathe may be preferable because cutting the thread is only one of a series of operations, and by doing this work in the lathe the piece is finished at one setting and the thread is accurately located with reference to other machined surfaces. Similar conditions may exist in connection with work done in turret lathes or screw machines. For example, when a part requiring an internal thread is turned and bored in a turret lathe, there is a decided advantage, in most cases, in finishing the part without removing it from the chuck and, ordinarily, some form of tap would be used, or a die in the case of external threads. In view of this close relationship between the method of cutting the thread and the work as a whole, it is apparent that any comparison between thread milling and other thread-cutting processes must be general and subject to modification. The classes of work for which the different thread-cutting methods are particularly adapted also merge into one another and there is no well-defined dividing line to serve as a guide.

Determining the relative merits of different thread-cutting processes is further complicated by the fact that a comparison between thread milling and the use of a lathe, die, or tap might be based either on the rate of production, accuracy of thread as to diameter and lead, smoothness of thread, or its location relative to other surfaces. The importance of these different features may vary considerably on different classes of work. As a general rule, the best method of cutting large screw threads of coarse pitch, or any form or size of thread requiring the removal of a relatively large amount of metal, is by means of a thread milling machine equipped with a single cutter. The milling process is particularly desirable if the pitch of the thread and size of the thread groove are large in proportion to the diameter of the screw, because the metal removed by each cutting edge around the circumference of the cutter during one revolution is small and the screw being milled is not subjected to any great torsional strain. When a die is used for work of this kind, the accuracy of the screw may be seriously affected by the torsional strain or the twisting of the screw blank when cutting the thread.

The single-cutter type of thread milling machine is superior to the lathe for cutting threads on lead-screws, worms, etc., because it gives a higher rate of production due to the fact that the action of the milling cutter is continuous and a single cut usually finishes the thread complete. The thread milling machine has little, if any, advantage over the lathe in regard to accuracy of lead since both machines duplicate the controlling lead-screw.

Are gear hobbing machines applicable to thread milling?

A hob is sometimes used in conjunction with a gear-hobbing machine for milling multiple screw threads. A hob used for this purpose has teeth which lie along a helical path, like a hob intended for cutting spur or helical gears, and it must be geared to revolve with the work at a definite speed ratio, the same as when hobbing a gear. The hobbing method is particularly efficient for cutting worms having several threads, because the hob finishes the different threads simultaneously. A hob having teeth of special form must be used for milling worms or other screw threads in order to generate threads having sides which are, at least, approximately straight. The threads are cut much more rapidly than when an ordinary milling cutter is employed. The sides of the threads are not exactly straight, but the curvature is slight. Multiple-threaded worms having four threads or more may often be milled to advantage with a single-threaded hob, but if the worm has only a single thread or a double thread, it should preferably be milled by using a single milling cutter, instead of employing a hob. When a single cutter is used on a gear-hobbing machine, the work-table is geared to the cutter feed so that a screw thread of the required lead will be milled, but without reference to the speed of the cutter which may be regulated to suit the thread milling operation. When a screw thread is milled in this way, the gear-hobbing machine is practically a thread milling machine, so far as the principle of its operation is concerned.

How are screw threads milled by the planetary method?

The planetary method of thread milling is similar in principle to planetary milling or "planamilling," as described in the section on Milling Flat, Curved and Irregular Surfaces. The part to be threaded is held stationary and the thread milling cutter, while revolving about its own axis, is given a planetary movement in order to mill the thread in one planetary revolution about the work. The machine spindle and the cutter which is held by it is moved longitudinally for thread milling an amount equal to the thread lead during the planetary movement. This operation is designated as "planathreading." It is applicable to both internal and external threads. For the latter operation, the thread milling cutter surrounds the work, the same as for planamilling. This thread milling is frequently accompanied by milling operations on other adjoining surfaces.

The accompanying diagram, Fig. 5, illustrates how a double-head type of machine is used for milling a screw thread and a cylindrical surface simultaneously in each side of the work. The cutters are shown in the neutral or non-cutting position. When the milling operation begins, the eccentrically mounted cutter-spindle feeds the cutter into the right depth and then the planetary movement begins, thus milling, in this case, the two threads and the two

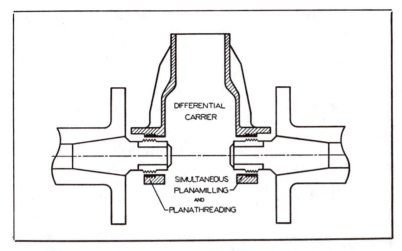

Fig. 5. Example of Milling and Threading Simultaneously
on Planetary Type of Machine

cylindrical surfaces indicated by the heavy lines, all simultaneously. The thin sharp starting edges are eliminated on threads milled by this planetary method and the thread begins with a smooth gradual approach. One design of machine will mill internal and external threads simultaneously. These threads may be of the same hand or one may be right hand and the other left hand. The threads may also be either of the same pitch or of a different pitch, and either straight or tapered.

How is the grinding process applied to screw threads?

Many screw threads are ground in order to correct whatever slight distortions may occur in connection with heat-treatment. These ground threads are very precise as to thread form and lead, assuming that precision grinding equipment is used. Thread grinding is applied both in the manufacture of duplicate parts and also in connection with precision thread work in the tool-room. Fig. 6 shows the grinding of a thread plug gage.

In grinding a thread, the general practice in the United

States is to use a large grinding wheel (for external threads) having a diameter of possibly 18 to 20 inches. The width may be 5/16 or ⅜ inch. The face or edge of this comparatively narrow wheel is accurately formed to the cross-sectional shape of the thread to be ground. The thread is ground to the correct shape and lead by traversing it past the grinding wheel. This traversing movement, which is equivalent to the lead of the screw thread for each of its revolutions, is obtained from a lead-screw. On one type of thread grinder, this lead-screw is attached directly to the work-spindle and has the same lead as the screw thread to be ground; hence, there is a separate lead-screw for each different lead of thread to be ground. On another design of

Fig. 6. Grinding External Thread on a Thread Plug Gage. Single Edge Type of Grinding Wheel Conforms to Shape of Thread Groove

machine, the lead-screw arrangement is similar to that on a lathe in that the required lead on the ground thread is obtained by selection of the proper change gears. The grinding wheel may have a surface speed of 7000 feet a minute, whereas the work speed may range from 3 to 10 feet per minute. The grinding wheel is inclined to suit the helix angle of the thread and either right- or left-hand threads may be ground. Provision is also made for grinding multiple threads and for relieving taps and hobs. The wheel shape is accurately maintained by means of diamond truing tools. On one type of machine, this truing is done automatically and the grinding wheel is also adjusted automatically to compensate for whatever slight reduction in wheel size may result from the truing operation. A single-edged wheel is used whenever the highest precision is required, grinding the work either from the solid or as a finishing operation.

Fig. 7 shows how an internal thread is ground. The operation is the same in principle as external thread grinding.

Fig. 7. Grinding an Internal Thread

Fig. 8. Grinding Threads on a Stud by Plunge-cut
Multi-edged Grinding Wheel

Fig. 9. Grinding a Long Thread with a Multiple-edged Wheel by
Shifting Wheel Axially after One-half of the Thread
Length has been Ground

Is it practicable to form threads entirely by grinding or without preliminary cutting?

On some classes of work, the entire thread is formed by grinding "from the solid," especially if the time required is less than would be needed for a rough thread-cutting operation followed by finish-grinding after hardening. In airplane engine manufacture, the threads of all such stressed parts as cylinder hold-down studs, adjusting screws, rocker hub bolts, and crankshaft clamp screws may be ground from the solid. The threads on such parts are ground after the parts have been hardened. Grinding has practically eliminated the minute cracks formerly found at the bottom of threads on parts hardened after the threads were cut. Other important advantages are the elimination of distorted threads due to hardening. Grinding threads from the solid is applied to the finer pitches. In some plants, threads with pitches up to about 1/16 inch are always ground by this method.

How is an entire screw thread ground in one revolution of the work?

An entire screw thread, if not too long, may be ground completely in one revolution by using a multiple- or multi-edged type of grinding wheel. The face of this wheel is formed of a series of annular thread-shaped ridges so that it is practically a number of wheels combined in one (see Fig. 8 and diagram A, Fig. 10). The principle is the same as that of milling screw threads by the multiple-cutter method. If the length of the thread to be ground is less than the width of the wheel, it is possible to complete the grinding in practically one work revolution as in thread milling. A grinding wheel having a width of, say, $2\frac{1}{2}$ inches, is provided with annular ridges or threads across its entire width. The wheel is fed in to the thread depth, and, while the work makes one single revolution, the wheel moves axially a distance equal to the thread lead along the face of the work. (Whenever a grinding wheel is fed straight in to the final depth, this is often called "plunge-cut" grinding.) Most threads which require grinding are not longer than the width of the wheel; hence, the thread is completed by one turn of the work.

In grinding long threads with a multi-edged wheel, what methods are employed?

There are two methods. One is to grind part of the thread and then shift the wheel axially one or more times for grinding the remaining part. For example, with a wheel 2½ inches in width, a thread approximately 12 inches long might be ground in five successive steps. Fig. 9 shows a job requiring one lateral adjustment of a multi-edged wheel. In this case, about half of the thread is ground during the first revolution and then the wheel is adjusted or indexed to the right position for "catching the thread" and grinding the remaining section.

The second method is that of using a multi-edged tapering wheel which is fed axially along the work. The taper is to distribute the work of grinding over the different edges or ridges as the wheel feeds along, as indicated by diagram B, Fig. 10. This method is used especially for grinding long threads on shafts or spindles, when it would be impracticable to use a plunge-cut multi-edged wheel and make several axial adjustments. While the tapered grinding wheel is practically a series of single-edged grinding disks, with each succeeding disk slightly larger in diameter than the previous one, all of these "disks" are actually part of one single, solid, wide-faced wheel. Each "disk" has only a comparatively light cut to take, and the method is, therefore, especially suitable for the grinding of long threads.

The tapered wheel is also used for internal grinding. When used for that purpose, it is frequently cylindrical to start with, but having the threads chamfered off at one end in a manner somewhat similar to that used for a tap or a thread chaser. Such a wheel has been found to give the best results for internal threads when grinding "from the solid." Where extreme accuracy is required, the thread would be finished by a single-edged grinding wheel.

How is a multi-edged wheel formed and trued when worn?

The multi-edged grinding wheel is dressed by means of an oil-hardened cylindrical carbon-steel roller having annular ridges of the exact thread profile required. This roller is

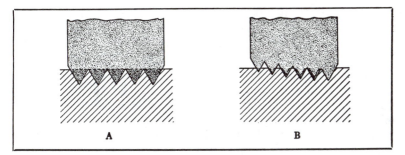

Fig. 10. (A) Method of Grinding Screw Thread with Multi-edged Wheel.
(B) Tapering Form of Multi-edged Wheel

provided with spiral flutes, somewhat similar to those of a gear-cutting hob. It is mounted in bearings in a fixture and produces the threads on the face of the grinding wheel by being pressed against the wheel while the latter is revolving slowly. This method of obtaining thread profiles on the grinding wheel might appear to be inaccurate, but the results are surprising, provided the wheels used are comparatively fine-grained with a dense texture. The grinding wheel while being dressed, revolves at about 150 revolutions per minute, driving the roller which is free to turn and is simply pressed against the wheel. Dressing requires, for ordinary pitches, from 2½ to 3 minutes to press the thread grooves into a new wheel. Redressing can be done in about one minute. This method of dressing a grinding wheel by means of a roller having the proper profile can be used not only for grinding threaded parts, but also when it is desired to shape the wheel for grinding profiles of various types. Many parts can be produced economically only in this manner.

How accurate are threads ground by different methods?

The following figures are based upon the experience of a German manufacturer: In grinding with a single-edged wheel, it is possible to hold the pitch diameter of the thread to an accuracy of plus or minus 0.0001 inch and to keep the accuracy of the pitch angle to plus or minus three minutes. Accuracy in lead can be maintained to within 0.0001 inch in 1 inch; to within 0.00015 inch in 4 inches; and to within 0.0003 inch in 20 inches of thread length.

The accuracy claimed for grinding with a multiple-edged wheel is approximately as follows: Pitch diameter, plus or minus 0.0004 inch; pitch angle, plus or minus 8 minutes; error in lead not over 0.0004 inch in 1½ inches. These tolerances are sufficient for most machine parts, including airplane motor parts.

The multi-edged wheel method is used for more rapid production when extreme accuracy is not required, but when a good thread surface is wanted and the material is difficult to cut by other means.

The accuracy obtained by the use of a tapered multi-edged wheel lies somewhere between that obtained by the use of a single-edged wheel and a multi-edged wheel.

What types of wheels are used for thread grinding?

The wheels used for steel have an aluminous abrasive and, ordinarily, either a resinoid bond or a vitrified bond. *Resinoid wheels,* as a rule, will hold a fine edge longer than a vitrified wheel but they are more flexible and, consequently, less suitable for accurate work, especially when there is lateral grinding pressure that causes wheel deflection. *Vitrified wheels* are utilized for obtaining extreme accuracy in thread form and lead because they are very rigid and not easily deflected by side pressure in grinding. This rigidity is especially important in grinding pre-cut threads on such work as gages, taps and lead-screws. Vitrified wheels are also recommended for internal grinding. *Diamond wheels* set in a rubber or plastic bond are also used for thread grinding, especially for grinding threads in carbide materials and in other hardened alloys.

Forming Screw Threads by Rolling Process

In forming screw threads by the rolling process, a die or roll having threads or ridges is forced into the metal and, by displacing it, produces a thread corresponding to the required shape and pitch. The plain blanks upon which threads are to be rolled are somewhat smaller in diameter than the finished thread, because the metal displaced by the rolling process is forced up above the original surface of the blank, thus producing a screw thread which is larger in diameter than the original blank. The increase in diameter is approximately equal to the depth of one thread. No material whatever is removed by the rolling process, the metal from the depression formed by the die simply being forced up on either side to form the upper half of the thread.

How is a screw thread formed on a bolt by rolling it between flat dies?

Most of the machines designed exclusively for rolling screw threads are equipped with hardened flat dies. There are two of these flat dies on a machine, as shown by the diagram, Fig. 1, which illustrates the general principle of the flat-die method of rolling threads. One die A is stationary and the other die B has a reciprocating movement. The face of each die has parallel grooves and ridges of practically the same cross-sectional shape as the thread to be rolled, and are spaced to correspond with the required pitch. These ridges, which represent a development of the thread the dies are intended to roll, incline at an angle equal to the lead angle of the thread, so that, as the screw blank rolls between the two dies, a screw thread of the same pitch and lead angle is reproduced on it. The thread is formed in one passage of the bolt, rod, or other part to be threaded,

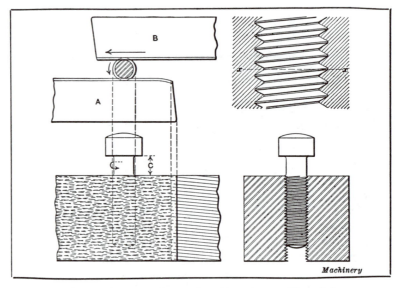

Fig. 1. Diagram Illustrating How Screw Thread is
Rolled between Flat Dies

the work being inserted at one end so that it simply rolls
between the die faces until it is ejected at the other end.
When the thread is not required along the entire body of
the bolt or screw, the work is started between the dies with
the head a distance C above the top edges of the dies. The
ridges on both dies incline in the same direction when
viewed from the rolling sides or faces, but when the dies
are in the thread-rolling machine, the ridges incline in the
opposite direction and are, therefore, in alignment with the
thread groove on the work at the two lines of contact. The
relation between the position of the dies and a screw thread
being rolled is such that the top of the thread-shaped ridge
of one die, at the point of contact with the screw thread,
is directly opposite the bottom of the thread groove in the
other die, at the point of contact, as indicated by the line
$x–x$ of the enlarged sectional view.

Thread-rolling machines are equipped with some form
of mechanism which insures starting the screw blank at
the right time and also square with the sides of the dies.
These machines differ in regard to the position of the dies

and the arrangement of the mechanisms which operate the moving die, the blank starting device, and other parts. Machines of this general class also vary in that some have automatic blank-feeding mechanisms, whereas others of simpler construction require the constant attention of an operator for feeding each blank between the dies by hand.

How are precision threads rolled in machines of cylindrical-die type?

With thread-rolling machines of this type, the blank is threaded while being rolled between either two or three cylindrical dies, depending upon type of machine. These dies are pressed into the blank at a rate of penetration adjusted to the hardness of the material, or wall thickness in the case of threading operations on tubing or hollow parts. The dies have ground or ground and lapped threads and a pitch diameter that is a multiple of the pitch diameter of the thread to be rolled. As the dies are much larger in diameter than the work, a multiple thread is required to obtain the same lead angle as that of the work. The thread may be formed in one die revolution or even less, or several revolutions may be required (as in rolling hard materials) to obtain a gradual rate of penetration equivalent to that obtained with flat or straight dies if extended to a length of possibly 15 or 20 feet. Provisions for accurately adjusting or matching the thread rolls to bring them into proper alignment with each other, are important features of these machines.

Two-Roll Type of Machine.—With a two-roll type of machine, the work is rotated between two horizontal power-driven threading rolls and is supported by a hardened rest bar on the lower side. One roll is fed inward by hydraulic pressure to a depth that is governed automatically.

Three-Roll Type of Machine.—With this machine the blank to be threaded is held in a "floating position" while being rolled between three cylindrical dies which, through toggle arms, are moved inward at a predetermined rate of penetration until the required pitch diameter is obtained. The die movement is governed by a cam driven through change gears selected to give the required cycle of squeeze, dwell and release.

When rolling a thread of given size, how is the blank diameter determined?

The diameter of the screw blank or cylindrical part upon which a thread is to be rolled should be less than the outside screw diameter by an amount that will just compensate for the metal that is displaced and raised above the original surface by the rolling process. If the screw blank is too large before rolling, there will be an excessive amount of metal and the screw will be larger than the standard size. On the contrary, if the blank is too small, either an incomplete thread will be formed or, if the dies are adjusted to roll a full thread, the diameter will be smaller than standard. Aside from the question of accuracy, the blank diameter is affected to some extent by the nature of the material of which the screw blanks are made; that is, whether it offers considerable resistance to displacement, or is easily formed by the threading roll. For instance, threads may be rolled in either brass or steel, but the action of brass is different from that of steel. The condition of the surface of a steel blank may also affect the diameter. When a thread is rolled on drawn stock, there is little, if any, compression of the metal as it is displaced to form a thread, because the surface is already quite dense as the result of the cold-drawing operation. According to the practice in different plants where thread rolling is done, there are three general classes of blank sizes, including: (1) Those which are a little larger than the pitch diameter; (2) those which are approximately equal to the pitch diameter; and (3) those which are slightly less than the pitch diameter.

The sizes in the first class are intended for screws which are to be rolled as accurately as possible. The blank diameters for screws in this class varying from $\frac{1}{4}$ to $\frac{1}{2}$ inch usually are from 0.002 to 0.0025 inch larger than the pitch diameter, and for screws varying from $\frac{1}{2}$ to 1 inch or larger, the blank diameters are from 0.0025 to 0.003 inch larger than the pitch diameter. Threads of the second class mentioned, or those rolled from blanks which are equal to the pitch diameter, are sufficiently accurate for many purposes. Blanks of the third class, or those which are slightly less than the pitch diameter, are intended for bolts, screws, etc., which are made to fit rather loosely, a comparatively free fit being desirable in many cases. Blanks for this grade of work, according to common practice, are from 0.002 to 0.003 inch less than the pitch diameters for screw threads varying from ¼ to ½ inch, whereas, for screw thread sizes larger than ½ inch, the blank diameters are frequently from 0.003 to 0.005 inch less than the pitch diameter. The blanks for screw threads smaller than ¼ inch are usually from 0.001 to 0.0015 inch less than the pitch diameter for ordinary grades of work, and about the same amount larger than the pitch diameter for more accurate screw threads.

What classes of screw threads are formed by rolling?

Machines designed especially for thread rolling are employed extensively for threading such parts as bolts, screws, studs, rods, etc., especially where such threaded parts are required in large quantities. Screw threads that are within the range of the rolling process may be produced more rapidly by this method than in any other way, which accounts for the use of thread-rolling machines in connection with bolt and screw manufacture and wherever thousands of duplicate threaded parts are required.

Precision Thread Rolling.—Both flat and cylindrical dies are used in aeronautical and other plants for precision work. The blank sizing may be by centerless grinding or by means of a die in conjunction with the heading operations. The blank should be round, and, as a general rule, the diameter tolerance should not exceed ½ to ⅔ the pitch diameter tolerance. The blank diameter should range from the correct size (which is close to the pitch diameter, but should be determined by actual trial), down to the allowable minimum, the tolerance being minus to insure a correct pitch diameter, even though the major diameter may vary slightly. Precision thread rolling has become an important method of threading alloy steel studs and other threaded parts, especially in aeronautical work where precision and high-fatigue resistance are required. Extensive experiments conducted at one of the leading universities showed that the average rolled thread tested had an elastic limit 13 per cent higher than the elastic limit of a cut thread of corresponding size and material.

What production rates are obtained when threads are formed by rolling?

Production rates in thread rolling depend upon the type of machine, the size of both machine and work, and whether the parts to be threaded are inserted by hand or automatically. A reciprocating flat die type of machine, applied to ordinary steels, may thread 30 or 40 parts per minute in diameters ranging from about ⅝ to 1⅛ inch, and 150 to 175 per minute in machine screw sizes from No. 10 (.190) to No. 6 (.138). In the case of heat-treated alloy steels in the usual hardness range of 26 to 32 Rockwell C, the production may be 30 or 40 per minute or less. With a cylindrical die type of machine, which is designed primarily for precision work and hard metals, 10 to 30 parts per minute are common production rates, the amount depending upon the hardness of material and allowable rate of die penetration per work revolution. These production rates are intended as a general guide only. The diameters of rolled threads usually range from the smallest machine screw sizes up to 1 or 1½ inches, depending upon the type and size of machine.

What steels are used for thread rolling?

Steels vary from soft low-carbon types for ordinary screws and bolts, to nickel, nickel-chromium and molybdenum steels for aircraft studs, bolts, etc., or for any work requiring exceptional strength and fatigue resistance. Typical SAE alloy steels are No. 2330, 3135, 3140, 4027, 4042 and 4640. The hardness of these steels after heat-treatment usually ranges from 26 to 32 Rockwell C, with tensile strengths varying from 130,000 to 150,000 pounds per square inch. While harder materials might be rolled, grinding is more practicable when the hardness exceeds 40 Rockwell C. Thread rolling is applicable not only to a wide range of steels but for non-ferrous materials, especially if there is difficulty in cutting due to "tearing" the threads.

Why are some screw threads rolled on parts turned in automatic screw machines?

Thread rolling may be done in automatic screw machines or turret lathes when a thread is required behind a shoulder where it would be impossible to cut it with a die. In this way, a second operation on the work is avoided. A circular threading roll or disk is used. This roll has a thread on its edge or periphery of the same shape and pitch as the thread to be rolled, and this roll is forced against the screw blank while the latter is revolving. Fig. 2 illustrates how a circu-

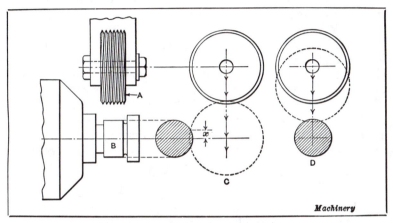

Fig. 2. Methods of Rolling Screw Threads on Automatic Screw Machines

lar roll may be applied. The unthreaded blank is held in the machine chuck and revolves about its axis. In screw machine practice, the threading roll A is usually mounted in a holder attached to the cross-slide of the machine. The roll is mounted on a pin or bolt so that it is free to revolve when brought into contact with the rotating piece to be threaded. A thread is formed (in this case on surface B) when the roll is forced against the work with sufficient pressure to leave an impression of its thread.

The roll may be presented to the work in either a tangential direction, as illustrated at C, or radially, as shown at D. When the roll is applied as shown at C, it gradually comes into contact with the periphery of the work and completes the thread as it passes across the blank surface in the direction indicated by the arrow. When the roll is presented in this way, the threaded piece should be cut off from the bar of stock before the roll is returned to its starting position, to prevent injury to the finished thread. This is done by simply mounting a cutting-off tool on the cross-slide in such a position that it severs the finished piece after the thread is rolled. The cam operating the cross-slide is arranged to quickly move the roll inward until it is nearly in contact with the surface of the blank; the cross-slide and roll are then given a slow feeding movement usually varying from about 0.002 to 0.004 inch per revolution of the work, while the roll is moving a distance x from the point of contact to the central position. The roll is then moved rapidly past the work in order to bring the cutting-off tool into position.

When the roll is in a radial position, as illustrated at D, it is simply forced against one side of the work until a complete thread is formed, the roll and work rotating together the same as two gears in mesh. Some roll-holders, instead of being held rigidly, are carried by a swinging arm which receives its motion from some form of cam. The exact method of applying the roll to the work and of holding it in position may depend, to some extent, upon the relation between the thread-rolling operation and other machining operations.

Should the diameter of the threading roll equal the screw thread diameter?

The diameter of the threading roll may be about the same as the diameter of the screw thread it is intended to roll, or some multiple of the screw thread diameter. If the thread to be rolled is larger in diameter than, say, ¾ inch, a roll may be used which is approximately equal to the diameter of the work, but if the required screw thread is much less than ¾ inch in diameter, the roll may be made twice as large, or some multiple of the screw diameter. When the roll and the screw thread are of about the same diameter, they are practically alike so far as the pitch and diameter of the thread are concerned. For instance, the roll for a ¾-inch screw thread might also be ¾ inch, minus a slight amount, in order to obtain a better rolling action. If the screw thread to be rolled were ½ inch in diameter, the roll might be two or three times as large.

Whatever the relation between the size of the screw thread and the roll, it is essential to use a roll having threads that mesh with or follow the threads on the work as the two revolve together. This means that the roll thread

must have the same pitch as the screw thread and about
the same helix angle. When the threads on the roll and
work are practically duplicates as to diameter, the roll for
a ¾-inch single screw thread would also have a single
thread, as shown at *A*, Fig 3. If the roll is twice as large
as the screw, it cannot have a single thread, because, for

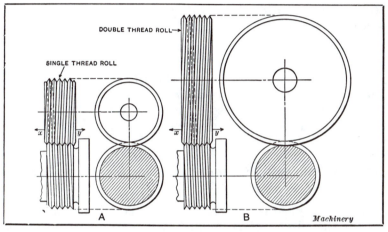

Fig. 3. (A) Threading Roll and Screw Thread of Approximately the
Same Diameter. (B) Threading Roll Nearly Twice as
Large in Diameter as Screw Thread

a given pitch, any increase of the roll diameter reduces the
angle between the thread and a plane perpendicular to the
axis; therefore, if the diameter of the roll is made twice as
large as the screw thread, it is necessary to use a roll having a double thread, as illustrated at *B*, instead of a single
thread. The pitch of the double thread remains the same,
but since the lead equals twice the pitch, the helix angle
of the thread is the same as for a single thread of one-half
the diameter. If a roll three times the size of the screw
thread were used, it would require a triple thread.

How is the diameter of the threading roll determined?

The following method of determining the pitch diameter
of the roll has been used successfully by a prominent manufacturer of automatic screw machines.

Rule.—Determine the pitch diameter of the thread to be rolled and deduct from it one-sixth of the double depth of the thread; then multiply by the number of threads or "starts" on the roll.

Ordinarily, the roll would be about two, three, or four times the diameter of the screw thread and have either a double, triple, or quadruple thread, in order to secure the proper thread helix angle. Expressing the preceding rule as a formula:

$$P = M \left(D - \frac{T}{2} \right) - T$$

in which P = pitch diameter of thread roll; M = multiple selected with reference to approximate diameter of roll desired (This number also equals the number of threads on the roll.); D = pitch diameter of screw thread; T = single depth of thread.

Why is the roll diameter not an exact multiple of the screw diameter?

If it is assumed that a roll is used having the same diameter as the thread after rolling, it will rotate at a slower rate than the work at the beginning of the rolling operation, because the roll is larger than the unthreaded screw blank and is driven by the blank. As the roll sinks into the blank, its speed gradually increases until, under ideal conditions, the speed is practically the same as that of the thread being rolled. It has been found in practice that better results are obtained when a roll is used that is slightly smaller in diameter than the diameter of the screw thread to be rolled. If the roll has a multiple thread, the same principle applies.

The reason for decreasing the diameter of the roll will be more apparent by considering the action of the roll and screw thread when a thread is being formed. As previously mentioned, the roll rotates at a slower speed than the screw blank, especially when the rolling operation begins, because the outside diameter of the roll is in contact with the plain blank which is about as large as the pitch diameter of the screw. As the rolling of the thread progresses, the roll speed might increase until it was practically the same as the work speed, but on account of frictional resistance between the roll and its holder, the natural tendency is for the roll to lag behind the screw thread. Now any retardation of the roll causes an increase of frictional resistance and the final result may be that the roll speed decreases to such an extent that it no longer meshes with the screw thread properly and, consequently, the thread is marred or spoiled.

The action between the roll and the screw thread will be more apparent by referring to Fig. 3. When the roll lags behind, the more rapidly revolving work, acting as a screw, forces the roll over against the side of the holder, as indicated by arrow x. The resulting increase of frictional resistance causes the roll to revolve still slower. On the contrary, if the roll is a little smaller in diameter than the work, so that it rotates faster, the roll tends to move in the opposite direction, as indicated by the arrow y. If this movement occurs and the roll presses against the side of the holder, thus increasing the frictional resistance, this tends to reduce the speed and cause the roll and work to rotate together properly; consequently, with the roll running a little faster than the screw, any difficulty from side thrust is cared for automatically, because then the driven roll tends to move axially in a direction opposite to that in which the screw thread on the work would force it.

Should the thread on the roll be an exact duplicate of the thread to be formed by rolling?

The thread (or threads) on the roll should be left-hand for rolling a right-hand thread, and *vice versa,* so that the thread on the roll and work will incline in the same direction at the point of contact. The thread on the roll should be sharp on the top for rolling American Standard threads. If the roll thread is made flat, the blunt edge will require too much pressure to force it into the metal and will not produce as smooth a thread. While a roll made this way will not form a correct American Standard thread, the fact that the thread has a fairly sharp bottom, instead of the standard flat form, may be of little importance as compared with the advantages of the rolling process. The bottom of the thread groove on the roll may also be left sharp or it may have a flat. If the thread groove is sharp in the bottom, the roll is only sunk far enough into the blank to form a thread having a flat top.

Planing Process and its General Application

The planing process is intended primarily for producing flat or plane surfaces. Such surfaces are required on a great many machine parts. Many of them, however, are machined by milling or by grinding. Flat surfaces on circular parts, such as the flat end of a shaft or the flat face of a cylinder flange, ordinarily are finished by turning. The particular method employed in any case depends upon various conditions. For example, the size and shape of the part, the number to be machined, the relation of the flat surface to other machined surfaces, and frequently the equipment available, are some of the factors influencing the selection of a method or type of machine. In brief, when two or more types of machines might be used for a certain class of work, usually the selection is based upon the relative machining costs, although accuracy may be, and often is, a deciding factor.

What is the principle of the planing process?

The planing process is illustrated by the diagrams, Fig. 1. The work or part to be planed is fastened securely to the planer table (or in a chuck or fixture attached to the table). This table has a reciprocating movement and the length of the stroke (which is adjustable) is a little longer than the length of the surface to be planed. During the cutting stroke, the table and work move in the direction of arrow *A* and the cutting edge of the tool removes a certain amount of metal. This tool remains stationary during the cutting stroke. When the work has traversed far enough to clear the tool, the table reverses or moves in the direction of arrow *B* preparatory to taking the next cut. This is known as the return stroke. Before the next cutting stroke begins, the tool is given a lateral feeding movement F (either by hand, or automatically by the feeding mechanism of the planer). When this intermitting feeding movement of the tool is parallel to the planer table, a horizontal surface will be planed. Vertical and angular flat surfaces may also be planed, as explained later. When parts are planed on a shaper, the principle is the same as described, but the tool is traversed and the work remains stationary excepting when the feeding movement occurs at the end of each stroke.

How much metal is removed in taking roughing cuts?

The amount of metal removed during each cutting stroke depends upon the depth D, Fig. 1, of the cut and the amount

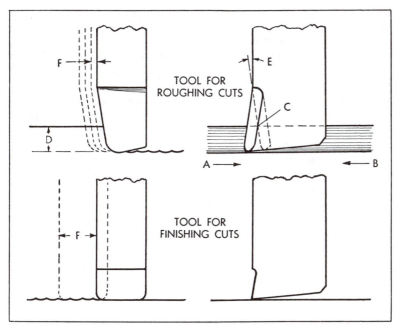

Fig. 1. Diagrams Illustrating Principle of Planing Process. Upper Diagram Shows a Roughing Cut and Lower Diagram a Finishing Cut in Cast Iron

of feed F per stroke. In planing, as in turning, there are two classes of cuts—namely, roughing and finishing. Since the object of the roughing cut is to remove surplus metal as quickly as possible, the depth of a roughing cut may be governed entirely by the amount of surplus metal. In some cases, for example, it may be necessary to remove an average of about ¼ inch to ½ inch from a casting, whereas, in other cases, there may be ¾ inch or more, especially if the casting is very large. As a general rule only one roughing cut is necessary. In taking roughing cuts, since the object is to remove surplus metal, the roughness of the planed surface is immaterial because it is finished later. This roughness will vary more or less according to the amount of feed per cut and the shape of the tool "nose" or end.

Roughing tools for planing vary more or less in shape. The face or surface against which the chips bear while being severed, has side rake and also back rake usually. This back rake, however, may be in the opposite direction to that shown, as indicated by the dotted lines and the angle E (upper right-hand diagram, Fig. 1). A tool ground, as indicated by these dotted lines, is said to have *negative back rake*, whereas the full lines represent *positive back rake*. The positive back rake results in a better cutting action at the point and is especially desirable when a tool must be fed downward into the metal. Roughing cuts, however, usually are started at one side or edge of a surface, the tool entering from the side after it has been adjusted downward to remove a required amount of metal or as much as can be taken in one cut. In such cases, the negative back rake indicated by the dotted lines, may be preferable, especially if the cut is deep and there is considerable shock when the cutting edge first strikes the work. The cutting edge of a tool having negative back rake first strikes the upper corner of the work as at C and enters to the full depth gradually because of its inclination; consequently, there is less shock at the beginning of the cut than occurs when using a tool without rake or even one with positive rake, because more of the cutting edge comes into action right at the beginning of the cut. The negative back rake angle E usually is about 8 degrees, and cemented-carbide tools for planers and shapers often are formed in this way.

Why are coarse feeds used for taking finishing cuts on a planer?

In taking finishing cuts, especially on cast-iron parts, the general practice is to use a broad tool having a straight cutting edge as indicated by the lower diagram, Fig. 1. This tool is merely used to remove the ridges left in rough planing with a round nose tool; consequently, the cuts are comparatively light and the tool can be given a large feeding movement F as indicated by the dotted lines. If the cutting edge is parallel with the direction of the feeding movement, the flat surface formed is smooth and accurate enough for most classes of work. The feeding movement must be somewhat less than the width of the flat cutting edge to provide some overlap and prevent the formation of ridges. This coarse finishing feed in conjunction with a broad flat-edged tool is used frequently in both turning and boring operations as well as for planing.

What are the main features of a planer?

There are several different types of planers which have been designed for planing certain classes of work to the best advantage. These various types, however, have certain main features which are similar, and the fundamental operating principle is the same in all cases. In general, a planer has a horizontal bed (see Fig. 2) upon which the platen or work-holding table slides with a reciprocating motion. This table, throughout its length, contains T-slots and holes for receiving bolts, stop-pins, etc. These are used for clamping either the work or a work-holding chuck or fixture. Adjustable dogs or stops at the sides of the platen serve to regulate the length of the stroke.

A horizontal cross-rail above the platen supports one or two tool-heads. The tool-holding slide forming part of the head is adjusted vertically by hand to set the tool for whatever depth of cut is required. The entire tool-head may be moved horizontally along the cross-rail. This lateral movement may be by hand for making adjustments; but when the planer is operating, the automatic feeding mechanism usually is employed. This mechanism provides means of adjusting the amount of feed. The range of feeding movements may vary, for example, from 1/32 inch up to 1 inch, in order to provide for various planing operations.

The cross-rail is adjustable on its supporting housing. This cross-rail must be high enough to clear the work, but too much clearance is undesirable because it reduces the rigidity of the tool support. When a planer has two tool-heads on the cross-rail, it is possible on many jobs to use them both simultaneously. Many planers, especially the larger sizes, also have supplementary tool-heads mounted on the sides of the housing. These heads are used for planing vertical surfaces and frequently are employed while the cross-rail heads are planing either horizontal or possibly other vertical or angular surfaces.

The particular planer shown in Fig. 2 is belt driven (the belts are not shown on the pulleys). Many planers are driven by motors.

How is the size of a planer designated?

The size of a planer is equivalent to the width and height of the largest part that will pass between the housings and under the cross-rail, when the latter is raised to its highest position. For instance, a 36- by 36-inch planer is one that will plane work 36 inches wide and 36 inches high. Sometimes the maximum length that can be planed is included when designating the planer size. Thus a 36-inch by 36-inch by 8-foot planer means that a piece 36 inches square will pass between the housings, and that a length of 8 feet can be planed.

In clamping work for planing, what are the essential principles?

Many planing operations are quite simple as far as the actual planing is concerned, but often considerable skill and ingenuity are required in setting parts on a planer and clamping them in the best manner. There are three important points that should be considered: First, the casting or forging must be held securely to prevent its being shifted by the thrust of the cut; second, the work should not be sprung out of shape by the clamping pressure; and third, the work must be held in such a position that it will be possible to finish the surfaces that require planing, in the

right relation with one another. Frequently a little planning before the "setting-up" operation will avoid considerable worry afterwards, to say nothing of spoiled work.
Much of the work done on a planer is clamped directly to
the platen, although in many cases planer chucks or special
fixtures are used.

How are errors caused and prevented in clamping parts to be planed?

When castings or forgings are set up on the planer for
taking the first cut, usually the side that is clamped against

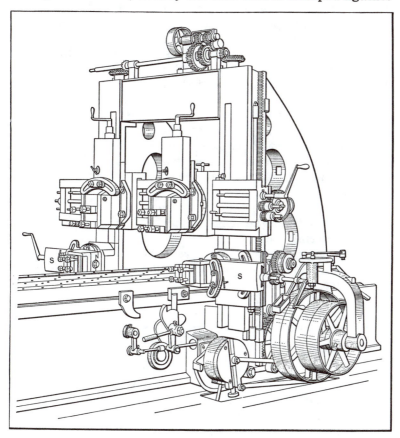

Fig. 2. Planer Equipped with Two Tool-heads on the
Cross-rail and Two Side-heads

the platen is rough and uneven, so that it bears on a few high spots. Assume that a casting does not touch the platen at the ends where the clamping is to be done. If the clamps are tightened without supporting the work at the unsupported ends, the entire casting is likely to be sprung out of shape more or less, depending on its rigidity; consequently the planed surface will not be true after the clamps are released, because the casting then resumes its natural shape. To prevent inaccurate work from this cause, there should always be a good bearing just beneath the clamps, which can be obtained by inserting pieces of sheet metal, or even paper when the unevenness is slight. Thin copper or iron wedges are also used for "packing" under the clamps. It is good practice when accuracy is required and the work is not very rigid to release the clamps slightly before taking the finishing cut. This allows the part to spring back to its normal shape and the finished surface remains true after the clamps are removed.

Very long castings or those which are rather frail but quite large and heavy sometimes bend by their own weight or are sprung out of shape by the pressure of the planing tool, unless supported at the weak points. When setting such castings on the planer, small jacks form a very convenient means of support. One type has a ball joint at the top which allows the end to bear evenly on the work, and the screw can be locked after adjustment to prevent it from jarring loose. These jacks, which are made in different heights, can be used in various ways for supporting work being planed. Hard-wood blocks cut to the right length are also used as supports.

Why are some castings distorted by internal stresses after planing?

Castings, even if properly clamped, are sometimes sprung out of shape by the internal stresses existing in the casting itself. These stresses are caused by the unequal cooling of the casting in the foundry. When a casting is made, the molten metal which comes in contact with the walls of the mold naturally cools first and, in cooling, contracts and becomes solid while the interior is still more or less molten.

The result is that, when the interior cools and contracts, the tendency is to distort the part which solidified first, and internal stresses are left in the casting. These stresses often act in opposite directions, and, when a roughing cut is taken from one side of such a casting, thus relieving the stress on that side, a slight distortion takes place. Suppose a casting is clamped so as to avoid all spring, and then a roughing cut is taken over one side, thus removing the hard outer surface. This might result in a slight change in shape. If a roughing cut were then taken from the opposite side, another change would probably occur because this would relieve the tension or stress of that side. The work would then assume what might be called its natural shape, and if both sides were then finished, they would tend to remain true, although slight changes might occur even then. Because of this tendency to distortion as the result of internal stresses, all precision work, especially if not rigid, should be rough-planed before any finishing cuts are taken. Such a change of shape does not always occur, because the stresses may be comparatively slight and the planed surface so small in proportion to the size of the casting that distortion is impossible. When great accuracy is required, it is the practice in some shops to rough plane the casting and then allow it to "season" for several weeks before taking the finishing cuts. This seasoning period is to allow the stresses to become adjusted so that little or no change will occur after all the surfaces are finished.

Why are work-holding fixtures often used on planers?

Some castings or forgings are so shaped that a great deal of time would be required for clamping them with ordinary means and, for such work, special fixtures are often used. These fixtures are designed to support the casting in the right position for planing and they often have clamps for holding it in place. Some work which could be clamped to the platen in the usual way is held in a fixture, because much less time is required for setting it up. Fixtures are especially useful when a large number of pieces have to be planed.

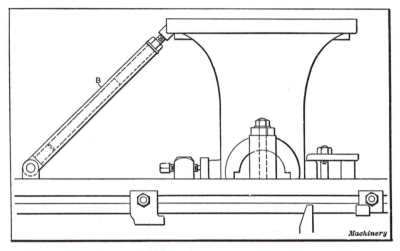

Fig. 3. High Casting Supported by Brace B which Takes Thrust
of Cut while Top Surface is Being Planed

Why are stop-pins and braces used
in conjunction with clamps and bolts?

Many parts which are planed cannot be held securely by
using only clamps and bolts, because the pressure of the
clamp is in a vertical direction, whereas the thrust of the
cut is in a horizontal direction, which tends to shift the
work along the platen. To prevent such a movement, practically all work that is clamped to the platen is further
secured by one or more stop-pins. These stop-pins ordinarily are placed at one end of the work to take the thrust
of the cut, but sometimes they also are needed along the
sides to prevent lateral movement.

Some castings have surfaces to be planed that are a considerable distance above the platen, as shown in Fig. 3,
which illustrates a large pillow-block set up for planing the
base. As will be seen, the end resting on the platen is comparatively small, and if the casting were simply clamped
at the lower end, it would tend to topple over when being
planed, because the thrust of the tool is so far above the
point of support. To prevent any such movement, braces *B*
are used. These braces serve practically the same purpose
as stop-pins. The style of brace shown has a hinged pin

in its lower end, which enters a hole in the platen, and the body of the brace is a piece of heavy pipe. At the upper end there is an adjustable fork-shaped piece which engages the work, and the hinged joint at the lower end enables the brace to be placed at any angle. In some shops, wooden blocks are used as braces. The arrangement of these braces and the number employed for any given case depend upon the shape and size of the casting, and this also applies to the use of stop-pins and clamps. The location of all braces and clamping appliances should be determined by considering the thrusts to which the part will be subjected during the planing operation.

When are angle-plates used for holding planer work?

Angle-plates may be used for holding parts while planing them, especially if a previously planed surface must be held in a vertical position while planing some other surface. The use of an angle-plate for holding an odd-shaped casting is shown in Fig. 4. The angle-plate A has two faces a and b which are square with each other, and the work W is bolted or clamped to the vertical face, as shown. The arrangement of the clamps or bolts depends, of course, on the shape of the work. The particular part illustrated,

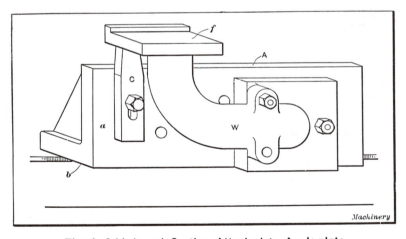

Fig. 4. Odd-shaped Casting Attached to Angle-plate

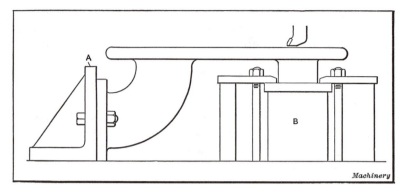

Fig. 5. Use of Angle-plate in Conjunction with Clamps for Holding Work

which is to be planed at f, is held by bolts inserted through previously drilled holes, and the left end is supported by a clamp C, set against the under side to act as a brace and take the downward thrust of the cut. Angle-plates are generally used for holding pieces, which, because of their odd shape, cannot very well be clamped directly to the platen.

Occasionally an angle-plate can be used in conjunction with clamps for holding castings, as illustrated in Fig. 5. In this example the angle-plate A is placed across the platen and serves as a stop for taking the thrust of the cut; it also holds the finished flange in a vertical position. The flange on the opposite end is supported by a block B against which the casting is clamped.

In holding cylindrical parts for planing, what equipment is used?

The planer is sometimes used for cutting keyways or splines in shafts, and, occasionally, other cylindrical work requires a planing operation. In order to hold and, at the same time, align round work with the platen, V-blocks (Fig. 6) are used. These blocks have a tongue t at the bottom which fits into a T-slot in the platen, and the upper part of the block is V-shaped as shown in the end view. This angular groove is parallel with the tongue so that it holds a round shaft in alignment with the travel of the platen. The diameter of a shaft held in one of these blocks can vary considerably, as indicated by the two circles, without affecting the alignment. In other words, the centers c and c_1 of the large and small circles, respectively, coincide with the vertical center-line.

Fig. 7 illustrates how V-blocks are used for holding a piston-rod while the cross-head, which is mounted on the end, is being planed. The bearing surfaces of the cross-head must be in line with the rod which fits a tapering hole in one end of the cross-head. By assembling the cross-head and rod and then mounting the latter in V-blocks, the bearing surfaces are planed in alignment with the rod. A work-supporting planer jack is inserted beneath the cross-head to take the downward thrust of the cut.

If duplicate V-blocks are to be planed, what method insures accuracy?

A good method of making a pair of accurate V-blocks is as follows: First plane the bottom of each block and form the tongue to fit the platen T-slots snugly. Then bolt both blocks in line on the platen and plane them at the same time so that they will be exact duplicates. A square slot or groove is first planed at the bottom of the vee, as shown in Fig. 6, to form a clearance space for the tool. The head is then set to an angle of 45 degrees and one side of the vee is rough-planed. The blocks are then reversed or turned

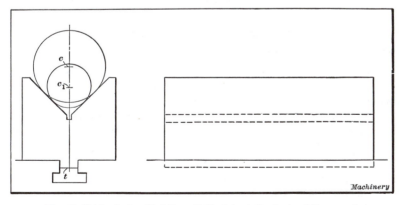

Fig. 6. V-block for Holding Cylindrical Parts in Alignment with Machine Table

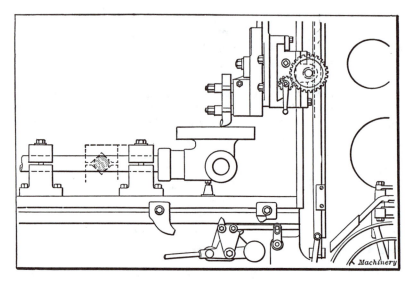

Fig. 7. Piston-rod and Attached Cross-head Mounted in V-blocks for
Planing Cross-head in Alignment with Piston-rod

"end for end" and the opposite side is rough-planed without disturbing the angular setting of the head. These operations are then repeated for the finishing cuts. This method
of reversing the work instead of setting the head to the
opposite angle insures equal angles for both sides and a vee
that is exactly central with the tongue on the bottom.

What types of tools are used for different planing operations?

The number and variety of cutting tools used on a planer
depend upon the character of the work which is done on
that particular machine. If the work varies considerably,
especially in its form, quite a number of tools of different
shapes will be needed, whereas, planers that are used principally for making duplicate parts do not need a large tool
equipment. Fig. 8 shows some typical tool forms and examples of the kind of planing for which the different tools
are adapted.

The tool shown at A is one form of roughing tool. This
form is particularly adapted for taking deep "roughing"

cuts in cast iron, when it is necessary to remove considerable metal. This style of tool is also made to the opposite hand, as at *B*, because it is sometimes desirable to feed the tool toward the operating side of the planer; ordinarily, however, horizontal surfaces are planed by feeding the tool *away* from the operator, the tool moving from right to left, as viewed from the front of the machine. This enables the workman to see just what depth of cut is being taken at the beginning of the cut.

The tool *C* with a broad cutting edge is used for taking finishing cuts in cast iron. Tools of this type are made in various widths, and when planing very large and rigid castings, wide cutting edges and coarse feeds are used. Wide finishing tools for cast iron are sometimes ground so that the cutting edge slopes back at an angle to give a shearing cut. A plain round-nose tool is shown at *D*. This style is often used for rough planing steel or wrought iron It can also be made into a finishing tool for the same metals by grinding the nose or tip end flat. The width of the flat cutting edge is much less, however, than for cast-iron finishing tools, because steel offers a greater resistance to cutting than cast iron. Tool *E*, which is known as a diamond point, is also used for rough-planing steel or iron and for taking light cuts.

The bent tools *F* and *G* are used for planing either vertical surfaces or those which are at a considerable angle with the platen. These are right- and left-side roughing tools, and they are adapted to either cast iron or steel. They can also be used for finishing steel. Finishing tools for vertical or angular cast-iron surfaces are shown at *H* and *I*. These have wide cutting edges to permit coarse finishing feeds. Vertical surfaces can often be planed to better advantage by using a straight tool in the side-head, when the planer is so equipped. Right and left angle tools are shown at *J* and *K*. This style of tool is for planing angular surfaces which, by reason of their relation to horizontal or other surfaces, can only be finished by a tool having a pointed form similar to that illustrated. A typical example of the kind of angular planing requiring the use of an angle tool is indicated in the illustration. After finishing side *a*, the horizontal surface *b* (from which a roughing cut should

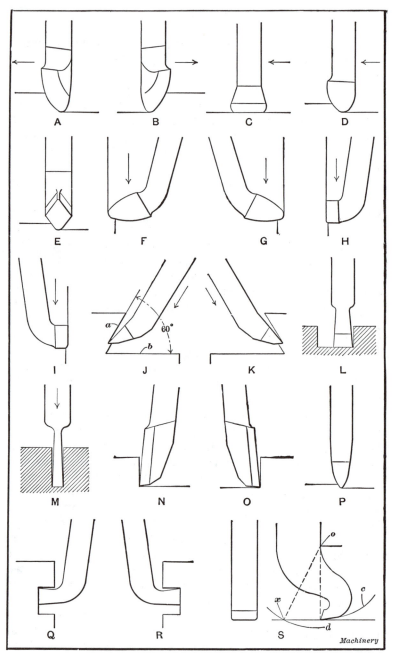

Fig. 8. Planing Tools of Different Shapes and Typical Application of Each Type

have been taken previously) could be planed by feeding the same tool horizontally.

A square-nose tool is shown at L. This is used for cutting slots and squaring corners, and the same style of tool is made in different widths. A narrow square nose or "parting" tool is shown at M. It is adapted to cutting narrow grooves, and can also be used for cutting a part in two, provided the depth does not exceed the length of the narrow cutting end. Right and left side-tools are shown at N and O. These can frequently be used to advantage on vertical or angular surfaces. A tool for planing brass is shown at P. It has a narrow rounded cutting edge and is very much like a brass turning tool. For finishing cuts in brass, tools having narrow flat ends are often used. Right and left, bent, square-nose tools are shown at Q and R. Such tools are used for cutting grooves or slots in vertical surfaces and for similar operations.

All of the tools shown are forged from a solid bar of steel, the cutting end being forged to about the right shape, after which the end is correctly formed by grinding. After the tool has been worn away considerably by repeated grindings, the end has to be reforged or "dressed" to bring it back to the original form. To eliminate this work, and also to reduce the amount of steel required, tools are often used on the planer and other machines, having shanks into which small cutters can be inserted. These tools are made in many different designs.

Why is the "goose-neck" type of tool used for some planing operations?

The peculiarly-shaped tool, shown by front and side views at S, Fig. 8, is especially adapted for finishing. This type is known as the "goose-neck" because of its shape, and it is intended to eliminate chattering and the tendency which a regular finishing tool has of gouging into the work, especially if the planer is not in good condition. By referring to the side view it will be seen that the cutting edge is on a line with the back of the tool shank, so that any backward spring of the tool while taking a cut would cause the cutting edge to move along an arc c or away from the work. When the cutting edge is in advance at some point x, as

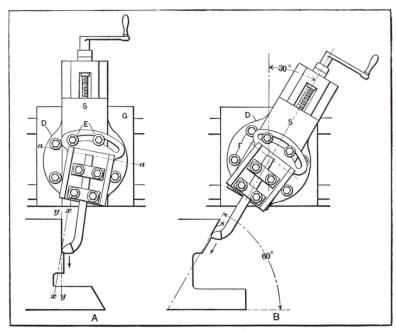

Fig. 9. Positions of Tool and Head for Planing Vertical
and Angular Surfaces

Fig. 10. Planing Two Vertical Surfaces Simultaneously
on a Double-head Planer

with a regular tool, it will move along an arc *d*, if the strain of the cut causes any springing action, and the cutting edge will gouge in below the finished surface. Ordinarily the tool and the parts of the planer which support it are rigid enough to prevent such a movement, so that the goose-neck tool is not always necessary.

What is the general procedure in planing vertical or angular surfaces?

When a vertical surface or one located at right angles to the platen is to be planed, the side-head can often be used to advantage. If the cross-rail tool head is used, the tool-block is set at an angle, as shown by the diagram at *A*, Fig. 9, by loosening bolts *E*, which permit it to be swiveled to the right or left from its vertical position. The tool-block is set over in this way to prevent the tool from dragging over the planed surface on the return stroke. The tool-block of a planer is free to swing forward upon a pin or pivot having an axis at *a-a*, so that the tool can lift slightly when returning for another cut. When a heavy cut is to be taken, the tool is sprung sidewise to some extent, as well as backward, and if it were held rigidly on the return stroke, the cutting edge would drag heavily over the work, and this would soon dull the edge. When a horizontal surface is being planed, the tool on its return tends to lift upward at right angles to the surface, because the tool-block is then set square with the platen. If, however, the tool-block were left in this position for vertical planing, the tool point would swing upward in a plane *y-y*, and drag over the finished surface, but by setting the block in an angular position, as shown, the tool-point swings in a plane *x-x*, or at right angles to the axis *a-a* of the pin on which the block swivels. As plane *x-x* is at an angle with the surface of the work, the tool-point moves away from the finished surface as soon as it swings upward. The angular position of the tool-block does not affect the direction of the movement of the tool, as this is governed by the position of slide *S* which is changed by swiveling the graduated base *D*.

A vertical surface is planed by adjusting the saddle *G* horizontally along the cross-rail until the tool is in position

for taking a cut. The tool is then fed down by hand, until
the cut is started, after which the vertical feed is engaged
(unless the surface is very narrow), thus causing slide S
and the tool to feed downward a certain amount for each
stroke, while the saddle remains stationary on the cross-
rail. The surface y-y will be planed square with the platen,
provided the swiveling base D is set in the proper position.
Before planing surfaces that are intended to be square with
the platen, the position of the tool-slide S should be noted
by referring to the graduation marks on the base D. When
the zero marks on the stationary and swiveling parts of
the base exactly coincide, the slide should be at right angles
to the platen. Its position, however, can be determined more
accurately by holding the blade of a square which rests on
the platen against one side of the tool-slide, as it is difficult to set graduation lines to exactly coincide, and even
if they were in line, errors might result from other causes.

The planing of an angular surface is illustrated at B,
Fig. 9. The tool-head is first set to the proper angle by
loosening bolts F and turning tne base D until the graduations show that it is moved the required number of degrees.
For example, if surface s were to be planed to an angle of
60 degrees with the base, as shown, the head should be set
over 30 degrees from the vertical or the difference between
90 and 60 degrees. The tool would then be fed downward,
as indicated by the arrow. The tool-block is also set at an
angle with slide S, for planing angular surfaces, so that
the tool will swing clear on the return stroke. The top of
the block should always be turned *away from the surface
to be planed*, to prevent the tool point from dragging during the return stroke. This rule applies to the planing of
either vertical or angular surfaces when using the cross-
rail head.

Fig. 10 shows how a double-head planer is used to plane
two vertical surfaces simultaneously. The tools in this case
are the inserted cutter type.

What is the general practice in planing with two or more tools simultaneously?

The use of both cross-rail tool-heads at the same time is
very common in connection with modern planer practice

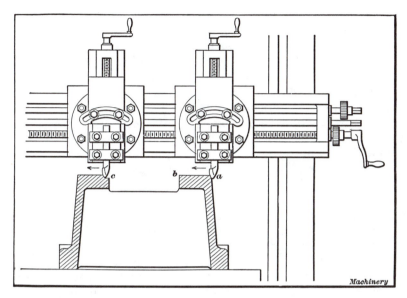

Fig. 11. Planing Two Horizontal Surfaces Simultaneously
with Two-head Planer

and on some jobs, two side-tools can also be used simultaneously. Whether it is feasible to use one tool or four at the same time, depends altogether on the shape of the work and the location and widths of the surfaces to be machined. Very often only one tool can be used, but in some cases the number may be increased to two, three or four, provided the planer is equipped with two or four heads. There are few fixed rules which can be applied generally to planer work, because the best way to set up and plane a certain part depends upon its shape, the relative location of the surfaces to be finished, the degree of accuracy necessary, and other things which vary for different kinds of work.

In using the two cross-rail tool-heads for a job like the one shown in Fig. 11, the tool at the left is started first, because it is the *leading* tool, as determined by the direction of the feed. This is a good rule to follow, especially when the tool-heads are quite close to each other on the cross-rail, as it prevents one head from feeding against the other, which might occur if the *following* tool were started first. The tools illustrated cut principally on the side and are intended for deep roughing cuts in cast iron. The surfaces should be finished with a broad tool with a wide feed. It is impossible to give any fixed rule for the amount of feed, as this is governed not only by the planer itself, but also by the rigidity of the work when set up for planing, the hardness of the metal, etc. The final cut should be taken by a single tool to insure finishing both sides to the same height. This tool should be fed by power from a to b, and then rapidly by hand (or by the rapid power traverse if the planer is so equipped) from b to c for finishing the opposite side. The use of two tools for rough-planing greatly reduces the time required for machining work of this kind.

If the planer is provided with side-heads, these have an automatic vertical feed and can often be used while the other tools are planing the top surface, the method being to start first the regular tools (which usually have the largest surfaces to plane) and then the side-heads. If the planing on the side requires hand manipulation, as in forming narrow grooves, etc., the planing would be done first on one side and then on the other, assuming that both sides required machining, but, when the surfaces are broad, the automatic feed enables both side-heads to be used at the same time, on some classes of work. These side-heads often greatly reduce the time required for planing and they also make it possible to finish some parts at one setting, thus saving time and insuring accuracy of alignment between all surfaces.

When is gang or multiple planing practicable?

When a number of duplicate parts have to be planed, much time can often be saved by arranging the castings in a straight row along the platen of the planer, so that they can all be planed at the same time. This method enables a number of parts to be finished more quickly than would be possible by machining them separately, and it also insures duplicate work. This method of planing cannot always be employed to advantage, as the shape of the work or location of the surfaces to be machined sometimes makes gang planing impracticable and even impossible. If the castings are so shaped that there will be considerable space between the surfaces to be planed, when they are placed in a row, so much time might be wasted while the tool was passing between the different surfaces that it would be better to plane each part separately. Some castings also have lugs or other projections which make it impossible for the tool to pass from one to the other without being raised to clear the obstruction. On the other hand, when castings are quite symmetrical in form and the surfaces are so located that the planing tool can pass from one to the other with a continuous stroke, the gang method of planing insures a uniform product and greatly reduces the time required for machining. Castings that are finished by gang planing are often held in special fixtures, especially if they are of irregular shape and not readily clamped directly to the planer table.

How are planing speeds regulated for planing different materials efficiently?

Many belt-driven planers can only be operated at one cutting speed, the driving pulleys being proportioned to give a speed that will be about right for average conditions. A change of two or more speeds, however, is very desirable, because the cutting speeds should be varied to suit the material to be planed, or the nature of the cut. Many modern planers have some form of variable-speed driving mechanism. There are several different methods of varying the planing speed. The speed-varying mechanisms which are applied to planers may be divided into four general classes: (1) Planer driving gears which can be shifted for varying the speed; (2) countershafts so arranged that two or more speeds can be obtained; (3) speed-changing mechanisms such as are applied to the top of planer housing; (4) variable-speed electric motor drives.

When planers are driven by electric motors, one method is to employ a non-reversing direct-current motor. This motor drives the planer through shifting belts which reverse the rotation for obtaining the forward and return strokes. Another method is to use a reversible type of motor to provide both the necessary speed changes and the reversal of rotation for the forward and return strokes.

In determining planer speeds what points are considered?

The cutting speed for planing usually varies from 30 to 60 feet per minute, whereas the return speed of the table is from two to four times as great, three times the cutting speed being a fair average. When a planer has a variable-speed drive, the cutting speed should not only be governed by the nature of the material to be planed, but also by other conditions. For instance, in planing a heavy cast-iron piece of considerable hardness, if there is plenty of stock to remove, the planer should run fairly slow on the cutting stroke and take heavy cuts. A tool will stand heavy cuts at a low speed better than when running at a comparatively high speed with light cuts; moreover, since the table and work travel a smaller total distance for the removal of a given quantity of stock, less power is required merely to move the table with its load. The return speed, in this case, should be moderate, because the rapid starting and moving of a heavy load subjects the planer driving mechanism to severe strains.

When the work is heavy, and there is comparatively little stock to remove, the cutting speed should be as high as the tool will stand, but the return speed should be kept within moderate limits for the same reason referred to in the preceding paragraphs. When the work is of light weight, but stocky, and considerable metal has to be removed, the cutting speed should be moderate, but the return speed may be high. When the work is light and there is little stock to be removed, both the cutting and return speeds may be comparatively high. In order to secure the best results, there are other factors which should be considered. For instance, it is of little benefit to have a high return speed on a short stroke, because this simply results in wearing the planer without accomplishing anything. In planing steel, a fairly high cutting speed may be used even if there is considerable metal to be removed, because most high-speed steels work better at a high speed on steel than on cast iron. When brass is to be planed, high cutting speeds may be used, but with some bronzes a low cutting speed is necessary in order to avoid tearing the work.

If the forward and return speeds are known, how is the net cutting speed calculated?

The net cutting speed of a planer is equal to the number of feet traversed by the tool in a given time while cutting or planing, and it is less than the speed of the table on the forward or cutting stroke, because of the idle or return period, when no work is being done.

Rule.—The net cutting speed equals the forward cutting speed divided by the total time required for the forward and return movements.

Example.—If the cutting speed is 40 feet per minute and the return speed 120 feet per minute, a forward movement, if continued for a distance of 40 feet, would require one minute and the return stroke one-third minute, or 1 1/3 minutes for forward and return strokes. Therefore the number of feet per minute traversed by the tool, while actually cutting, equals 40 ÷ 1 1/3 = 30 feet per minute or 10 feet per minute less than the nominal cutting speed.

In testing the alignment of a planer cross-rail, what simple methods are applicable?

The cross-rail of a planer which is in good condition is parallel with the upper surface of the platen, so that the planing tool, as it feeds horizontally, moves in a line parallel with this surface. If accurate work is to be done, especially on a planer that has been in use a long time, it is well to test the position of the cross-rail. One method of making this test is as follows: An ordinary micrometer is fastened to the tool-head in a vertical position either by clamping it to the butt end of a tool or in any convenient way, and the head is lowered until the end of the micrometer thimble is slightly above the platen. The thimble is then screwed down until the end just touches the surface to be tested, and its position is noted by referring to the regular graduations. The thimble is then screwed up slightly for clearance, and, after the tool-head is moved to the opposite side, it is again brought into contact with the platen. The second reading will show in thousandths of an inch any variations in the position of the cross-rail. The parallelism of the cross-rail can also be tested accurately by clamping an ordinary dial indicator to the tool-head, and noting the

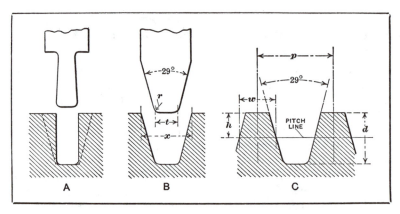

Fig. 12. Diagrams Showing Rack Planing Tools and
Formation of Rack Teeth

readings when the indicator is moved from one side of the
platen to the other. If the cross-rail is not parallel, the low
side should be raised by adjusting the elevating screw.

When a planer is used for rack cutting, how are the rack teeth formed?

Rack cutting generally is done by milling as explained at
the end of the section on cutting spur gears by milling. If
a planer is used, the usual method is to first rough out the
teeth by cutting plain rectangular grooves with a square-
nosed tool as indicated at A, Fig. 12. A form tool is then
used to finish the teeth to the required angle. The included
angle between the sides of the teeth for involute gears of
$14\frac{1}{2}$-degree pressure angle equals 29 degrees, as shown at
C. The proportions of a rack tooth are the same as for a
gear of corresponding pitch. The whole depth d of a full-
depth tooth equals the linear pitch p (or the circular pitch
of the meshing gear) multiplied by 0.6866. The width w at
the pitch line equals one-half the linear pitch. The distance
h from the top of the tooth to the pitch line (which would
be required when setting a vernier gear-tooth caliper for
measuring width w) equals the linear pitch multiplied by
0.3183. The width of the point or end of the rack tool (before the corners are rounded) equals the linear pitch multi-

plied by 0.31. The radius r of the corners of the tool equals 0.066 multiplied by the pitch p.

How is the rack planing tool indexed for spacing the rack teeth uniformly?

When planing the teeth, the tool is indexed a distance equal to the linear pitch p, Fig. 12, either by means of a micrometer dial on the cross-feed screw or by the use of a positive locating device attached to the cross-rail. If the planer has a micrometer dial for adjusting the tool by direct measurement, care should be taken to avoid errors due to lost motion between the screw and feed-nut, by moving the tool continuously in one direction when making an adjustment. In using the micrometer dial, it is well to set it to zero before making each adjustment. The distance that the tool is indexed can then be read direct from the zero mark, which tends to prevent mistakes that might otherwise be made.

To avoid the possibility of error when measuring the indexing movement of the tool by using the cross-feed screw, a special indexing bar is sometimes attached to the cross-rail, especially when a number of duplicate racks are to be planed. This bar has notches or teeth of the same pitch as the rack teeth to be cut, and the planer tool-head is located positively by a pin or plunger on the saddle which engages the notches in the fixed indexing bar.

Another very simple but positive method of indexing the tool is to clamp a stop on the cross-rail in such a position that when the saddle is against the stop the tool will be set for planing the first tooth space; when the first space is cut, the stop is set ahead of the saddle a distance equal to the pitch of the rack, by using a gage. The saddle is then moved up to this stop for planing the next tooth space, and this operation of adjusting the stop by the gage and locating the saddle against it is repeated for each successive tooth.

How is a planer equipped for machining circular segments?

In some cases, it is either necessary or desirable to use a planer for circular work. Various types of attachments are used to guide the planing tool along a circular path. A comparatively inexpensive radius planing device that requires very few parts is shown in Fig. 13 (upper diagram). The upper ends of the two tool-slides are coupled together by a link or bar. This bar has holes at each end which fit pins set in the tool-slides. The clamping bolts of the swivel slide at the left are loosened just enough to permit the slide to be rotated. Thus, when the tool-slide to the right is fed along the cross-slide as indicated by arrow F, the tool in

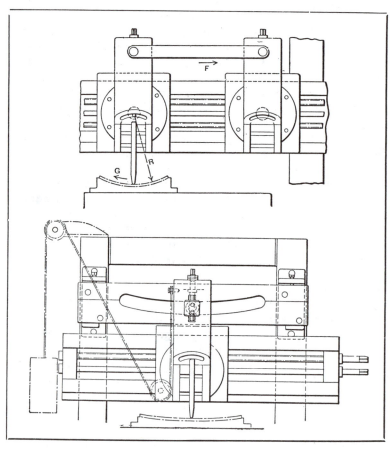

Fig. 13. (Upper Diagram) Simple Arrangement of Planer for Radius Work. (Lower Diagram) Planer with Former Plate Attachment for Machining Curved Surfaces of Large Radii or Irregular Surfaces

the swiveling slide to the left rotates in a clockwise direction G, so that it cuts along an arc of radius R as indicated by the dot-and-dash lines.

For planing curved surfaces of large radii or irregular surfaces, a former plate attachment (see lower diagram, Fig. 13) may be used. The same former plate can be used for planing both concave and convex parts by reversing its position. The roller which travels in the curved slot of the former plate is carried on a pin fixed to the tool-slide. The regular down-feed screw is removed to allow the tool to follow the curvature of the slot in the former plate when it is traversed across the work. The former plate may be split at the ends of the slot to provide for taking up wear in the slot and to facilitate machining. A weight may be employed, as shown, to keep the roller in contact with the lower surface of the slot in the former plate.

How does a radius-bar type of circular planing attachment operate?

There are different methods of using a pivoted bar for controlling the movement either of the tool or work in planing to a given radius. A concave planing attachment of the radius-bar type is shown in Fig. 14. The radius-bar, which can have holes for various radii, is secured to the tool-slide by a pin at its lower end. The upper end is pivoted on a combined nut and pin carried in a slide which is supported by two cross-beams bolted to the uprights of the machine. Vertical feed can be applied by means of a feed-screw which moves the combined nut and pin that is connected to the upper end of the radius-bar. When the feed is not being used, the nut can be locked in position. With this attachment, the vertical screw in the regular feed-slide is removed. Radius R is always equal to the center distance between the holes used in the radius-bar. The tool is traversed in the usual way.

The diagram, Fig. 15, illustrates an attachment employing a special table mounted on the main table of the planer. The auxiliary table is carried on an adapter plate which, with its hold-down plates, is bolted to the main table. The table is arranged to slide on the adapter plate at right angles to the main table and it is pivoted to a radius-bar.

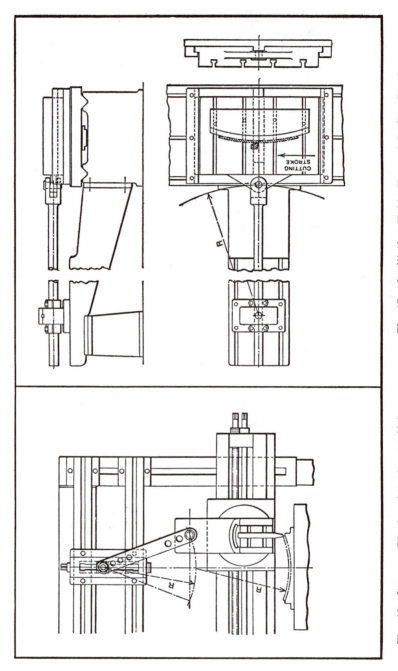

Fig. 15. Oscillating Table Type of Radius Planing Attachment

Fig. 14. Concave Planing Attachment Using Radius-bar

The outer end of the radius-bar is gripped in a pivot block supported by a housing. This housing is carried on an arm bolted to the planer bed at one end and supported from the floor by a column at the other end. The arm has T-slots in its top face for setting the pivot-block housing to the required radius.

The radius machined is always equal to the center distance R between the swivel pin and the pivot pin of the radius-bar. In operation, the main table of the planer is run in the usual way, the radius-bar causing the auxiliary table to oscillate across the main table and so reproduce the radius R on the work shown in dotted cross-section. A pointer that indicates when the radius-bar is exactly at right angles to the table facilitates setting up the work. One disadvantage of this attachment, particularly on work of small radius, is the necessity for a large side clearance on the cutting tool.

It should be noted that the work does not move radially past the tool, but always at right angles to the longitudinal movement of the planer bed. It is at the end of this cutting stroke that the extra side clearance is required. At the beginning of the stroke the side clearance is excessive. This attachment is easily applied to open-side planers. On the planer with two uprights, care must be taken to see that the swinging radius-bar does not interfere with the adjacent upright. The center line through the pivot bearing of the radius-bar at right angles to the machine table, should be in line with the cutting edge of the tool. The arm that supports the radius-bar must, therefore, be located the proper distance from the uprights. An extension head can be fitted to the tool-slide, in order to permit the attachment to be brought farther away from the planer uprights.

What principles govern the grinding of planing tools?

While the action of a planer is entirely different from that of a lathe, many of the principles which govern the shape of turning tools also apply in the grinding of tools for planing. Front and side views of one form of planer roughing tool are shown at A, Fig. 16. As the cutting is done by the curved edge e, the front surface b is ground

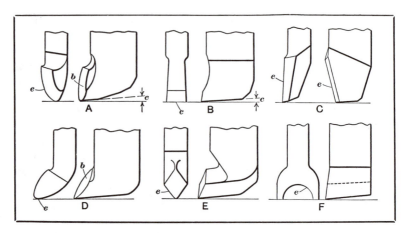

Fig. 16. Different Types of Planing Tools

to slope backward from this edge to give the tool keenness, the slope being away from the *working part* of the cutting edge. The end or flank of the tool is also ground to slope inwards to provide clearance. The angle of clearance c is about 4 or 5 degrees for planer tools, which is much less than that for lathe tools. This small clearance is allowable because a planer tool is held about square with the platen, whereas a lathe tool, the height of which may be varied, is not always clamped in the same position. A lathe tool also requires more clearance because it has a continuous feeding movement, whereas a planer tool is stationary during the cut, the feed taking place just before the cut begins. This point should be considered when grinding planer tools, because the clearance of any tools should not be greater than is necessary to permit the tool to cut freely, as excessive clearance weakens a tool. The slope of the top surface b depends upon the hardness of the metal to be planed, the slope angle being less for hard material, to make the cutting edge more blunt and stronger. When tools are ground by hand, the angles of slope and clearance are not ordinarily measured, the workman being guided by experience.

The edge of a broad cast-iron finishing tool should be ground straight by testing it with a small straight-edge or scale. The corners should be rounded slightly, because a square corner on the leading side will dull quickly. The square-nose tool *B* cuts along its lower edge *e*, and is given clearance *c* on the end and sides, as shown in the two views. The lower edge is the widest part of the cutting end, the sides sloping inward in both a vertical and horizontal direction, which prevents the tool from binding as it moves through a narrow slot.

The side-tool *C* cuts along edge *e*, which, as the side view shows, slopes backward. Planer side-tools are not always made in this way, but it is a good form, as the sloping edge starts a cut gradually, whereas a vertical edge takes the full width of the cut suddenly, thus producing a greater shock. As tool *D* is used for vertical planing, the cutting is done by edge *e*; hence, face *b* should slope back from edge *e*. The diamond point *E* is ground with a narrow rounded point; this type of tool is useful when a light cut is necessary, either because the work cannot be held securely for planing or for some other reason. The form tool *F* is used for rounding edges. Tools of this type are sharpened by grinding on the front face only, in order to retain the curved edge. When sharpening other than formed tools, the grinding is done both on the face and end, because a sharp edge can be secured more quickly by this method.

Reference has been made to the grinding of these few types of tools merely to point out some of the principles connected with the grinding of planing tools. When the principle of tool grinding is understood, the various tools required, whether regular or special in form, can be ground without difficulty.

Why are shapers used for planing operations?

The shaper, like the planer, is used principally for producing flat or plane surfaces, but it is intended for smaller work than is ordinarily done on a planer. The shaper is preferable to the planer for work within its capacity because it is less cumbersome to handle and quicker in its movement. The action of a standard shaper, when in use, is quite different from the planer; in fact, its operation is just the reverse, as the tool moves across the work, which remains stationary, except for a slight feeding movement for each stroke.

There are several different types of shapers, but the design generally used consists of the following main features: A column or bed supports a ram or tool-holding slide and a work-holding table (see Fig. 17). The ram operates in ways or guiding surfaces at the top of the column. The tool-holding head is at the front end of this ram. This head, with its tool-slide, is used in adjusting the tool for obtaining a given depth of cut; it is also used for feeding a tool in planing vertical surfaces. Angular surfaces may also be planed by setting the tool-head at whatever angle is required. The stroke of the ram and tool is adjusted to suit the length of surface to be planed.

The work-holding table is held by a cross-rail attached to the front of the bed. This cross-rail is adjustable vertically to suit the height of the work. In planing a horizontal surface, the work-table is given an intermittent feeding movement along the cross-rail. This movement, which

Fig. 17. Typical Application of a Shaper

follows each cutting stroke, may be controlled by hand or automatically by the shaper feeding mechanism. The planing of horizontal, vertical, or angular surfaces in a shaper is practically the same as when a planer is used. The tools used in connection with shaper work are similar in form to planer tools, although smaller.

Why are some shapers designated as "universal" types?

If a shaper has a work table which can be swiveled about an axis that is parallel to the line of motion, and an auxiliary tilting side which has angular adjustment with reference to the axis about which the main table swivels, the shaper is sometimes known as a *universal* type. A shaper designed in this way is especially adapted for tool and die work, owing to the universal adjustment. The range of such a machine may be still further increased by means of extra attachments.

What is a draw-cut shaper?

A shaper of the draw-cut type differs from the ordinary design in that the tool cuts when it is moving towards the column of the machine. In other words, the tool is pulled or drawn through the metal on the cutting stroke instead of being pushed. For this reason, the name "draw-cut" is applied to a shaper of this type. The planing tool is set with the cutting edge reversed. The object in designing a shaper to take a draw cut is to secure greater rigidity. The thrust of the cut is toward the column and this tends to relieve the cross-rail and other bearings from excessive strains, especially when taking deep cuts. As the ram is subjected to a tensile stress, it is claimed that vibrations are practically eliminated.

How is the cutting speed of a shaper changed?

The cutting speed of a shaper or the number of strokes which the ram makes in a given time may be varied. There are several different ways of effecting these speed changes, the method depending upon the type of driving mechanism. On some shapers, there is simply a cone-pulley, and the

Fig. 18. Vertical Shaper Cutting Keyway in Shaft Coupling

speeds are varied by shifting the belts from one step on the pulley to another of different size. Shapers are sometimes equipped with a single constant-speed belt pulley (instead of a cone-pulley) and a gear-box, by which the necessary speed changes are obtained. Many modern shapers are driven by direct-connected motors instead of transmitting the power by belt from an overhead countershaft to a pulley on the machine. If a constant-speed motor is used for driving, the speed changes are obtained by gearing. Shapers are also driven by adjustable-speed motors.

What is a vertical shaper?

The ram which carries the cutting tool of a vertical shaper moves vertically. The ram can be placed perpendicular to the table or at an angle for slotting dies, etc. It is mounted in an independent bearing, the upper part of which is pivoted, so that both the bearing and ram can be adjusted to an angular position, which is indicated by degree graduations. The work table of this shaper can be given a transverse, longitudinal, or rotary movement. Work can often be completed at one setting by a shaper of this type, as it may be used for machining either straight, curved, or irregular surfaces.

Fig. 18 shows a vertical shaper cutting keyways in a marine shaft coupling. The sizing tool seen in the ram is used to take a final light cut on the two sides and bottom of the keyways after they have been machined to approximate size with other tools. Four keyways are cut around the coupling. The work-table is rotated for locating the coupling in the four positions.

What is a slotter?

The slotting machine or "slotter," as it is commonly called, operates on the same general principle as an ordinary shaper, except that the ram which carries the planing tool moves in a vertical direction (like a vertical shaper) and at right angles to the work-table. Slotters are used for finishing slots or other enclosed parts which could not be planed by the tool of a horizontal machine like a planer or shaper. The slotter is also used for various other classes of work which can be machined to better advantage by a tool which moves vertically.

A slotter is arranged to machine either flat or curved surfaces. The work-holding table or platen can be moved crosswise along a saddle and the latter can be traversed at right angles along the bed. In addition, the platen can be rotated about its center for slotting circular surfaces. These three movements can be effected either by hand or power.

A typical example of the kind of work done on the slotter is shown in Fig. 19 which illustrates, diagrammatically, the slotting of a locomotive driving-wheel box. The side and top views at A indicate how the inner parallel sides of the box are finished. The work is set on parallel strips s to provide clearance for the tool at the lower end of the stroke, and it is secured to the platen by four clamps. The stroke of the ram R should be about one inch greater than the width of the surface to be slotted and most of the clearance between the tool and the work should be at the top of the stroke where the feeding movement takes place. When the stroke is adjusted, the ram is placed in its lowest position and it is lowered until the end is a little above the top of the work. The tool is extended below the end of the ram far enough to allow the cutter c to reach through the box when at the bottom of the stroke. The line previously scribed on the work to show the location of the finished surface is next set parallel to the cross travel of the platen. This can be done by comparing the movement of the line with relation to the stationary tool point while the work is fed laterally by hand. If adjustments are necessary, these can be made by swiveling the platen one way or the other as required. The cut is started at one end as shown in the plan view and the side is planed by the vertical movement of the tool combined with the lateral feeding movement of the platen and work after each cutting stroke. The opposite side is slotted without disturbing the position of the work by simply turning the tool halfway around.

The sketch at B indicates how the curved seat for the

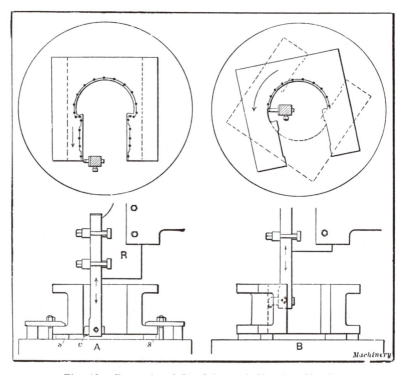

Fig. 19. Example of Straight and Circular Slotting

brass bearing is finished. The radius of the seat is shown by a scribed line which must be set concentric with the center or axis about which the platen rotates. The platen must also be adjusted laterally and longitudinally, if necessary, until the tool will follow the finish line as the work feeds around. The position of the work soon after the cut is started is shown in the plan view by the full lines, and the dotted lines indicate how the box feeds around as the tool moves up and down, thus machining a circular seat into which the bearing brass is afterwards inserted. After the slotter is set in motion, the cut is started by hand and then the power feed is engaged. The finish lines on work of this kind usually serve merely as a guide, and the final measurements are determined by calipers or special gages.

Two or more duplicate parts sometimes can be slotted simultaneously by clamping one piece above the other in a stack or pile. The tool then planes the entire lot to the same shape.

How are castings seasoned to prevent warpage after machining?

Castings, such as machine beds, etc., may warp out of shape gradually after planing or milling, especially if considerable metal is removed from one side only, thus relieving the internal strains unequally. There are several methods of "seasoning" castings to avoid excessive warpage.

1. Leave castings in the mold until thoroughly cold.

2. If equipment is available for heating casting, heat to 1050 degrees F. (low red) for a period of one hour per inch of wall thickness; then cool slowly in furnace. Comparatively small additional improvement may be obtained by re-heating, but ordinarily this is not required.

3. Machine extra surfaces (as on the under side of a bed) in order to neutralize the strains.

4. Suspend casting from crane and pound with wooden blocks, thus setting up strain-relieving vibrations (metal hammers would cause peening).

5. Store castings several months after rough-machining. This method may be objectionable because of the time required.

Milling Flat, Curved and Irregular Surfaces

Surfaces are milled by using a circular type of cutter having a number of teeth or cutting edges which successively mill away the metal as the cutter rotates. These cutting edges may be straight and parallel to the axis of the cutter for milling flat surfaces; they may be inclined to it for forming an angular-shaped groove or surface, or they may have an irregular outline corresponding to the shape or profile of the parts which are to be milled by them. The milling machine rotates the cutter (or cutters) and is equipped with mechanism for regulating the cutting speed and the rate of feed. Milling machines are used for a great variety of operations, and many types have been designed for handling certain classes of work to the best advantage. The milling machine was originally developed for manufacturing the small irregular-shaped parts used in the construction of fire-arms, and the milling process is still employed very extensively in the production of similar work, especially when intricate profiles are required and the parts must be interchangeable. Milling machines are also widely used at the present time for milling many large castings or forgings.

How are flat surfaces produced by the milling process?

Fig. 1 illustrates, diagrammatically, one method of producing a flat surface by milling. The cutter C rotates as shown by the arrow, but rotates in one position, while the work W, which is held on the table of the milling machine and is adjusted vertically to give the required depth of cut, slowly feeds to the left in a horizontal direction. Each tooth on the periphery of the cutter removes a chip every revolu-

100

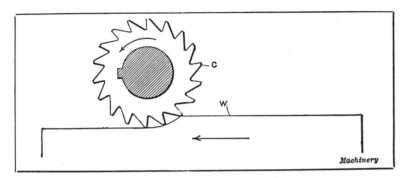

Fig. 1. End View of Cylindrical Cutter Milling Flat Surface

tion, and, as the work moves along, a flat surface is formed.

Another method of producing flat surfaces is by *face milling*. This term, as generally used, means the production of a plane surface by the teeth of a milling cutter which operate in a plane that is at right angles to the axis of the cutter. (See Fig. 2.) The particular machine shown has two cutters so that one may be used for the roughing cut and the other for finishing.

What are the chief functions of a milling machine?

The function of the milling machine, in all cases, is to rotate the cutter and, at the same time, automatically feed the work in the required direction. As it is necessary to vary the feeding movement and the speed of the cutter, in accordance with the material being milled and the depth of the cut, the milling machine must be equipped with feed- and speed-changing mechanisms and other features to facilitate its operation.

When a milling machine of the general type shown in Fig. 3 is in operation, the work-table and work which is attached to it feeds beneath a revolving cutter which may be mounted on the arbor *H* or attached directly to the spindle *G*. The form and size of the cutter, and the rate of the cutter speed and table feed depend upon the nature of the milling operation. Power is transmitted to the main spindle *G* through gearing enclosed in the column of the machine, and, by varying the combinations of this gearing (which is done by changing the positions of levers Q on this particular machine) the required speed changes are obtained. The knee B may be adjusted vertically on the front face of the column for varying the distance between the cutter arbor and the work-table, in accordance with the height of the work. The saddle D has a cross movement on the knee, and the table can be traversed at right angles to the axis of the spindle.

Any of the three feeding movements for the table, that is, the longitudinal, cross, and vertical movement, may be effected by hand or by power. The hand movements are used principally for adjusting the table and the work to the required position when starting a cut, whereas the automatic power feed is employed when milling, except on very short surfaces. The hand-wheel M is used for lowering or raising the knee with its attached parts, handwheel L is for the cross-feed of the saddle and table, and crank N is for the longitudinal adjustment of the table. The work to be milled is held either in a vise or is attached to the

Fig. 2. An Example of Face Milling

table by other means. When duplicate parts are to be milled in quantities, they are usually held in some form of special fixture which is bolted to the table. When an arbor is used to hold the cutter, it is rigidly supported by the bearing K and arbor brace J, which is attached to the knee and over-arm. Many of the machines do not have the extra bearing K, but this is desirable for many classes of work as it can be adjusted on the over-hanging arm and provides support for the arbor close up to the cutter.

The power feed mechanism at O transmits its movement to the front of the machine by a shaft having universal joints and telescopic connection to permit of raising or lowering the knee on the column. The power feed may be transmitted to the table, saddle, or knee as may be required.

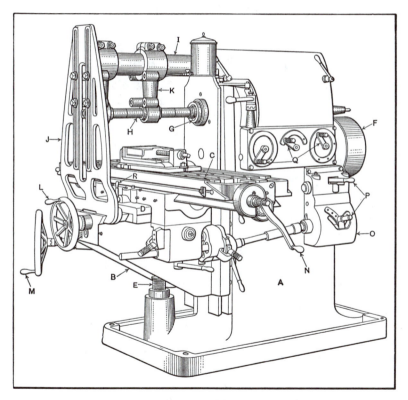

Fig. 3. Plain Type of Milling Machine

The rate or amount of feed per revolution of the cutter can be varied by levers *P*. An index plate shows what the rate of feed is for any position of the levers. The feeding movements of the table can be automatically stopped at any predetermined point by means of dogs on the front side of the table, which may be adjusted so that the automatic trip for the feed will operate after the cut is completed. This automatic tripping mechanism is a very convenient feature, as it prevents feeding too far and makes the machine more independent of the operator.

This machine is driven by a belt on pulley *F*. Many milling machines of the same general type as the one illustrated are driven by a motor. While machines of other makes have the speed- and feed-changing mechanisms, automatic trips, etc., the arrangement of these parts varies more or less in different designs.

What are the distinguishing features of different types of milling machines?

The names used to designate different classes of milling machines may indicate some constructional feature that is characteristic, or they may relate to the nature of the work for which the machine is intended. There are a few exceptions to this method of classification, however, as special names are used to some extent; moreover, in some cases, manufacturers of milling machines do not use the same name for similar types of machines. The constructional features which are generally indicated by the name are the position of the spindle, the design of the bed or frame, and the arrangement of the work-table. A great many milling machines have horizontal spindles, and some of them are known as *horizontal* types, but, in most cases, the name indicates some other constructional feature in order to distinguish between different classes having horizontal spindles. For instance, there are *plain* machines and *universal* machines; both of these types have horizontal spindles, but one is simpler in construction than the other and is not adapted to such a wide range of work. A vertical milling machine has a vertical spindle; some special vertical milling machines, however, are named according to the class of work for which they are intended, as for example, die-sinking and profiling machines, which are, in reality, special types of vertical spindle milling machines.

The frames of milling machines are usually of two general forms; the most common design is a vertical column which supports the horizontal cutter spindle, and has on the front face or side an adjustable knee upon which the work-table is mounted as shown in Fig. 3. This is known as the *column-and-knee* construction; several different types of milling machines are designed in this way. Other milling machines have beds or frames which extend horizontally like the bed of a planer, instead of vertically. The design of the bed in any case is governed partly by the position of the spindle, and it is also affected largely by the general requirements of the work for which it is intended; for instance, milling machines which are used for milling long surfaces and for doing the same general class of work which is done on planers must have a long horizontal bed and work-table, with the necessary feeding movements. Such machines are often known as the *horizontal* or *planer* type.

Why are some milling machines classed as "plain" types and others as "universal" types?

In general, milling machines of the *plain type* (Fig. 3) are adapted to a smaller range of work than the universal type, although many modern plain machines have attachments which practically make them universal so far as adaptability to different classes of work is concerned. The "plain" type differs from the "universal" type, in that the work-table cannot be set at an angle relative to the spindle. Plain milling machines are commonly used for milling operations which can be performed by feeding the work in a straight line, although in modern practice there are many exceptions to this rule. Ordinarily, plain machines are more rigid and heavier in construction than universal designs for a given size of machine, and are intended for heavier milling operations. The plain type is used principally for manufacturing operations, whereas the universal machine is intended more for tool-rooms and for a diversified line of work.

The *universal type* of milling machine is so named be-

cause it is adapted to a very wide range of milling operations. The general construction is similar to that of a plain
milling machine, although the universal type has certain
attachments which plain machines do not ordinarily have,
and which make it possible to mill a comparatively large
variety of work. The table of a universal machine can be
fed automatically at an angle to the spindle by swiveling
the saddle on a lower base or "clamp-bed" which is interposed between the saddle and the knee. The circular swiveling base of the saddle has degree graduations which show
the angular position of the table. This angular adjustment
makes it possible to do work, such as helical milling, which
could not be done on a plain machine unless a spiral milling attachment were used that provided the required angular adjustment for the cutter. Practically all universal machines are equipped with auxiliary appliances, such as the
dividing or indexing head, vertical milling attachments,
etc. Many of these same attachments are also used on plain
milling machines which are thus converted, to some extent,
into universal types.

How are parts held for a milling operation?

The proper method of holding work for milling is governed by the size of the work, its shape, and the nature
of the milling operation. The number of duplicate parts
required should also be taken into consideration. Some
pieces are clamped directly to the machine table which has
T-slots for receiving the clamping bolts. Vises are frequently used for holding small pieces, but are not suitable
for many classes of work. When large quantities of duplicate pieces are to be milled, they are usually held in special
fixtures which are so designed that the work can quickly
be clamped in the correct position for milling. The arrangement or form of a fixture depends upon the shape of the
part for which it is intended and the nature of the milling
operation. The work must be held securely enough to prevent its shifting when a cut is being taken, and it is equally
important to support the part so as to overcome any springing action due either to its own weight or to the pressure
of the cut. Some parts are sprung out of shape by applying the clamps improperly or by omitting to place supports

under some weak or flexible section; as a result, the milled surface is not true after the clamps are removed and the casting springs back to its natural shape.

What form of milling cutter is known as the "plain" type?

The *plain milling cutter* is cylindrical in form and is used for milling flat surfaces. A plain milling cutter is illustrated at *A*, Fig. 4. The teeth on the cylindrical surface may be either "straight," that is, parallel with the axis of the cutter, or helical (as at *A*) and form an angle with the axis of the cutter. Cutters having helical teeth are generally used in preference to the type with straight or parallel teeth, especially for milling comparatively wide surfaces because the former cut more smoothly. When the teeth are parallel to the axis, each tooth begins to cut along its entire width at the same time; hence, if a wide surface is being milled, a shock is produced as each tooth engages the metal. This difficulty is not experienced with helical teeth which, being at an angle, begin to cut at one side and continue across the work with a smooth shaving action. Helical cutters also require less power for driving and produce smoother surfaces.

The plain type of cutter is made in a large variety of diameters and lengths. Another cutter of the plain cylindrical type is shown at *B*. This differs from cutter *A* in that the teeth are "nicked" at intervals along the cutting edges, as shown. The object of nicking the teeth is to break up the chips. As the processes of milling can be applied to an almost unlimited range of work, the cutters used are of many different types. Some of these can be used for general work of a certain class, while others are made especially for milling one particular part.

What is a side milling cutter?

A side milling cutter (also known as a *straddle* milling cutter when cutters of this type are used in pairs for straddle milling) is illustrated at *C*, Fig. 4. This type is provided with teeth on both sides as well as on the periphery of the cylindrical surface. Side milling cutters are

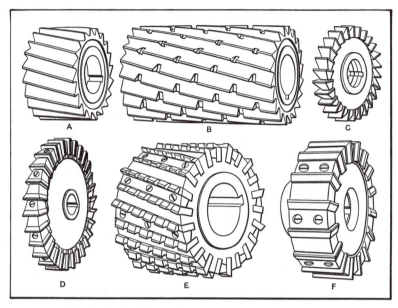

Fig. 4. Cylindrical, Side, and Face Milling Cutters

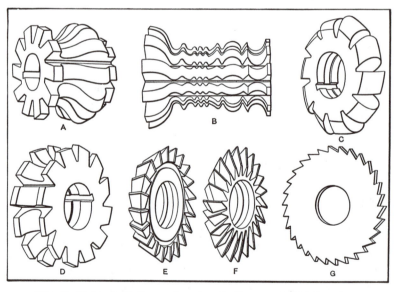

Fig. 5. Formed Cutters, Angular Cutters, and Slitting Saw

used for cutting grooves or slots, as well as for many other operations. They are often used in conjunction with other forms of cutters for milling special shapes in a single operation.

Why are milling cutters of the inserted-tooth type used extensively?

The inserted-tooth construction is ordinarily used for large cutters, in preference to the solid form, because it is more economical, and the inserted teeth can readily be replaced when necessary; moreover when solid cutters are made in large sizes, there is danger of their cracking while being hardened, but with the inserted-tooth type, this is eliminated.

An inserted-tooth side milling cutter is shown at *D*, Fig. 4. This mill, instead of being made of one solid piece of steel, has a machinery steel body into which high-speed tool steel teeth are inserted. These teeth fit into slots and they are held in place by flat-sided bushings which are forced against them by the screws shown; but there are many different methods of holding teeth in cutters of this type.

A large cylindrical cutter with inserted teeth is shown at *E*. The cutter illustrated at *F* also has inserted teeth and is called a face milling cutter. This form is especially adapted to end or face milling operations for producing flat surfaces. When in use, the cutter is mounted on a short arbor which is inserted in the milling machine spindle. Some of the larger face milling cutters are mounted directly on the machine spindle "nose" without the use of an arbor.

Why is a "formed" cutter so named?

The type of cutter used for milling an irregular outline, or a curved surface, is known as a *formed* or *form* cutter because the cutting edges conform to the profile of the work. There is a distinction between a *form* cutter and a *formed* cutter, which, according to the common use of these terms, is as follows: A formed cutter has teeth which are so relieved or "backed off" that they can be sharpened by grinding, without changing the tooth outline, whereas the term "form cutter" may be applied to any cutter for

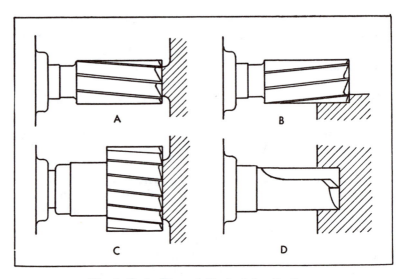

Fig. 6. End-mills and Typical Applications

form milling, regardless of the manner in which the teeth are relieved. Examples of formed milling cutters are shown at *A*, *B*, *C*, and *D*, Fig. 5. The simplest types of formed cutters are the convex cutter shown at *C* and the concave cutter shown at *D*, the outlines of the form in each case being a half-circle. The invention of the formed milling cutter in 1864 by J. R. Brown marked a great step in advance, as the contour of the cutting edge was not affected by successive sharpenings. This type of cutter is used for cutting gear teeth by the milling process.

What is the difference between a right-hand and a left-hand milling cutter?

A cutter which rotates to the right (clockwise), as viewed from the spindle or rear side, is said to be right-hand, and, inversely, a left-hand cutter is one that turns to the left (counter-clockwise) when viewed from the spindle of the machine.

A cutter does not always rotate in the direction illustrated in Fig. 1. If this cutter were turned end for end on the arbor, thus reversing the position of the teeth, the rotation would have to be reversed and a right-hand cutter would be changed to left-hand because the "hand" is determined by the direction of rotation only.

Why are end-mills so named and what are the common types?

End-mills are so named because they have end cutting teeth. Diagram *A*, Fig. 6, shows how an end-mill might be used for milling a raised pad or surface on the side of a casting. As the diagram shows, end-mills have cutting teeth along the cylindrical body as well as at the end, and, in many cases, both the body and end teeth are used as illustrated, for example, by the diagram *B*.

End-mills may either be right-hand as at *A* or left-hand as at *B*, and most end-mills have helical or spiral teeth so that the end cutting teeth have positive rake. Surfaces frequently are milled with end-mills because it would be impractical to use a cutter mounted upon an arbor.

Diagram *C* illustrates a typical application of what is known as a shell end-mill. This is merely a larger size, and it is mounted upon an arbor instead of having its own shank like the other end-mills referred to. Commercial shell end-mills usually have minimum sizes of 1 inch or 1¼ inches, and the range of sizes may extend to 6 inches or more.

For milling grooves, what type of end-mill is preferable?

The two-lipped type of end-mill is especially adapted to milling grooves, as illustrated by diagram *D*, Fig. 6. It has two cutting edges only, and, in starting a groove, can be sunk directly into the metal without preliminary drilling. According to one manufacturer, a depth of cut equal to one-half the diameter of the mill usually can be taken in solid metal. The diameters of commercial two-lipped end-mills may range from ¼ to 1½ inches, or more.

How are end-mills held and driven?

The smaller end-mills represented by diagrams *A*, *B* and *D*, Fig. 6, usually range in diameter from ¼ inch to 1¼ inches, and these sizes have their own shanks. These shanks usually conform to Brown & Sharpe tapers Nos. 5, 7 and 9, although the Morse tapers Nos. 1, 2 and 3 are also used. These taper shanks are much smaller than the standard holes in milling machine spindles; consequently, the shank of the end-mill is inserted in an adapter which tapers on the outside to fit the standard taper of a milling machine spindle, as illustrated by the diagram *A*, Fig. 7. The inner end of this adapter is threaded to receive a draw-in rod which passes through the machine spindle. The adapter also

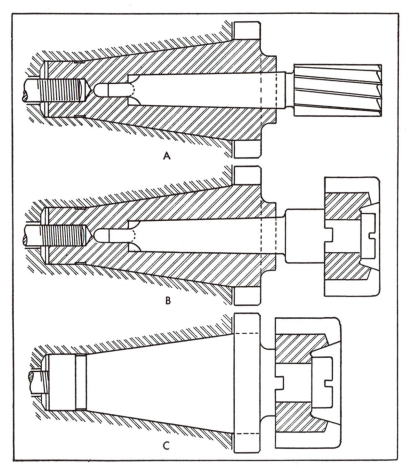

Fig. 7. Methods of Holding End-mills

has an outer flange in which there are two notches that engage driving keys on the spindle.

Diagram *B* shows one method of holding and driving a shell end-mill. An arbor is used. A tongue on the arbor engages a cross-slot at the rear of the end-mill to insure a positive drive. The arbor shank in this case is also inserted in an adapter which fits the machine spindle.

Diagram *C* represents another method of driving a shell end-mill. In this case, the end-mill is mounted directly upon the end of an adapter which has a driving tongue. This is, of course, a more direct and rigid drive than the one represented at *B* and is preferable especially for the larger end-mills and for roughing cuts.

What is "gang milling"?

A great deal of the milling done in different types of milling machines is effected by a combination or "gang" of two or more cutters mounted on one arbor. This is known as *gang milling*. It is not only a rapid method, but one conducive to uniformity when milling duplicate parts. Gang milling is applied in various ways and to many different classes of work. A simple example is shown in Fig. 8. The operation is that of milling 13 equally-spaced slots in rectangular strips (samples of the work may be seen on the machine table). The gang of thirteen cutters on the cutter-arbor not only mills these slots simultaneously, but also very accurately. The slots are $\frac{1}{2}$ inch wide and the width tolerance on this particular job is minus 0.0000 plus 0.0003 inch. The spacing is held within a plus or minus tolerance of 0.0001 inch.

Fig. 9 shows another example of gang milling. In this case, seven standard cutters are so located as to mill various flat surfaces at different levels and one concave or half-round surface.

Gang milling is usually employed when duplicate pieces are milled in large quantities, and the application of this method is almost unlimited. Obviously, the form of a gang-cutter and the number of cutters used depend altogether upon the shape of the part to be milled. Gang-cutters are sometimes made by combining cylindrical and formed cutters, for producing an irregular or intricate profile. When

Fig. 8. Gang Cutter Arranged for Milling Simultaneously
Equally Spaced Notches in Plates

Fig. 9. Gang Cutter for Milling Several Surfaces Simultaneously

more than one helical type of cylindrical cutter is mounted on one arbor, for forming a gang-mill, cutters having both right- and left-hand helical teeth should be used: that is, the teeth of one cutter should form a right-hand helix and the teeth of the other cutter, a left-hand helix. The reason why cutters of opposite hand are used is to equalize the end-thrust, the axial pressure caused by the angular position of the teeth of one cutter being counteracted by a pressure in the opposite direction from the other cutter.

Why is the cutter diameter important in gang milling?

In selecting the cutters for a gang milling operation, it is advisable to use fairly large cutters, if possible, in order to reduce the disproportion of the diameters. For instance, when a 3-inch mill is working in the same gang with a 6-inch mill, the peripheral speed has to be kept within the maximum speed capacity of the larger cutter, which means that the 3-inch size is running much below its maximum peripheral speed. On the other hand, the feed is limited to the capacity of the 3-inch mill; consequently, one cutter may be working at one-half its maximum peripheral cutting speed, whereas the other is operating at one-half its feed capacity. Now, if the diameter of the small mill were six inches and that of the larger one nine inches, the same form would be produced, because the difference between the radii of the two cutters is the same, and the output would also be greatly increased. The same principle applies in the use of dovetail cutters, when the width of the dovetail affords plenty of room for a cutter of large diameter.

What are the general features of a horizontal milling machine?

Any milling machine equipped with a horizontal spindle might be classified as a horizontal type, but this name is generally applied to designs which have a horizontal work-table like that of a planer. There are exceptions, however, to this classification. When a machine has a horizontal work-table and bed, but is equipped only with vertical spindles, it is termed by some manufacturers a "vertical-spindle, horizontal machine," whereas others classify such a machine as a "vertical" type. There are various other names in common use for milling machines, which, so far as the work-table and bed are concerned, are horizontal types such, for example, as *plano* milling machines.

The horizontal design is generally used for heavy milling operations and often for the same general classes of work to which the planer is adapted. Horizontal machines are often used for gang milling. (See Fig. 10.) They are built in many different designs which are modified to suit different classes of work. Some large machines, instead of having a single horizontal cutter arbor, are equipped with four heads. Two of these heads are carried by the cross-rail and the other two are attached to the right and left housings. The cross-rail heads have vertical spindles and the side-heads, horizontal spindles, so that the sides and top surfaces of castings can be milled simultaneously. The side-heads can be adjusted vertically on the housings, and the vertical heads laterally along the cross-rail. Machines of the same general design are also built with three heads, one being located on the cross-rail and two on the housings, and

Fig. 10. Gang Milling with Cutters of Inserted-blade Type

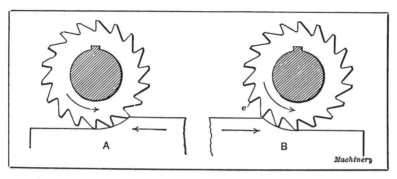

Fig. 11. (A) Work Feeding Against Rotation of Cutter. (B) Work Feeding with Rotation of Cutter

there are various other modifications. With the multiple-spindle machines, the number of spindles used at one time depends upon the nature of the work. For some work, it is necessary to use the horizontal spindles, whereas other parts are milled by using the horizontal and vertical spindles in combination. This type of machine is very efficient for certain kinds of milling.

Should a milling cutter rotate in the same direction as the feeding movement?

When a cylindrical milling cutter rotates as shown at *A*, Fig. 11, the part being milled feeds *against* the direction of rotation, whereas, at *B*, the movement is *with* the cutter rotation. In the first case, the cutter tends to push the work away, but, when the relative movements are as at *B*, the cutter tends to draw the part forward, and if there is any backlash or lost motion between the table feed-screw and nut, this actually occurs when starting a cut; consequently, the cutter teeth which happen to be in engagement take deeper cuts than they should, which may result in breaking the cutter or damaging the work. As a general rule, the relation between cutter rotation and work-feeding movement is as shown at *A*. In other words, the feeding movement and cutting movement are in *opposite* directions. This is sometimes known as the "normal" or "conventional" method of milling to distinguish it from the method referred to in the next paragraph.

What is the meaning of the term "in-cut" or "climb" milling?

The term *in-cut* or *climb* milling means that the feeding movement of the work and the cutting movement are in the same direction, as illustrated by diagram *B*, Fig. 11. Several advantages are claimed for in-cut milling, assuming that conditions are favorable to its application. One important advantage cited is that the cutter life is increased and at the same time higher speeds and feeds may be employed. When milling as shown by diagram *A*, the cutting edge of each tooth rubs against the work or rides upon it momentarily before beginning to cut, which results in greater dulling of the cutter than when each cutting edge enters the metal at the top of the cut or at the point of greatest chip thickness. The advantage in this respect is said to be even greater for the harder materials, although there may be an exception when castings have a hard sandy scale.

In-cut milling has another important advantage in that it enables pieces that are difficult to clamp securely in a fixture or on the machine table to be milled efficiently. The downward action of the cutter teeth in in-cut milling such pieces tends to seat them firmly in the holding devices. Cutters used in the conventional manner would tend to lift the pieces from their seats and might make the operation impractical.

It is evident that a machine used for in-cut milling must be in good condition and be so constructed that the machine table will resist the cutting forces in either direction. Any play or lost motion which would permit the cutter to climb into the work faster than intended would, of course, be objectionable.

Are the conventional and in-cut methods of milling ever combined?

Yes, in some cases the conventional method of milling illustrated by diagram *A*, Fig. 11, is combined with the in-cut method shown by diagram *B*. For example, parts may be rough-milled by the in-cut method *B*, and then be finish-milled while the table is moving in the opposite direction as at *A*. This combining of conventional and in-cut milling may be used in connection with duplicate work-holding fixtures as illustrated by the example in Fig. 12. This particular operation is milling of teeth in special chaser blanks of tool steel. The machine table reverses automatically, and, while a blank is being milled in one fixture, the other fixture is reloaded so that the machine operates without interruption. In-cut milling is applied to pieces held in one fixture and conventional milling to pieces held in the other fixture.

When duplicate parts of irregular form are required, why is the milling process effective?

One of the great advantages of the milling process is that duplicate parts having intricate shapes can be finished

Fig. 12. Climb and Conventional Milling Used in Combination

Fig. 13. An Example of Form Milling

Fig. 14. Another Example of Form Milling

within such close limits as to be interchangeable. Because of this fact, milling machines are widely used for manufacturing a great variety of small machine parts having an irregular outline. The improved cutters now used, and the powerful machines which have been developed for driving these cutters, also make it possible to machine many heavy parts more rapidly by milling than in any other way. Examples of milling with formed cutters are illustrated in Figs. 13 and 14. As these views show, the shape of the milled part is a duplicate of the cutter profile. These particular cutters have a maximum diameter of about 5 inches and they remove about 1/16 inch of material in this form-milling operation. Directly back of each cutter is a flattened spout which directs a stream of cutting compound upon the cutter and work.

What are the principal features of a vertical-spindle milling machine?

The vertical milling machine, as the name implies, has a cutter spindle which revolves about a vertical axis. (See Fig. 15.) While there are a number of different types which might be classified as vertical, the term is usually applied to machines intended for general work. Most special types usually have names which indicate the nature of the work for which they are adapted. The table has longitudinal, crosswise and vertical movements, and the spindle can also be fed vertically within certain limits. Vertical milling machines do not always have vertical adjustments for both the spindle and the table. In some designs, the table, instead of being carried by a sliding knee, is mounted on a fixed part of the base which extends forward beneath it, whereas other machines have a table that can be moved vertically, and a spindle which remains fixed so far as vertical movement is concerned. Some vertical milling machines are also constructed along the lines of a planer, there being a horizontal table, a cross-rail supported on housings, and a vertical spindle mounted upon the cross-rail. Machines of this type are also equipped with two or three spindles, and are known by some manufacturers as vertical-spindle, horizontal milling machines. The double, vertical-spindle and the triple, vertical-spindle machines are especially adapted for milling parallel rows of duplicate parts, or surfaces located in different planes.

Fig. 15 shows a close-up view of a vertical milling machine. The particular operation in this case is milling the opposite sides of a forging which is held between the index centers.

What is the function of a vertical milling attachment?

When an end-mill is driven directly by inserting it in the spindle of a milling machine having a horizontal spindle, it is difficult to mill some surfaces, especially if much hand manipulation is required, because the mill operates on the rear side of the work, where it cannot readily be

Fig. 15. Using a Vertical Milling Machine for Finishing
Opposite Sides of a Forging

seen when one is in the required position for controlling the machine; moreover, it is frequently necessary to clamp the work against an angle-plate to locate it in a vertical position or at right angles to the end-mill, when the latter is driven by a horizontal spindle. In order to overcome these objectionable features, special vertical milling attachments are used in some shops to convert a horizontal machine temporarily into a vertical type. These vertical attachments are very useful, especially when the shop equipment is comparatively small and a horizontal machine must be employed for milling a great many different parts. Where there is a great deal of work that requires end milling, it is better to use a machine having a vertical spindle.

There are several types of vertical attachments designed for different classes of work. The principal difference between these designs, aside from minor details, is in the adjustment of the cutter spindle. There is a compound type of vertical spindle attachment which is adjustable in two vertical planes, one being parallel with the axis of the spindle, and the other being at right angles to the spindle. The universal milling attachment is so named because the spindle can be set at any angle in both horizontal and vertical planes.

When are high-speed milling attachments used?

For some milling operations on such work as diemaking, etc., it is necessary to use a small cutter which should be run more rapidly than the fastest spindle speed available, and, in order to obtain these high speeds, special attachments are sometimes used. These attachments consist principally of an auxiliary spindle that holds and drives the cutter, and suitable gearing connecting with the main spindle and so proportioned as to give the necessary increase of speed. The gearing and spindle are carried by a housing which is attached to the machine. High-speed attachments are used on both horizontal- and vertical-spindle machines.

If a milling operation requires a circular feeding movement, how is this obtained?

If a circular feeding movement is required, a circular milling attachment may be used. This attachment has a

Fig. 16. Milling Operation Requiring Use of Circular
Milling Attachment

Fig. 17. Rotary Milling Machine Employed for Machining the
Joint Faces of Connecting-rods and Their Caps, which
are Loaded Alternately around the Fixture

round work-table which can be rotated for milling circular surfaces or slots. It is placed upon the main table of the machine and is either used on a vertical-spindle machine or in connection with a vertical-spindle attachment on a horizontal-spindle machine. A circular attachment is also used in connection with a slotting attachment for machining curved surfaces. Some of the simpler designs are so arranged that the table is rotated by a handwheel, whereas others have, in addition to the hand-feeding movement, a power feed for the table which can be disengaged automatically at any point by means of adjustable stops which are bolted to the periphery of the table in the required position.

A universal milling machine is seen in Fig. 16 finishing a narrow slot in a rifle part. In this operation, the table is fed toward the cutter for milling one side of the slot until the closed end of the slot is reached, at which point the circular work-table is revolved through 180 degrees for milling the closed end of the slot. Then the regular table of the machine is once more fed in a straight direction to feed the second side of the slot past the cutter. A dial gage on the head of the machine is used to position the cutter accurately for height relative to the work fixture, while the dial gage seen at the left-hand end of the table indicates the point at which the table movement is to be stopped for the rotation of the circular table.

What is rotary milling?

The term *rotary milling* is applied to a machine having a work-holding table which operates with a rotary feeding movement. Castings or forgings which are so shaped as to be readily clamped or released from a fixture are sometimes milled by a continuous rotary milling operation. The rotary milling machine may be designed along the lines of a vertical milling machine, but it is intended for milling large quantities of duplicate parts. The castings or forgings to be milled are held in a fixture near the edge of the table and, as the latter revolves, one piece after another passes beneath the revolving cutter and is milled or faced. As the finished parts come around to the front of the machine, they are removed by the operator and replaced by rough pieces without stopping the machine, so that the milling operation is practically continuous. A fixture for continuous rotary milling must be designed so that the work can be removed quickly and without stopping the rotation of the table.

Fig. 17 shows a rotary type of milling machine arranged for finishing the joint surfaces of connecting-rods and caps. This machine is equipped with a fixture on which rods and caps are mounted alternately with the joint faces upward, so that they are finished as they pass under the shell end-mill attached to the vertical spindle. The connecting-rods are located lengthwise from the previously reamed piston-pin hole, and sidewise from the milled boss faces. The caps are also located from previously milled surfaces.

Some machines for rotary milling have two spindles so that the rough-milling operation is followed by finish-milling. Machines of the rotary type are only used for milling duplicate parts in large quantities.

What is die-sinking and how is it done?

Dies for making "drop forgings" have impressions for giving forgings the required shape. Dies for die-casting have impressions for forming castings, and some sheet-metal forming dies also have impressions or cavities. Cutting out these impressions is known as *die-sinking* (see Figs. 18 and 19). A simple or plain type of "die-sinker" is similar, in its general arrangement, to a vertical-spindle milling machine and in milling out an impression, cutters are used which will assist, as far as possible, in obtaining whatever shape is required for the die-cavity. Such a machine is largely manipulated by hand, and, in finishing the cavity, it is often necessary to do more or less chipping, filing, scraping, and "typing" by hand. Many die-cavities have circular or spherical impressions or sections. The term "cherrying" relates to the milling of these circular or spherical impressions in dies, as, for example, when a milling cutter is sunk to one-half its depth in milling out a circular recess. Many devices and attachments for both the milling machine and die-sinking machine have been devised to eliminate chipping and the difficult hand work. For example, die-sinking machines have been equipped with a

Fig. 18. Universal Die-sinker Used for Cutting Irregular Contours in Forging Dies and Similar Work

cherrying head or attachment to facilitate milling circular impressions in dies.

Why are some die-sinking machines classed as universal types?

A die-sinking machine known as a universal type, is so designed that both cherrying and straight die-sinking operations can be performed without special attachments or special cutters. The machine shown in Fig. 18 is a universal type. The principal feature of the machine is an oscillating head by means of which an ordinary die-sinking cutter can be moved through a circular path, so that both roughing and finishing cherrying operations can be performed. A binder provides for locking the entire oscillating head solidly to the column when the machine is to be used for ordinary die-sinking cuts in which the table elevating and transverse movements are employed. The oscillating head is equipped with power feed and it may also be moved by hand, through a handwheel on the front of the head. This machine will perform many types of cherrying cuts that are impossible on a plain machine. For instance, by combining the rotary table feed and the oscillating cutter movement, it is possible to sink a spherical cut in the surface of a die. The oscillating movement without the rotary motion produces a cylindrical cavity.

Some types of machines for milling cavities in dies reproduce the shape of a model or master form. These are described in the section on "Milling Irregular Contours by Reproducing Shape of Model."

In die-sinking, what cutter shapes are used?

In using a die-sinking machine of either the plain or universal types previously described, the shape of the milled cavity is obtained partly from the shape of the cutter, although, in most cases, the shape is due chiefly to proper

Fig. 19. Typical Die-sinking Operation

manipulation of the machine by the operator, excepting cases where the form is obtained by a cherrying operation, as previously described. A close-up view of a die-sinking operation is shown in Fig. 19 and some typical cutter shapes have been placed on the table to illustrate the types commonly used. A ball or round-end cutter is one common form. The cylindrical cutter having a square end similar to an end-mill is another common shape. A third form that is often required is tapering and has a ball end or round point. With these three simple shapes, a great variety of cavities can be formed by proper manipulation of the machine. The cutters may vary in size from ⅛ inch to 1¼ inches diameter.

What is a milling machine slotting attachment?

A slotting attachment, as the name indicates, is used for converting a milling machine into a slotter. One of these attachments is illustrated in Fig. 20. The tool-slide, which has a reciprocating movement like the ram of a slotter, is

Fig. 20. Slotting Attachment

driven from the main spindle of the machine by an eccentric. Some attachments are driven by an adjustable crank, which enables the stroke to be varied. These varying strokes are indicated by a graduated scale. When the attachment is in use, a slotting tool of the required shape is clamped to the end of the slide. The slide swivels about the machine spindle and can be set at any angle required. The setting is indicated by graduations. These angular adjustments are especially desirable when slotting out dies, in order to obtain the necessary clearance. Attachments of this type are largely used in connection with diemaking and toolmaking for slotting out small blanking dies, box tools for screw machines, etc.

How is the planetary principle applied in milling internal and external surfaces?

In milling parts on a Hall planetary type of machine, the work is held stationary and the cutter is given a planetary movement. For example, in milling a hole (or any internal surfaces) the work is securely held in a chuck and the milling cutter rotates about its own axis, while at the same time it follows a circular path so that the milling operation extends around the entire circumference and is completed in one planetary revolution. The cutter rotation about its axis is rapid in order to obtain a suitable cutting speed, whereas the planetary movement is relatively slow or at the proper rate for feeding the cutter.

The diagrams, Fig. 21, illustrate the principle of planetary milling or "planamilling" as it is called. While these diagrams represent internal milling, the same general principle applies to milling external surfaces by this method. The work represented by the circular section remains stationary. The machine spindle is carried in two eccentric sleeves. The function of these sleeves is to control the cutter's inward and outward movements. The cutter is represented by diagram A in the starting position. Diagram B illustrates how it is moved outward to obtain the required depth of cut; then, as the cutter continues to revolve, it also has a planetary movement as indicated by diagram C. In other words, the cutter moves around in a circle, thus finishing a complete planetary revolution and the milling of the surface as indicated by diagram *D*. The cutter then returns automatically to the inner or neutral position.

In milling external surfaces, the same general principle is applied. The work is held stationary in all cases and the cutter, which for external milling completely surrounds the work, rotates about its axis and has, in addition, a planetary movement so that the milling of the entire circumference is completed in one planetary revolution. This process is known as "planamilling," unless the operation is that of thread milling, in which case it is designated as "planathreading." In milling internal surfaces, the cutters

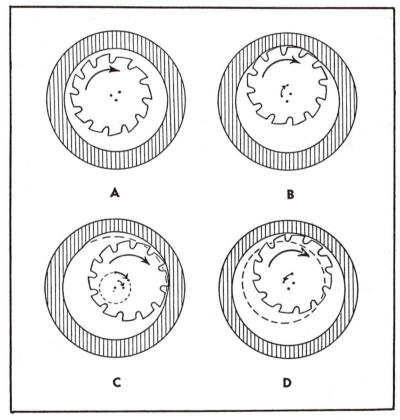

Fig. 21. Diagrams Showing How the Planetary Principle
is Applied to Internal Milling

are only about 20 per cent smaller in diameter than the machined surfaces, and for external surfaces the diameter across the cutting edges is only about 20 per cent greater than the diameter of the milled surface; consequently, in both cases several cutter teeth are always in contact with the work because there is not much difference between the cutter and work diameters. This method of milling is efficient, accurate, and applicable to a great many different classes of work, especially when it is desirable to form-mill simultaneously various internal or external surfaces, shoulders, offsets, or irregular forms of circular cross-section. Threads may be milled at the same time that other milling cuts are taken, and in many cases it would be difficult or impossible to machine parts satisfactorily by any other method. The part to be milled is held in a chuck especially designed to hold it accurately and rigidly. Some machines of the planetary type have double heads in order to perform these annular milling operations at both ends of a casting or forging.

Why are some machines designed for boring, drilling and milling?

While machines of the boring, drilling and milling class (which are of the horizontal type) are designed primarily for boring and drilling operations, they are also used frequently for milling. These milling operations, however, are usually of a secondary nature. For instance, after a casting has been drilled and bored, it may also be necessary to finish some surface, such as a pad, boss, or flange and this may be done readily in machines of this type, by simply attaching a face-milling cutter directly to the main spindle. These machines are very efficient for certain classes of work, because they enable all the machining operations on some parts to be completed at one setting. To illustrate, a casting which requires drilling, boring, and milling at different places can often be finished without changing its position on the machine table after it is once clamped in place. Frequently, a comparatively small surface needs to be milled after a boring operation. If this milling can be performed while the work is set up for boring, accurate results will be obtained (provided the machine is in good

Fig. 22. Application of Boring, Drilling and Milling
Machine to Face Milling

condition), and the time saved that would otherwise be required for re-setting the part on another machine. Fig. 22 shows a large boring, drilling, and milling machine face-milling bosses on the end of a turbine casing.

If both planing and milling processes are applicable, which method should be employed?

In some instances, it may be difficult to decide whether the planer or a milling machine should be used for a given class of work. The number of parts required and general character of the work must be considered. To illustrate, it might be possible to finish a casting by milling much more rapidly than by planing. It does not necessarily follow, however, that milling will be more economical than planing. In the first place, milling cutters are much more expensive than the single-pointed planer tools, and more time is also required to set up a milling machine than a planer, especially when a gang of cutters must be arranged for milling several surfaces simultaneously; hence, if only a few parts are required and especially if the necessary milling cutters are not in stock, the cost of the cutters and the time for arranging the machine might much more than offset the time gained by the milling process. On the other hand, when a large number of duplicate parts are required, milling is often much more economical than planing. It must not be inferred from this that the planer should always be used for small quantities of work, and the milling machine when there is a large number of parts, although the quantity of work to be done frequently decides the question.

Dividing Circumferences into Equal Parts by Indexing

In producing certain classes of machine parts and tools of circular form, it is necessary to machine equally spaced grooves across the periphery or circumference. Work of this kind usually is done by milling, and the equal spaces or divisions are obtained mechanically by using a dividing-head or index-head. For example, if some part of circular form requires 28 equally spaced grooves, the dividing-head is used to rotate the work 1/28 revolution after cutting each groove. A dividing-head ordinarily is used for such work as gear cutting whenever this is done on a milling machine, milling flutes in reamers or taps, milling teeth in milling cutters, or for holding any circular part requiring accurate spacing.

What are the principal mechanical features of an index- or dividing-head?

Index-heads of different makes vary more or less in design but the principle of operation is the same. The general arrangement of a design which has been extensively used is shown in Fig. 1. The main spindle S has attached to it a worm-gear B (see cross-sectional view) which meshes with the worm A on shaft O, and the outer end of this shaft carries a crank J which is used for rotating the spindle when indexing. Worm-wheel B has 40 teeth and a single-threaded worm A is used, so that 40 turns of the crank are required to turn spindle S one complete revolution; hence, the required number of turns to index a fractional part of a revolution is found simply by dividing 40 by the number of divisions desired. In order to turn crank J a definite amount, a plate I is used, having several concentric rows of holes that are spaced equidistant in each separate row. When indexing, spring-plunger P is withdrawn by pulling out knob *J* and the crank is rotated as many holes as may be required. The number of holes in each circle of the index plate varies, and the plunger is set in line with any circle by adjusting the crank radially. One index plate can be replaced by another having a different series of holes, when this is necessary in order to obtain a certain division. A dividing-head forms a part of the equipment of all milling machines of the universal type, and of many plain machines. When in use it is bolted to the table of the machine and is employed in connection

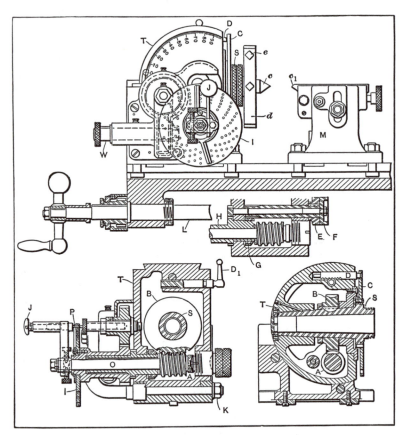

Fig. 1. Dividing-head which is Used for Holding and Indexing Circular Parts as in Milling Equally Spaced Grooves. It is also Used for Helical Milling

with a footstock for milling work that must be supported between the centers c and c_1. The dividing-head is also used independently or without the footstock, in which case the work is usually held in a chuck attached to the spindle. By means of the dividing-head, the circumference of a cylindrical part can be divided into almost any number of equal spaces. It is also used for imparting a rotary motion to work (in addition to the longitudinal feeding movement of the table) for milling helical or spiral grooves, and is sometimes called a "spiral head." The term "index centers" is also used because so much work requiring indexing is held between centers. A great deal of the work done in a universal milling machine requires an attachment of this kind.

How is an index-head used in obtaining an equal number of divisions?

As an example of the work that requires indexing between successive cuts, suppose we have a cylindrical milling cutter blank which requires 18 equally-spaced teeth to be cut across the circumference parallel to the axis and with the front face of each tooth on a radial line. The first step would be to press the blank on an arbor, assuming that it has previously been bored and turned to the proper diameter. The arbor and work is then placed between the centers of the index head and footstock, as shown in Fig. 2. After attaching a dog to the left-hand end, set-screw e is set against the dog to take up any play between these parts, and the footstock center is adjusted rather tightly into the center of the arbor to hold the latter securely.

The cutter is mounted on arbor b and, in this case, the straight side or vertical face is set in line with the center of the arbor as shown by the detail end-view A. The next step is to set the cutter to the right depth for milling the grooves. The depth is regulated according to the width which the tooth must have at the top, this width being known as the "land." The usual method is to raise the knee, table and blank far enough to take a cut, which is known to be somewhat less than the required depth. The blank is then indexed or turned 1/18 of a revolution (as there are to be 18 teeth) in the direction shown by arrow a, and a second groove is started as at *B*. Before taking this cut, the blank is raised until the required width of land is obtained. The second groove is then milled, after which the blank is again indexed 1/18 of a revolution, thus locating it as at *C*. This operation of cutting a groove and indexing is repeated, without disturbing the position of the cutter, until all the teeth are formed as shown at *D*.

What is the rapid or direct method of indexing to obtain a given number of divisions?

There are several different methods of indexing. The method employed may depend upon the number of divisions required and it may also be determined by the type or design of index-head used. The simplest method of indexing is known as the rapid or *direct method*. This will be described first.

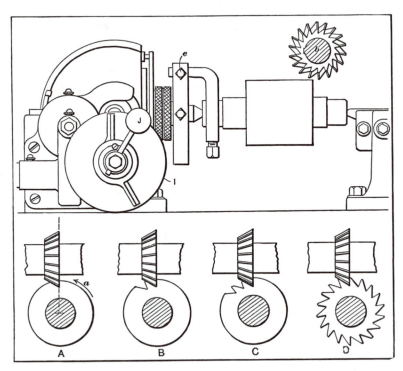

Fig. 2. Views Illustrating Use of Dividing-head for Indexing

The index-head shown in Fig. 1 has a plate C attached to its main spindle. This plate has a circle of 24 equally spaced holes. Any one of these holes may be engaged by pin D which is operated by a small hand-lever D_1. Before the index-head spindle can be rotated for rapid indexing, worm gearing inside of the head must be disengaged. Crank J turns the spindle by means of this gearing when the divisions are not obtainable by the direct method. A knob E at the side of the index-head is used to throw the worm out of mesh with its wheel and this permits indexing the spindle rapidly by hand. The direct method is employed whenever the number of divisions required can be obtained by using plate C. If this plate C has 24 holes, only the numbers 2, 3, 4, 6, 8, 12, and 24 can be indexed by the direct method, because these are the only numbers that will divide evenly into 24. Hence, the following rule applies to the direct method.

Rule for Direct Method.—To find the index movement, divide the total number of holes in the direct index plate by the number of divisions required.

Example.—If 8 divisions are required and the plate has 24 holes, the indexing movement equals 3 holes (24 ÷ 8 = 3), not counting the hole occupied by the locking pin D.

What is plain or simple indexing?

If the required division or movement can be obtained by simply turning the index-crank of the dividing-head the required amount, and engaging it with one of the holes in the index plate, this is known as *plain* or *simple* indexing, because only one indexing movement is necessary, instead of two movements, as with compound indexing. The work is rotated whatever part of a revolution is required, by turning crank J, Fig. 1. If the shaft carrying this crank has a single-threaded worm and there are 40 teeth in the worm-wheel, 40 turns of the crank are necessary to rotate the spindle one complete revolution. If only a half revolution were wanted, the number of turns would equal 40 ÷ 2, or 20, and for 1/12 of a revolution, the turns would equal 40 ÷ 12, or 3 1/3, and so on. In each case, the number of turns the index-crank must make is obtained by dividing the number of turns required for one revolution of the dividing-head spindle by the number of divisions wanted. As the number of turns for one revolution is 40 (with rare exceptions), the rule then is as follows:

Rule.—Divide 40 by the number of divisions into which the periphery of the work is to be divided to obtain the number of turns for the index-crank.

Example.—A gear is to have 18 teeth. What indexing movement is required?

By applying the rule, we find that crank J, Fig. 1, must be turned 2 2/9 times to rotate the gear blank 1/18 revolution, because there are 18 teeth, or divisions, and 40 ÷ 18 = 2 2/9.

How is the index crank rotated a fractional part of a turn to obtain the indexing movement?

When the index crank must be rotated a fractional part of a turn this is done by means of the index plate I, Fig. 1, which has concentric rows or circles of holes. It will be assumed that the numbers of holes in the different circles of this particular plate are 33, 31, 29, 27, 23, and 21. In order to turn crank J two-ninths of a revolution (as required in connection with the preceding example) it is first necessary to adjust the crank radially until the latch-pin is opposite a circle having a number of holes exactly divisible by the denominator of the fraction (when reduced to its lowest terms) representing the part of a turn required. As the denominator of the fraction in this case is 9, there is only one circle on this plate that can be used, namely, the 27-hole circle. In case none of the circles have a number which is exactly divisible by the denominator of the fractional turn required, the index plate is replaced by another having a different series of holes. The number of holes that the latch-pin would have to move for two-ninths of a turn equals 27 × 2/9, or 6 holes. After the latch-pin is adjusted to the 27-hole circle, the indexing of the work 1/18 of a revolution is accomplished by pulling out the latch-pin and turning the crank two complete turns, and then two-ninths of a turn, or 6 holes in a 27-hole circle. After each groove is milled in the work, this indexing operation is repeated, the latch-pin being moved each time 2 2/9 of a turn from the position it last occupied, until the work has been indexed one complete revolution and all of the teeth are milled.

What is the function of a dividing-head sector?

After withdrawing the latch-pin of a dividing-head for indexing, one might easily forget which hole it occupied, or become confused when counting the number of holes for the fractional turn, and, to avoid mistakes of this kind as well as to make it unnecessary to count, a device called a *sector* is used. The sector has two radial arms *A* and *B* (Fig. 3), which have an independent angular adjustment for varying the distance between them. The sector is used by so adjusting these arms that when the latch-pin of the index-crank is moved from one arm to the other, it will traverse the required number of holes for whatever fractional turn is necessary.

When setting the sector, arm *A* is first set against the

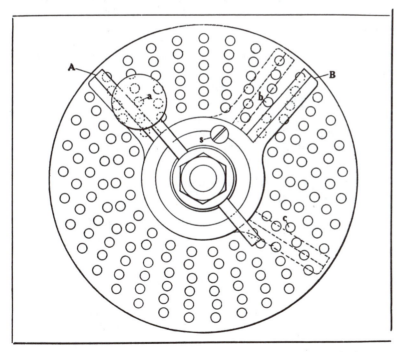

Fig. 3. Diagram Showing How Sector is Used when Indexing

left side of the latch-pin, and then arm *B* is shifted to the right until the number of holes between arm *B* and the latch-pin represent the required fractional part of a turn. In indexing, the latch-pin is withdrawn from hole *a* and the crank is first given as many complete turns as may be necessary, and then the fractional part of a turn by moving the crank until the latch-pin enters hole *b* adjacent to the arm *B* of the sector. The sector is then revolved until arm *A* again rests against the pin, as shown by the dotted lines. When another indexing movement is necessary, the crank is again given the required number of complete revolutions as before, with hole *b* as a starting point, and then the fractional part of a revolution, by swinging the latch-pin around to arm *B* and into engagement with hole *c*. This operation of indexing and then moving the sector is repeated after each space or groove is milled, until the work has made one complete revolution.

When setting the sector arms, the hole occupied by the latch-pin should not be counted or, in other words, the arms should span one more hole than the number needed to give the required fractional turn. Thus, if six holes in the 27-hole circle are required, the sector arms should be adjusted to span seven holes or six spaces, as shown in the illustration. The sectors of some dividing-heads have graduations that make it unnecessary to count the holes when adjusting the sector arms. The setting is taken directly from the index table accompanying the machine, the sector being adjusted to whatever number is given in the table.

In actual practice are indexing movements calculated in the shop?

In actual practice, the number of turns of the index-crank for obtaining different divisions is usually determined by referring to indexing tables. These tables give the numbers of divisions and show what circle of holes in the index plate should be used, and also the turns or fractional part of a turn (when less than one revolution is necessary) for the index-crank. The fractional part of a turn is usually given as a fraction having a denominator which equals the number of holes in the index circle to be used, whereas the numerator denotes the number of holes the latch-pin should be moved, in addition to the complete revolutions, if one or more whole turns are required. For example, the movement for indexing 24 divisions would be given as 1 26/39 of a turn, instead of 1 2/3, the denominator 39 representing the number of holes in the index circle, and 26 the number of holes that the crank must be moved for obtaining two-thirds of a revolution, after making one complete turn.

What is compound indexing?

Ordinarily, the index-crank of a dividing-head must be rotated a fractional part of a revolution, when indexing, even if one or more complete turns are required. This fractional part of a turn is measured by moving the latch-pin a certain number of holes in one of the index circles; but, occasionally, none of the index plates furnished with the machine has circles of holes containing the necessary number for obtaining a certain division. One method of indexing for divisions which are beyond the range of those secured by the plain or simple method is to first turn the crank a definite amount in the regular way, and then the index plate itself, in order to locate the crank in the proper position. This is known as *compound* indexing, because there are two separate movements which are, in reality, two simple indexing operations. The index plate is normally kept from turning by a stationary stop-pin at the rear, which engages one of the index holes. When this stop-pin is withdrawn, the index plate can be turned.

To illustrate the principle of the compound method, suppose the latch-pin is turned one hole in the 19-hole circle and the index plate is also moved one hole in the 20-hole circle and in the same direction that the crank is turned. These combined movements will cause the worm (which engages the worm-wheel on the dividing-head spindle) to rotate a distance equal to $\frac{1}{19} + \frac{1}{20} = \frac{39}{380}$ of a revolution. On the other hand, if the crank is moved one hole in the 19-hole circle, as before, and the index plate is moved one hole in the 20-hole circle, *but in the opposite direction*, the rotation of the worm will equal $\frac{1}{19} - \frac{1}{20} = \frac{1}{380}$ rev-

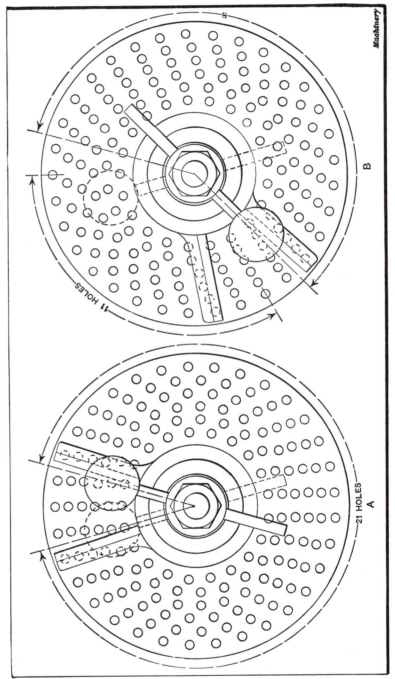

Fig. 4. Diagrams Illustrating the Principle of Compound Indexing

olution. By the simple method of indexing, it would be necessary to use a circle having 380 holes in order to obtain these movements, but by rotating both the index plate and crank the proper amount, either in the same or opposite directions, as may be required, it is possible to secure divisions beyond the range of the simple or direct system.

How is a given number of divisions obtained by compound indexing?

To illustrate the use of the compound method, suppose 69 divisions were required. In order to index the work 1/69 revolution, it is necessary to move the crank 40/69 of a turn (40 ÷ 69 = 40/1 × 1/69 = 40/69), and this would require a circle having 69 holes, if the simple method of indexing were employed, but suppose a 69-hole circle is not available; then, by the compound system, this division could be obtained by using the 23- and 33-hole circles. The method of indexing 1/69 revolution by the compound system, assuming that the 23- and 33-hole circles were used, is as follows: The crank is first moved to the right 21 holes in the 23-hole circle, as indicated at A in Fig. 4, and it is left in this position; then the stop-pin at the rear, which engages the 33-hole circle of the index plate, is withdrawn, and the plate is turned backward, or to the left, 11 holes in the 33-hole circle. This rotation of the plate also carries the crank to the left, or from the position shown by the dotted lines at B, to that shown by the full lines, so that, after turning the plate backward, the crank is moved from its original position a distance x which is equal to $\dfrac{21}{23} - \dfrac{11}{33} = \dfrac{40}{69}$ which is the fractional part of a turn the crank must make, in order to index the work 1/69 of a revolution.

The actual movement required to index or rotate the index-head spindle and work a given fractional part of a revolution, depends upon the particular index-plates or circles available. With one set of plates, a certain number of divisions may require compound indexing; with another set supplied by a different manufacturer, it may be possible to obtain the same number of divisions by simple indexing. The compound indexing movements in the accom-

Compound Indexing

In obtaining certain divisions, such as 69, 77, 87, etc., each index movement equals the division or spacing required. This is not the case in obtaining divisions such, for example, as 53, 67, 71, etc. For these, and most of the divisions listed in the table below, the actual index movement is several times greater than the required spacing, but the work is indexed around a number of times (as shown by the third column) and the required divisions finally are obtained.

No. of Divisions	Indexing Movements	No. of Times Around	No. of Divisions	Indexing Movements	No. of Times Around	No. of Divisions	Indexing Movements	No. of Times Around
51	8 41/47 − 12/49	11	133	3 23/29 − 16/43	11	198*	3/47 + 3/43	...
53	6 43/47 − 9/49	9	134	3 27/47 + 15/49	13	199	2 13/41 − 5/49	11
57	4 4/47 + 3/49	7	137	3 17/43 − 9/49	11	201	2 15/47 + 19/49	13
59	7 19/47 + 12/49	11	138*	1 1/43 − 1/23	...	202	3 19/41 + 9/49	17
61	3 43/47 + 3/49	6	139	2 25/47 + 24/49	11	203	1 23/29 + 9/49	9
63	4 19/29 + 14/63	8	141	1 32/29 + 22/49	8	204	2 29/41 + 3/49	13
67	2 27/41 + 19/49	5	142	4 1/47 + 19/49	15	206	2 33/29 + 3/49	15
69*	2 1/23 − 1/23	...	143	1 39/47 − 18/49	5	207	3 5/41 − 24/49	14
71	3 33/41 − 23/49	6	146	2 3/47 − 9/49	7	208	1 19/47 + 19/49	9
73	6 28/47 − 1/49	12	147*	1 3/29 − 3/49	...	209	8/49 + 9/41	2
77*	9/41 + 3/43	...	149	3 5/43 − 9/49	11	211	1 29/29 + 18/49	11
79	2 43/43 + 3/49	6	151	1 43/43 − 9/49	7	212	3 5/47 + 9/49	17
81	5 3/41 − 9/49	10	153	2 45/47 − 5/49	11	213	1 19/29 + 9/49	8
83	3 45/47 − 5/49	8	154*	3/41 − 5/43	...	214	3 9/47 − 19/49	15
87*	23/29 − 11/43	...	157	2 33/41 + 5/43	11	217	2 3/43 + 19/49	13
89	3 28/29 − 9/49	8	158	5 5/43 − 15/49	19	218	1 27/47 − 9/49	7
91*	9/49 + 1/49	...	159	2 7/47 + 19/49	10	219	3 29/43 − 19/49	19
93*	3/41 + 1/43	...	161	2 19/29 − 1/49	9	221	1 5/47 − 1/49	6
96*	3/18 + 5/20	...	162	1 30/29 − 3/49	7	222	2 5/43 − 19/49	11
97	4 27/41 − 9/49	11	163	3 7/47 − 24/49	11	223	2 29/43 + 13/49	16
99*	15/29 − 5/43	...	166	1 19/43 + 13/49	7	224	2 5/23 + 5/43	13
101	4 23/43 − 19/49	11	167	2 1/29 + 5/43	9	225*	5/18 − 5/20	...
102	4 17/43 − 5/49	11	169	1 32/47 + 13/49	9	226	1 38/29 + 19/49	13
103	1 5/43 + 15/49	4	171	1 29/47 + 1/49	7	227	3 3/43 + 5/49	18
106	2 38/41 + 23/49	9	173	1 7/43 + 1/49	6	228	2 5/41 − 13/49	11
107	2 21/41 − 2/43	7	174*	1 1/43 − 3/29	...	229	2 19/41 − 18/49	12
109	2 19/29 + 4/49	7	175	1 41/41 + 5/43	6	231*	3/41 + 1/43	...
111	3 29/47 + 17/49	11	176	1 14/43 + 13/49	7	233	1 38/47 + 9/49	11
112	4 19/41 − 13/43	11	177	2 19/47 + 5/49	11	234	2 21/29 + 9/43	17
113	3 29/47 − 19/49	9	178	3 25/47 + 11/49	17	236	2 39/43 + 9/49	17
114	1 35/47 + 25/49	7	179	2 34/47 − 13/49	11	237	2 12/47 − 3/49	13
117	7 1/47 − 9/49	20	181	2 5/43 + 12/49	11	238	2 3/41 + 11/43	15
118	1 5/49 + 24/49	5	182*	3/49 + 7/49	...	239	1 23/43 + 15/49	11
119	3 5/43 − 16/43	8	183	1 2/47 + 9/49	8	241	1 1/41 + 23/49	9
121	1 14/47 − 15/49	3	186*	17/31 − 11/33	...	242	2 23/41 − 5/49	15
122	3 41/43 − 17/49	11	187	1 29/47 + 14/49	8	243	1 29/41 − 3/49	10
123	1 34/43 + 17/49	5	189	2 29/41 − 15/49	11	244	2 15/31 + 19/43	17
125	2 33/41 − 3/49	8	191	1 39/47 − 15/49	10	246	1 9/43 − 15/49	5
126	3 19/19 − 7/20	11	192	2 23/41 − 12/49	11	247	2 15/43 − 5/49	14
127	2 23/29 + 13/49	9	193	1 5/47 − 15/49	4	249	3 5/43 − 3/49	19
129	5 24/41 + 15/49	19	194	2 23/27 − 16/49	11	250	2 9/27 − 9/49	13
131	2 40/43 + 21/49	11	197	1 39/43 + 15/49	11	...	....	...

* The indexing movements are exact for the divisions marked with an asterisk (*); the errors of the other divisions are so slight as to be negligible for all ordinary classes of work, such as gear-cutting, etc.

panying table apply to index-plates containing the following numbers of holes in the different circles: 18, 19, 20, 21, 23, 27, 29, 31, 33, 37, 39, 41, 43, 47, and 49.

Is there more than one method of compound indexing?

There are two general methods of indexing by the compound method. In obtaining certain divisions, such as 69, 77, 87, etc., each final indexing movement equals 1/69, 1/77, and 1/87 revolution, respectively. In other words, the movement equals the division or spacing required. This, however, is not the case in obtaining divisions such, for example, as 53, 67, 71, etc. For these, and, in fact, most of the divisions listed in the table "Compound Indexing," the actual index movement of the work is several times greater than the required spacing, but the work is indexed around a number of times (as shown by the third column) and the required divisions finally are obtained.

To illustrate, the table shows that to obtain 53 divisions, the work is indexed around 9 times. This means that each indexing turns the work 9 times farther than the indexing movement equivalent to 1/53 revolution. Just how divisions 1/53 turn apart are obtained when the actual indexing movements of the work are equivalent to 9/53 revolution, will be explained later. First, we shall consider the compound indexing that requires indexing the work around only once, so that for 69 divisions, for example, the indexing movement is that required to obtain 1/69 turn of the index-head spindle and work.

When work is indexed around one turn, how are compound indexing movements determined?

Ordinarily, the number of circles to use and the required number of movements to make when indexing is determined by referring to a table, as this eliminates all calculations and lessens the chance of error. The table shows that 69 divisions require a movement of 21/23 minus 11/33. This movement equals 1/69 revolution of the work; hence, it is only necessary to index around once. In such cases a rule for determining what index circles can be used for indexing by the compound method is as follows:

Rule.— Resolve into its factors the number of divisions required; then choose at random two circles of holes, subtract one from the other, and factor the difference. Place the two sets of factors thus obtained above a horizontal line. Next factor the number of turns of the crank required for one revolution of the spindle (or 40) and also the number of holes in each of the chosen circles. Place the three sets of factors thus obtained below the horizontal line. If all the factors above the line can be cancelled by those below, the two circles chosen will give the required number of divisions; if not, other circles are chosen and another trial made.

To illustrate this rule by using the example given in the foregoing:

$$\frac{\begin{array}{r} 69 = \cancel{3} \times \cancel{23} \\ 33 - 23 = 10 = \cancel{2} \times \cancel{5} \end{array}}{\begin{array}{l} 40 = \cancel{2} \times 2 \times 2 \times \cancel{5} \\ 33 = \cancel{3} \times 11 \\ 23 = \cancel{23} \times 1 \end{array}}$$

As all the factors above the line cancel, an index plate having 23- and 33-hole circles can be used. The next thing to determine is how far to move the crank and the index plate. This is found by multiplying together all the uncancelled factors below the line; thus: $2 \times 2 \times 11 = 44$. This means that to index 1/69 of a revolution the crank is turned *forward* 44 holes in the 23-hole circle, and the index-plate is move *backward* 44 holes in the 33-hole circle. The movement can also be forward 44 holes in the 33-hole circle and backward 44 holes in the 23-hole circle, without affecting the result. The movements obtained by the foregoing rule are expressed in compound indexing tables in the form of fractions, as for example: $+ 44/23 - 44/33$. The numerators represent the number of holes indexed and the denominators the circles used, whereas, the $+$ and $-$ signs show that the movements of the crank and index plate are opposite in direction.

These fractions can often be reduced and simplified so that it will not be necessary to move so many holes, by adding some number to them algebraically. (The smaller value is *subtracted* from the greater and the sign of the greater is prefixed to the result.) The number is chosen by trial, and its sign should be opposite that of the fraction to which it is added. Suppose, for example, a fraction is added, representing one complete turn, to each of the fractions referred to; then:

$$+\frac{44}{23} - \frac{44}{33}$$

$$-\frac{23}{23} + \frac{33}{33}$$

$$+\frac{21}{23} - \frac{11}{33}$$

If the indexing is governed by these simplified fractions, the crank is moved forward 21 holes in the 23-hole circle and the plate is turned backward 11 holes in the 33-hole circle, instead of moving 44 holes, as stated. The result is the same in each case, but the smaller movements are desirable, especially for the index plate, because it is easier to count 11 holes than 44 holes. For this reason, the fractions given in index tables are simplified in this way.

When the work is indexed around several times, how are the required divisions obtained?

With the particular set of index circles listed in the table "Compound Indexing," the work is indexed around from 1 to 20 times. For example, to obtain 53 divisions, the part being milled would be indexed around 9 times. If the work were indexed around once, each movement of the index-crank would equal $\frac{40}{53} = 0.754717$ of a revolution, but this requires a 53-hole index circle which, we shall assume, is not available. With the index circles previously referred to, the most accurate dividing will be done by indexing the work around 9 times, as the table shows. This means that the index-crank, instead of being moved 40/53 of a turn (as required for indexing around once) must be moved 9 times this amount; hence, the movement equals $\frac{40}{53} \times \frac{9}{1} =$ 6.79245 turns of the index-crank; consequently, the actual rotation of the index-head spindle and work for each indexing movement of 6.79245 turns of the crank equals $\frac{1}{53} \times \frac{9}{1} = \frac{9}{53}$ revolution.

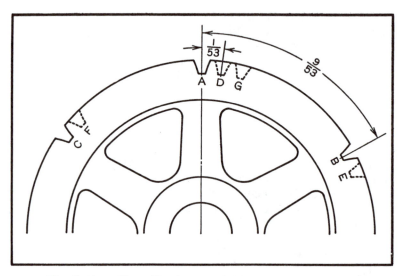

Fig. 5. How Given Number of Divisions is Obtained when
Work is Indexed Around Several Times

But, suppose we are milling a spur gear having 53 teeth.
How are tooth spaces equal to 1/53 of the circumference to
be obtained when the gear blank actually moves 9/53 revolution every time it is indexed? These 1/53 divisions are
obtained as follows: After milling the first tooth space *A*
(see diagram, Fig. 5), the gear blank is indexed 9/53 turn
and then space *B* is milled, and so on around the gear ring.
The fifth indexing movement in this case locates the gear
for milling space *C* which is 8/53 from the starting point
at *A*, measuring from *C* to *A*. After the sixth indexing
movement, the work has turned $\frac{9}{53} \times \frac{6}{1} = \frac{54}{53}$ turn or 1/53
revolution past point *A*; hence, space *D* is 1/53 turn from *A*
so that one tooth has been formed. As the gear is indexed
around a second time, space *E* is milled 1/53 turn past
space *B*, thus forming a tooth here. As the indexing continues, the gear moves around to the positions for milling
spaces *F* and *G*, and this filling-in process is continued until
the entire circumference has spaces milled 1/53 turn apart.

Most of the divisions listed in the "Compound Indexing"
table require indexing the work around more than once to

obtain the most accurate divisions; however, in some cases it will be possible to obtain practically the same degree of accuracy by indexing around once. This is true, for example, of 67 divisions. For 67 divisions, the accurate indexing movement equals $\frac{40}{67} = 0.597015$ turn of the index-crank.

If the 67 divisions are to be obtained in *one* revolution of the work, the index movement can be obtained as follows:

$$\frac{12}{21} + \frac{1}{39} = \frac{163}{273} = 0.597069 \text{ turn of index-crank.}$$

In this case, the error in the indexing movement is only $0.597069 - 0.597015 = 0.000054$, which is negligible. However, it is possible, with the set of index-plates represented by the table, to obtain — theoretically, at least — a more accurate result by increasing the indexing movement to $0.597015 \times 5 = 2.98507$ turns of the crank. This requires indexing around 5 times.

If indexing around more than once is required, how is the number of work turns calculated?

When work must be indexed around more than one turn to secure greater accuracy, the number of turns depends upon available index circles, and there are so many variable factors in the problem that a general rule or formula is impracticable. Suppose the indexing movement for 53 divisions is required and a table is not available. Further, assume that indexing the work around once is not accurate enough, as shown by calculations. The problem then is to determine how many turns will result in the greatest accuracy. This number happens to be 9, but suppose there is no table to show this.

Several numbers are selected for trial, as, for example, 5, 6, 7, 8, 9, 10. The indexing movements equivalent to these 6 numbers may be obtained by multiplying 40/53 by each number. For example, if the work is indexed around 5 times, the movement equals $\frac{40}{53} \times \frac{5}{1} = 3.77358$ turns for the index-crank. If the work is indexed around 6 times, the index-crank movement equals 4.52830 turns; if indexed 7 times, crank movement equals 5.28302 turns; if indexed

8 times, crank movement equals 6.03773; and if indexed 9 times, crank movement equals 6.79245 turns.

The problem now is to determine which fractional part of the movement can be obtained with the greatest accuracy. To illustrate, if the work is indexed around 5 times, will available circles make it possible to obtain the fractional part of the movement or 0.77358 of a turn? By repeated trials, it has been found that the fractional movement or 0.79245 required for indexing around 9 times can be obtained almost exactly by using the 47- and 49-hole circles which are available in the set previously referred to. The compound indexing table shows that the fractional movements for 53 divisions are $\dfrac{43}{47} - \dfrac{6}{49}$. The decimal equivalent of $\dfrac{43}{47}$ is 0.914894, and the decimal equivalent of $\dfrac{6}{49}$ is 0.122449. The difference between these values or 0.914894 — 0.122449 = 0.792445, which is almost the exact fractional movement of 0.79245 required when the work is to be indexed around 9 times in obtaining 53 divisions. By similar trial solutions, it has been found, for example, that the work should be indexed around 7 times for 57 divisions, 6 times for 61 divisions, 3 times for 121 divisions, and so on as shown by the table.

Are the simple and compound systems ever combined to simplify indexing?

Sometimes the simple method of indexing can be used to advantage in conjunction with the compound system. For example, if it is desired to cut a 96-tooth gear, every other tooth can be cut first by using the simple method and indexing for 48 teeth, which would require a movement of 15 holes in an 18-hole circle. When half of the tooth spaces have been cut, the work is indexed 1/96 of a revolution by the compound method, for locating the cutter midway between the spaces previously milled. The remaining spaces are then finished by again indexing for 48 divisions by the simple system. Compound indexing should only be used when necessary, because of the chances of error, owing to the fact that the holes must be counted when moving the index plate. This counting also requires considerable time and, because of these disadvantages, the compound system is avoided as far as possible.

Is compound indexing similar in principle to differential indexing?

Differential indexing is the same in principle as compound indexing, but differs from the latter in that the index-plate is rotated by gearing connecting with the dividing-head spindle, as shown in Fig. 6. This rotation or differential motion of the index-plate takes place when the crank is turned, the plate moving either in the same direction as the crank or opposite to it, as may be required. The result is that the *actual* movement of the crank, at every indexing, is automatically increased or decreased as may be required to obtain the required movement of the index-head spindle. As the hand movement of the plate and the counting of holes is eliminated, the chances of error are also reduced. The proper gear sizes to use for moving the index-plate the required amount would ordinarily be determined by referring to a table or by calculation as shown later. (A table of this kind is included in MACHINERY'S HANDBOOK.)

To illustrate the principle of the differential indexing system, suppose the dividing-head is geared for 271 divisions. For this number the gear on the worm-shaft has 56 teeth, and the spindle gear, 72 teeth. The sector should be set for giving the crank a movement of 7 holes in the 49-hole circle or 3 holes in the 21-hole circle, either of which equals one-seventh of a turn. If an index-plate having a 49-hole circle happens to be on the spindle head, this would be used. Now if the spindle and index-plate were not connected through gearing, 280 divisions would be obtained by successively moving the crank 7 holes in the 49-hole circle, but the two gears (which are connected by an idler) cause the index-plate to turn in the same direction as the crank at such a rate that when 271 indexings have been made the work is turned one complete revolution; therefore, 271 divisions are obtained instead of 280, the number being reduced because the total movement of the crank, for each indexing, is equal to its movement relative to the index-plate, *plus* the movement of the plate itself when (as in this

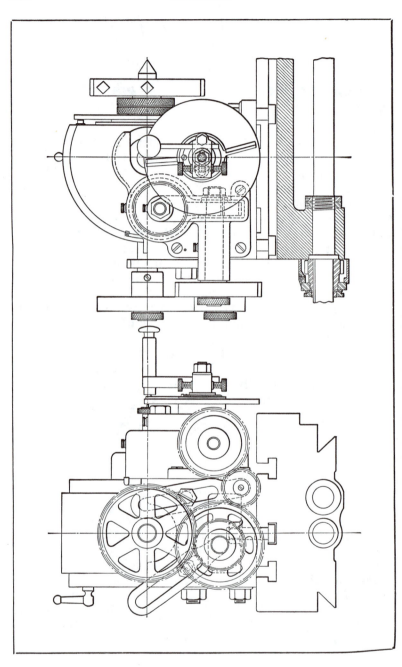

Fig. 6. Dividing Head Geared for Differential Indexing

case) the crank and plate rotate in the same direction. If they were rotated in opposite directions, the crank would have a total movement equal to the amount it turned relative to the plate, *minus* the movement of the plate. In each case, there is a *differential* movement between the crank and the index-plate. Sometimes it is necessary to use compound gearing, in order to move the index-plate the required amount for each turn of the crank. Fig. 6 shows a dividing-head equipped with compound gearing.

How is the gearing ratio calculated for differential indexing?

To find the gearing ratio for differential indexing, first select some approximate number A of divisions either greater or less than the required number N. To illustrate, if the required number N is 67, the approximate number A might be 70; then if 40 turns of the index-crank are required for 1 revolution of the spindle,

$$\text{Gearing ratio } R = (A - N) \times \frac{40}{A}$$

If the approximate number A is less than N, the formula is the same as above except that $A - N$ is replaced by $N - A$.

Example.—Find the gearing ratio and indexing movement for 67 divisions. If $A = 70$,

$$\frac{\text{Gearing}}{\text{Ratio}} = (70 - 67) \times \frac{40}{70} = \frac{12}{7} = \frac{\text{Gear on spindle (driver)}}{\text{Gear on worm (driven)}}$$

The fraction $\frac{12}{7}$ is raised to obtain a numerator and denominator equivalent to available gears. For example,

$$\frac{12}{7} = \frac{48}{28}.$$

Various combinations of gearing and index circles are possible for a given number of divisions. The index movements and gear combinations as given in a table apply to a given series of index circles and gear-tooth numbers. The approximate number A upon which any combination is based may be determined by dividing 40 by the fraction representing the index movement. For example, the approximate number used for 109 divisions equals $40 \div \dfrac{6}{16}$

or $40 \times \dfrac{16}{6} = 106\dfrac{2}{3}$. If this approximate number is inserted in the preceding formula, it will be found that the
gear ratio is 7/8, as shown in the HANDBOOK table.

Second Method of Determining Gear Ratio.—In illustrating a somewhat different method of determining the gear
ratio, 67 divisions will again be used. If 70 is selected as

the approximate number, then $\dfrac{4\cap}{70} = \dfrac{4}{7}$ or $\dfrac{12}{21}$ turn of the

index crank will be required. If the crank is indexed four-

sevenths of a turn sixty-seven times, it will make $\dfrac{4}{7} \times 67$

$= 38\dfrac{2}{7}$ revolutions. This is $1\dfrac{5}{7}$ turns less than the forty

required for one revolution of the work (indicating that
the gearing should be arranged to rotate the index-plate in
the same direction as the index-crank to increase the index-

ing movement) ; hence, the gear ratio $= 1\dfrac{5}{7} = \dfrac{12}{7}$.

When is compound gearing required for differential indexing?

In some cases, as will be noted by referring to a table,
it is necessary to use a train of four gears in order to obtain the required ratio with gear-tooth numbers in the
available series.

Example.—Find the gear combination and indexing
movement for 99 divisions, assuming that an approximate
number A of 100 is used.

$$\text{Ratio} = (100 - 99) \times \frac{40}{100} = \frac{4}{10} = \frac{4 \times 1}{5 \times 2} = \frac{32}{40} \times \frac{28}{56}$$

These final numbers conform to available gear sizes. The
gears having 32 and 28 teeth are the drivers (gear on spindle and first gear on stud), and gears having 40 and 56
teeth are driven (second gear on stud and gear on worm).
The indexing movement is represented by the fraction
40/100 which is reduced to 8/20, the 20-hole index circle
being used in this case.

Example.—Determine the gear combination to use for indexing 53 divisions. If 56 is used as an approximate number (possibly after one or more trial solutions to find an approximate number and resulting gear ratio coinciding with available gears).

$$\text{Gearing ratio} = (56 - 53) \times \frac{40}{56} = \frac{15}{7} = \frac{3 \times 5}{1 \times 7} = \frac{72 \times 40}{24 \times 56}$$

The tooth numbers above the line represent *gear on spindle* and *first gear on stud*. The numbers below the line represent *second gear on stud* and *gear on worm*.

$$\text{Indexing movement} = \frac{40}{56} = \frac{5}{7} = \frac{5 \times 7}{7 \times 7} = \frac{35 \text{ holes}}{49\text{-hole circle}}$$

In setting sector arms, do not count the hole containing the index-crank pin.

How is the movement of the index-crank determined for differential indexing?

The indexing movement of the crank is represented by the fraction $\frac{40}{A}$. For example, if 70 is the approximate number A used in calculating the gear ratio for 67 divisions, then, to find the required movement of the index-crank, reduce 40/70 to any fraction of equal value and having as denominator any number equal to the number of holes available in an index circle. To illustrate,

$$\frac{40}{70} = \frac{4}{7} = \frac{12}{21} = \frac{\text{number of holes indexed}}{\text{number of holes in index circle}}$$

When are idler gears used in the gear train for differential indexing?

In differential indexing, idler gears are used (1) to rotate the index-plate in the same direction as the index-crank, thus *increasing* the actual indexing movement, or (2) to rotate the index-plate in the opposite direction, thus *reducing* the actual indexing movement.

Case 1.—If the approximate number A is *greater* than the actual number of divisions N, simple gearing will require one idler, and compound gearing no idler. Index-plate and crank rotate in the same direction.

Case 2.—If the approximate number A is *less* than the actual number of divisions N, simple gearing requires two idlers, and compound gearing one idler. Indexing-plate and crank rotate in opposite directions.

What number of divisions will be obtained with a given gear ratio and index movement?

Invert the fraction representing the indexing movement and let C equal this inverted fraction. R = gearing ratio.

Case 1.—If simple gearing is used with one idler or compound gearing with no idler,

Number of divisions $N = 40\,C - RC$

Case 2.—If simple gearing is used with two idlers or compound gearing with one idler,

Number of divisions $N = 40\,C + RC$

Example.—The gear ratio is 12/7; there is simple gearing and one idler (Case 1), and the indexing movement is 12/21, making the inverted fraction $C = 21/12$; find the number of divisions N.

$$N = \left(40 \times \frac{21}{12}\right) - \left(\frac{12}{7} \times \frac{21}{12}\right) = 70 - \frac{21}{7} = 67 \text{ divisions}$$

Example.—The gear ratio is 7/8; two idlers are used with simple gearing (Case 2) and the indexing movement is 6 holes in the 16-hole circle. Then

$$N = \left(40 \times \frac{16}{6}\right) + \left(\frac{7}{8} \times \frac{16}{6}\right) = 109 \text{ divisions}$$

What is angular indexing?

Angular indexing is utilized when it is desirable to index a certain number of degrees instead of a fractional part of a revolution. As there are 360 degrees in a circle and 40 turns of the index-crank are required for one revolution of the dividing-head spindle, one turn of the crank must equal $\frac{360}{40} = 9$ degrees. Therefore, two holes in an 18-hole circle, or three holes in a 27-hole circle, are equivalent to a one-degree movement, as this is one-ninth of a turn. If it is required to index 35 degrees, the number of turns the crank must make equals $35 \div 9 = 3\,8/9$, or three complete turns and 8 degrees. As a movement of two holes in an 18-hole circle equals one degree, a movement of 16 holes is required for 8 degrees. If it is required to index 11½ degrees, the one-half degree movement is obtained by turning the crank one hole in the 18-hole circle, after the 11 degrees have been indexed by making one complete revolution (9 degrees), and four holes (2 degrees). Similarly, one and one-third degrees can be indexed by using the 27-hole circle, three holes being required to index one degree, and one hole, one-third degree.

When it is necessary to index to minutes, the required movement can be determined by dividing the total number of minutes represented by one turn of the index-crank, or 540 ($9 \times 60 = 540$), by the number of minutes to be indexed. For example, to index 16 minutes requires approximately 1/34 turn ($540 \div 16 = 34$, nearly), or a movement of one hole in a 34-hole circle. As the 33-hole circle is the one nearest to 34, this could be used and the error would be very small.

Rule for Angular Indexing.—The following is a general rule for the approximate indexing of angles, assuming that forty revolutions of the index-crank are required for one turn of the dividing-head spindle: Divide 540 by the number of minutes to be indexed. If the quotient is nearly equal to the number of holes in any index circle available, the angular movement is obtained by turning the crank one hole in this circle; but, if the quotient is not approximately equal, multiply it by any trial number which will give a product equal to the number of holes in one of the index circles, and move the crank in the circle as many holes as are represented by the trial number. If the quotient of 540 divided by the number of minutes to be indexed is greater than the largest indexing circle, it is not possible to obtain the movement by the ordinary method of simple indexing.

Is angular indexing applicable to very small angles?

One method of indexing small angles is by moving the indexing pin one hole in the 27-hole circle, which gives an indexing movement of 20 minutes, or one hole in the 18-hole circle, which gives an indexing movement of 30 minutes. There is another simple method which makes it possible to index small angles accurately. This consists of using the rear pin which is used for holding the index-plate in a fixed position while indexing in the usual way. This pin may also be used to index in an opposite direction to that of the regular indexing-arm pin, in a different circle, or to index in the same direction, thus adding to or subtracting from the movement made by the indexing-arm pin.

If the back pin is in the 20-hole circle and the regular indexing pin in the 18-hole circle, then by simply withdrawing the rear pin from the 20-hole circle and turning the plate backward or counter-clockwise one hole, again inserting the pin, and then withdrawing the regular indexing pin from the 18-hole circle and moving it one hole in the 18-hole circle clockwise, the work will be moved through an arc of exactly three minutes, and, therefore, any angle may be indexed within a maximum error of 1½ minutes. Any part of an angle greater than 30 minutes may be indexed by first moving one hole in the 18-hole circle for each 30 minutes and then indexing the remainder within 1½ minutes by moving one hole backward in the 20-hole circle and one hole forward in the 18-hole circle for every three minutes.

A still more accurate result may be obtained by placing the plate having the 27-hole circle on the outside of the 20-hole plate and pinning the two together through any two holes which happen to come opposite each other. This is equivalent to having a plate with both a 20- and a 27-hole circle. With this arrangement an exact movement of 1 minute may be made by moving the back pin three holes in the 20-hole circle in a clockwise direction and then the regular index-arm pin four holes in the 27-hole circle in a counter-clockwise direction. This gives a movement to the dividing-head of 1 minute in a clockwise direction, so that all angles may be thus indexed with an error of 30 seconds.

Duplex Dividing Mechanism: Dividing heads are available which are especially adapted to angular indexing even when very small angles are required. For example, one universal dividing head is equipped with a "wide range divider" which may be used for indexing at intervals of 6 seconds without the use of change gears or additional index plates.

The large or main index plate is equipped with a smaller auxiliary plate. Each of these plates has its own index crank and sector. With this arrangement the usual 40 to 1 ratio is supplemented by 100 to 1 ratio of the smaller plate; hence, any angle likely to be required may be obtained readily.

Why should the axis of a tapering part be in alignment with axis of dividing-head spindle?

When milling flutes in taper reamers, milling cutters, etc., which are held between centers, the axis of the divid-

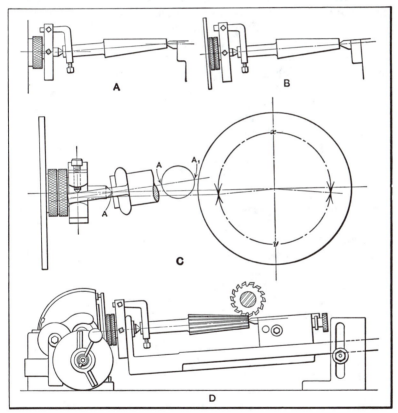

Fig. 7. Tapering Tools Arranged for Fluting

ing-head spindle and the axis of the work should coincide or be in line with each other so as to prevent errors in indexing. Sketch A, Fig. 7, shows an incorrect method of holding a taper reamer mounted between centers for fluting. When a "plain" indexing head is used that does not have angular adjustment like a "universal" head, taper work is sometimes held in this way; that is, the tailstock is blocked up on parallels to hold the reamer or cutter blank at the required angle. The axis of the work, however, is at an angle with the dividing-head spindle and, consequently, the indexing is inaccurate. The cause of this inaccuracy is illustrated at C. If the work is indexed exactly one-half revolution from the position shown, point A will move to position A_1, but, owing to the angularity of the driving dog and work, the dividing-head will move through an angle x or y (depending upon the direction of the indexing movement) which is either greater or less than 180 degrees or one-half revolution. The result is that all the flutes will not be milled to the same width, because a given angular movement of the dividing-head spindle rotates the work farther when the driving dog is passing through the lower half y of the circle than when passing through the upper half x. Another objection to setting taper parts, as shown by sketch A, is that the dog moves in and out through the slot of the driver plate; hence, the dog must be clamped rather loosely to permit this movement and prevent any binding action. If a taper part must be held as shown at A, a compensating dog should be used.

Sketch B illustrates the proper method of holding a taper blank for fluting. The dividing-head spindle is set at whatever angle is required to hold the blank in the proper position, and the center of the tailstock (which is adjustable on modern designs) is also aligned with the axis of the work. As the axis of the dividing-head spindle and blank coincide, they rotate the same amount for each indexing and equal flutes are milled; moreover, there is no movement of the dog relative to the driver plate.

Sketch D shows a special attachment for taper milling. With this attachment, alignment between the dividing-head spindle, work, and tailstock center is assured, and the desired angular setting can be obtained easily. One end of the attachment is secured to the spindle of the dividing-head and the opposite end is bolted to a slotted knee clamped to the table of the machine. This knee is graduated to correspond with the graduations on the dividing-head. The footstock of the attachment can be adjusted along a T-slot to suit the length of the work. This attachment is especially desirable when there is considerable taper milling.

How is an index-head used in graduating flat scales?

A milling machine equipped with an index- or dividing-head is sometimes used for graduating verniers, flat scales, and other parts requiring odd fractional divisions or graduations. The index-head spindle should be geared to the table feed-screw so that a longitudinal movement of the table is secured by turning the indexing crank. In using a Brown & Sharpe machine, the gear for the index-head is mounted on the differential index center inserted in the main index-head spindle. By varying the indexing movement, graduations can be spaced with considerable accuracy. The graduation lines are cut by a sharp-pointed tool held either in a fly cutter arbor mounted in the spindle, or between the collars of a regular milling cutter arbor. The lines are drawn by feeding the table laterally by hand, and the lengths of the lines representing different divisions and subdivisions can be varied by noting the graduations on the cross-feed screw. The gearing between the index-head spindle and feed-screw should be equal or of such a ratio that the feed-screw and spindle rotate at the same speed.

In flat-scale graduating, how are the indexing movements determined?

Assuming that the lead of the feed-screw thread is 0.25 inch and that 40 turns are required for one revolution of the index-head spindle; then one turn of the index-crank will cause the table to move longitudinally a distance equal to one-fortieth of 0.25 or 0.00625 inch.

$$\left(\frac{1}{40} \times \frac{25}{100} = 0.00625 \right)$$

Suppose graduation lines 0.03125 or 1/32 inch apart were required on a scale. Then the number of turns of the index-crank for moving the table 0.03125 inch equals 0.03125 ÷ 0.00625 = 5 turns. If the divisions on a vernier reading to thousandths of an inch are to be 0.024 inch apart, the indexing movement would equal 0.024 ÷ 0.00625 = 3.84 turns. This fractional movement of 0.84 turn can be obtained within very close limits by indexing 26 holes in the 31-hole circle; thus, three complete turns will move the work 0.00625 × 3 = 0.01875 inch, and $\frac{26}{31}$ turn will give a longitudinal movement equal to 0.00524+; therefore, a movement of $3\frac{26}{31}$ turns = 0.01875 + 0.00524 = 0.02399 inch, which is only 0.00001 inch less than the required amount—a theoretical error of no practical importance. The index-crank should always be turned in one direction after beginning the graduating operation, in order to prevent errors from any play or backlash that might exist between the table feed-screw and nut.

The accompanying table "Indexing Movements for Graduating on Milling Machine" was compiled by the Brown & Sharpe Mfg. Co., and will be found very convenient for determining what circle to use and the number of holes to index for given dimensions. The whole number of turns required is first determined and then the indexing movement for the remaining distance is taken from the table. Thus, for graduation lines 0.0218 inch apart, three complete turns give a movement of 0.00625 × 3 = 0.01875 inch. 0.0218 — 0.01875 = 0.00305 inch. By referring to the table it will be seen that a movement of 21 holes in the 43-hole circle equals 0.00305+. Therefore, to index the work 0.0218 inch, the crank should be given $3\frac{21}{43}$ turns. When graduating, the index-crank should always be turned in one direction after beginning the graduating operation, in order to prevent errors from any play or backlash that might exist between the table feed-screw and nut.

Why is irregular spacing sometimes required, in connection with indexing?

The flutes of reamers may be spaced unevenly to prevent the reamer from chattering. The flutes of half of the reamer

Indexing Movements for Graduating on Milling Machine

Movement of table.	Holes.	Circle.	Movement of table.	Holes.	Circle.	Movement of table.	Holes.	Circle.
0.0001275	1	49	0.0006377	5	49	0.0011479	9	49
0.0001330	1	47	0.0006410	4	39	0.0011574	5	27
0.0001454	1	43	0.0006465	3	29	0.0011628	8	43
0.0001524	1	41	0.0006579	2	19	0.0011718	3	16
0.0001603	1	39	0.0006649	5	47	0.0011824	7	37
0.0001689	1	37	0.0006757	4	37	0.0011905	4	21
0.0001894	1	33	0.0006944	3	27	0.0011968	9	47
0.0002016	1	31	0.0006944	2	18	0.0012096	6	31
0.0002155	1	29	0.0007268	5	43	0.0012195	8	41
0.0002315	1	27	0.0007353	2	17	0.0012500	4	20
0.0002551	2	49	0.0007576	4	33	0.0012500	3	15
0.0002660	2	47	0.0007622	5	41	0.0012755	10	49
0.0002717	1	23	0.0007653	6	49	0.0012820	8	39
0.0002907	2	43	0.0007813	2	16	0.0012930	6	29
0.0002976	1	21	0.0007979	6	47	0.0013081	9	43
0.0003049	2	41	0.0008012	5	39	0.0013158	4	19
0.0003125	1	20	0.0008064	4	31	0.0013257	7	33
0.0003205	2	39	0.0008152	3	23	0.0013298	10	47
0.0003289	1	19	0.0008333	2	15	0.0013513	8	37
0.0003378	2	37	0.0008446	5	37	0.0013587	5	23
0.0003472	1	18	0.0008621	4	29	0.0013722	9	41
0.0003676	1	17	0.0008721	6	43	0.0013888	6	27
0.0003788	2	33	0.0008929	7	49	0.0013888	4	18
0.0003826	3	49	0.0008929	3	21	0.0014031	11	49
0.0003906	1	16	0.0009146	6	41	0.0014113	7	31
0.0003989	3	47	0.0009259	4	27	0.0014422	9	39
0.0004032	2	31	0.0009308	7	47	0.0014535	10	43
0.0004167	1	15	0.0009375	3	20	0.0014628	11	47
0.0004310	2	29	0.0009469	5	33	0.0014706	4	17
0.0004361	3	43	0.0009616	6	39	0.0014881	5	21
0.0004573	3	41	0.0009869	3	19	0.0015086	7	29
0.0004630	2	27	0.0010081	5	31	0.0015152	8	33
0.0004808	3	39	0.0010136	6	37	0.0015202	9	37
0.0005068	3	37	0.0010174	7	43	0.0015244	10	41
0.0005102	4	49	0.0010204	8	49	0.0015306	12	49
0.0005319	4	47	0.0010417	3	18	0.0015625	5	20
0.0005435	2	23	0.0010638	8	47	0.0015625	4	16
0.0005682	3	33	0.0010671	7	41	0.0015957	12	47
0.0005814	4	43	0.0010776	5	29	0.0015989	11	43
0.0005952	2	21	0.0010869	4	23	0.0016026	10	39
0.0006048	3	31	0.0011029	3	17	0.0016128	8	31
0.0006098	4	41	0.0011218	7	39	0.0016204	7	27
0.0006250	2	20	0.0011363	6	33	0.0016303	6	23

Indexing Movements for Graduating on Milling Machine — *Continued*

Movement of table.	Holes.	Circle.	Movement of table.	Holes.	Circle.	Movement of table.	Holes.	Circle.
0.0016447	5	19	0.0021682	17	49	0.0026785	9	21
0.0016581	13	49	0.0021738	8	23	0.0026785	21	49
0.0016666	4	15	0.0021802	15	43	0.0027028	16	37
0.0016768	11	41	0.0021875	7	20	0.0027174	10	23
0.0016892	10	37	0.0021960	13	37	0.0027243	17	39
0.0017045	9	33	0.0022059	6	17	0.0027344	7	16
0.0017241	8	29	0.0022176	11	31	0.0027440	18	41
0.0017288	13	47	0.0022436	14	39	0.0027618	19	43
0.0017361	5	18	0.0022607	17	47	0.0027777	8	18
0.0017442	12	43	0.0022728	12	33	0.0027777	12	27
0.0017628	11	39	0.0022866	15	41	0.0027925	21	47
0.0017857	6	21	0.0022959	18	49	0.0028017	13	29
0.0017857	14	49	0.0023027	7	19	0.0028060	22	49
0.0018144	9	31	0.0023148	10	27	0.0028125	9	20
0.0018292	12	41	0.0023257	16	43	0.0028225	14	31
0.0018382	5	17	0.0023438	6	16	0.0028409	15	33
0.0018518	8	27	0.0023649	14	37	0.0028717	17	37
0.0018581	11	37	0.0023706	11	29	0.0028846	18	39
0.0018617	14	47	0.0023809	8	21	0.0028963	19	41
0.0018750	6	20	0.0023937	18	47	0.0029070	20	43
0.0018896	13	43	0.0024038	15	39	0.0029167	7	15
0.0018939	10	33	0.0024192	12	31	0.0029256	22	47
0.0019021	7	23	0.0024235	19	49	0.0029337	23	49
0.0019132	15	49	0.0024306	7	18	0.0029412	8	17
0.0019231	12	39	0.0024390	16	41	0.0029605	9	19
0.0019396	9	29	0.0024455	9	23	0.0029762	10	21
0.0019532	5	16	0.0024622	13	33	0.0029890	11	23
0.0019737	6	19	0.0024710	17	43	0.0030094	13	27
0.0019818	13	41	0.0025000	8	20	0.0030172	14	29
0.0019947	15	47	0.0025000	6	15	0.0030241	15	31
0.0020161	10	31	0.0025266	19	47	0.0030303	16	33
0.0020271	12	37	0.0025339	15	37	0.0030406	18	37
0.0020350	14	43	0.0025463	11	27	0.0030448	19	39
0.0020485	16	49	0.0025510	20	49	0.0030488	20	41
0.0020833	13	39	0.0025640	16	39	0.0030524	21	43
0.0020833	5	15	0.0025736	7	17	0.0030586	23	47
0.0020833	11	33	0.0025862	12	29	0.0030611	24	49
0.0020833	9	27	0.0025915	17	41	0.0031250	9	18
0.0020833	7	21	0.0026164	18	43	0.0031250	10	20
0.0020833	6	18	0.0026209	13	31	0.0031250	8	16
0.0021277	16	47	0.0026316	8	19	0.0031889	25	49
0.0021342	14	41	0.0026515	14	33	0.0031915	24	47
0.0021552	10	29	0.0026596	20	47	0.0031978	22	43

Indexing Movements for Graduating on Milling Machine — *Continued*

Movement of table.	Holes.	Circle.	Movement of table.	Holes.	Circle.	Movement of table.	Holes.	Circle.
0.0032014	21	41	0.0036990	29	49	0.0042091	33	49
0.0032050	20	39	0.0037038	16	27	0.0042152	29	43
0.0032095	19	37	0.0037163	22	37	0.0042232	25	37
0.0032197	17	33	0.0037234	28	47	0.0042338	21	31
0.0032257	16	31	0.0037500	12	20	0.0042553	32	47
0.0032327	15	29	0.0037500	9	15	0.0042685	28	41
0.0032408	14	27	0.0037793	26	43	0.0042765	13	19
0.0032607	12	23	0.0037878	20	33	0.0042971	11	16
0.0032738	11	21	0.0038043	14	23	0.0043104	20	29
0.0032895	10	19	0.0038112	25	41	0.0043268	27	39
0.0033088	9	17	0.0038195	11	18	0.0043368	34	49
0.0033164	26	49	0.0038265	30	49	0.0043477	16	23
0.0033245	25	47	0.0038305	19	31	0.0043562	23	33
0.0033333	8	15	0.0038460	24	39	0.0043605	30	43
0.0033431	23	43	0.0038564	29	47	0.0043750	14	20
0.0033538	22	41	0.0038692	13	21	0.0043883	33	47
0.0033654	21	39	0.0038794	18	29	0.0043922	26	37
0.0033784	20	37	0.0038853	23	37	0.0043980	19	27
0.0034091	18	33	0.0039063	10	16	0.0044119	12	17
0.0034273	17	31	0.0039246	27	43	0.0044210	29	41
0.0034375	11	20	0.0039352	17	27	0.0044354	22	31
0.0034439	27	49	0.0039475	12	19	0.0044643	15	21
0.0034482	16	29	0.0039540	31	49	0.0044643	35	49
0.0034574	26	47	0.0039636	26	41	0.0044871	28	39
0.0034722	10	18	0.0039773	21	33	0.0045060	31	43
0.0034722	15	27	0.0039894	30	47	0.0045140	13	18
0.0034885	24	43	0.0040064	25	39	0.0045213	34	47
0.0035063	23	41	0.0040322	20	31	0.0045259	21	29
0.0035156	9	16	0.0040443	11	17	0.0045452	24	33
0.0035255	22	39	0.0040541	24	37	0.0045610	27	37
0.0035325	13	23	0.0040625	13	20	0.0045732	30	41
0.0035474	21	37	0.0040700	28	43	0.0045835	11	15
0.0035714	12	21	0.0040759	15	23	0.0045920	36	49
0.0035714	28	49	0.0040817	32	49	0.0046055	14	19
0.0035904	27	47	0.0040948	19	29	0.0046194	17	23
0.0035984	19	33	0.0041160	27	41	0.0046296	20	27
0.0036186	11	19	0.0041223	31	47	0.0046371	23	31
0.0036289	18	31	0.0041666	22	33	0.0046473	29	39
0.0036339	25	43	0.0041666	14	21	0.0046512	32	43
0.0036585	24	41	0.0041666	18	27	0.0046543	35	47
0.0036637	17	29	0.0041666	12	18	0.0046875	15	20
0.0036765	10	17	0.0041666	10	15	0.0046875	12	16
0.0036858	23	39	0.0041666	26	39	0.0047195	37	49

Indexing Movements for Graduating on Milling Machine — *Continued*.

Movement of table.	Holes.	Circle.	Movement of table.	Holes.	Circle.	Movement of table.	Holes.	Circle.
0.0047256	31	41	0.0052296	41	49	0.0057180	43	47
0.0047299	28	37	0.0052327	36	43	0.0057400	45	49
0.0047349	25	33	0.0052365	31	37	0.0057433	34	37
0.0047414	22	29	0.0052419	26	31	0.0057692	36	39
0.0047620	16	21	0.0052635	16	19	0.0057874	25	27
0.0047796	13	17	0.0052884	33	39	0.0057927	38	41
0.0047873	36	47	0.0053030	28	33	0.0058142	40	43
0.0047968	33	43	0.0053125	17	20	0.0058187	27	29
0.0048074	30	39	0.0053194	40	47	0.0058336	14	15
0.0048384	24	31	0.0053242	23	27	0.0058466	29	31
0.0048470	38	49	0.0053364	35	41	0.0058512	44	47
0.0048613	14	18	0.0053572	42	49	0.0058599	15	16
0.0048613	21	27	0.0053572	18	21	0.0058674	46	49
0.0048782	32	41	0.0053781	37	43	0.0058710	31	33
0.0048912	18	23	0.0053880	25	29	0.0058825	16	17
0.0048989	29	37	0.0054057	32	37	0.0059027	17	18
0.0049202	37	47	0.0054170	13	15	0.0059122	35	37
0.0049244	26	33	0.0054348	20	23	0.0059215	18	19
0.0049345	15	19	0.0054434	27	31	0.0059294	37	39
0.0049420	34	43	0.0054486	34	39	0.0059375	19	20
0.0049569	23	29	0.0054522	41	47	0.0059455	39	41
0.0049677	31	39	0.0054690	14	16	0.0059524	20	21
0.0049745	39	49	0.0054848	43	49	0.0059598	41	43
0.0050000	16	20	0.0054878	36	41	0.0059782	22	23
0.0050000	12	15	0.0054924	29	33	0.0059841	45	47
0.0050308	33	41	0.0055148	15	17	0.0059951	47	49
0.0050402	25	31	0.0055238	38	43	0.0060188	26	27
0.0050532	38	47	0.0055555	24	27	0.0060346	28	29
0.0050596	17	21	0.0055555	16	18	0.0060480	30	31
0.0050676	30	37	0.0055746	33	37	0.0060607	32	33
0.0050785	13	16	0.0055852	42	47	0.0060812	36	37
0.0050876	35	43	0.0055925	17	19	0.0060898	38	39
0.0050928	22	27	0.0056035	26	29	0.0060980	40	41
0.0051022	40	49	0.0056088	35	39	0.0061052	42	43
0.0051136	27	33	0.0056123	44	49	0.0061171	46	47
0.0051281	32	39	0.0056250	18	20	0.0061224	48	49
0.0051474	14	17	0.0056403	37	41	0.0062500	...	1
0.0051627	19	23	0.0056450	28	31	....	...	..
0.0051721	24	29	0.0056546	19	21	....	...	..
0.0051830	34	41	0.0056690	39	43	....	...	..
0.0051861	39	47	0.0056816	30	33	....	...	..
0.0052083	15	18	0.0057065	21	23	....	...	..

may be spaced irregularly but made to correspond with the other half of the reamer, opposite cutting edges being exactly diametrically opposite. Another method is to space the cutting edges around the whole reamer irregularly so that no two cutting edges are diametrically opposite. The advantages obtained by having the two halves of the reamer identical are that the reamer can be exactly measured, and that an equal width of the land of all the cutting edges can be more easily obtained if desired, as the milling machine table on which the reamer is mounted while fluting would have to be raised or lowered only half the number of times for obtaining this result, than would be the case if every tooth were irregularly spaced. It is the opinion of experienced reamer makers, however, that these advantages do not outweigh the disadvantages resulting from this method and that when the flutes are not irregularly spaced around the whole reamer, the tool is liable to chatter. When all the cutting edges are diametrically opposite each other, the positions of each opposite pair coincide after half a revolution, whereas, if all the cutting edges are irregularly spaced there is no coincidence of position of any two cutting edges until the reamer has been turned around a complete revolution.

The error in measuring a reamer when all the cutting edges are irregularly spaced, and when no two are diametrically opposite, is very slight, provided the variation in spacing is not excessive. The irregularity may be so small that the error in measuring will not exceed 0.0003 inch. The difference in the widths of the lands of the reamer teeth, which is the inevitable result of "breaking up" the flutes if the table or the cutter is not raised or lowered between consecutive cuts so as to make up for the difference in spacing, ordinarily is not considered important. The only reason for lands of even width would be for the sake of appearance, as the unequal width of the lands in no way interferes with the efficiency of the reamer.

How are irregular indexing movements determined?

When the flutes of a reamer have unequal spacing which is alike on each half so that opposite cutting edges are in line, the indexing may be done conveniently by milling the

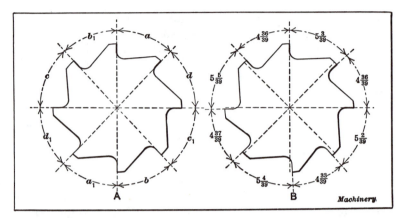

Fig. 8. (A) Irregular Spacing with each Half Uniform and Cutting
Edges Opposite. (B) Flutes so Spaced that Cutting
Edges are not Exactly Opposite

flutes in pairs; that is, after milling a flute, the index-head is turned half a revolution and the corresponding flute on the opposite side of the reamer is cut. Then, after milling the adjacent flute, the index-head is again turned half a revolution, and so on. If the depths of the flutes are to be varied to secure lands of equal width, milling the flutes in pairs saves time as the cutter is only set once for opposite flutes. The flutes the first time around should be milled somewhat less than the required depth; then when milling them the second time, the cutter can be sunk to depth by noting the width of the land. After finishing one pair of opposite flutes, the cutter is re-set for the adjoining pair, etc.

Example.—To illustrate how the indexing movements for irregular spacing are determined, suppose a reamer is to have eight flutes with the spacing of each half corresponding. Assuming that the 20-hole circle in the index-plate is to be used, a complete revolution expressed as a total number of holes equals $20 \times 40 = 800$ holes. The number for 8 equal divisions equals $800 \div 8 = 100$ holes. The next thing to decide is the amount of irregularity in the spacing. The difference should be slight and need not exceed 2 degrees, although it is often made 3 or 4 degrees. Assuming that it is to be 2 degrees, the movement of the index-crank necessary to give this variation must be determined. As

Irregular Spacing of Teeth in Reamers

Number of flutes in reamer......	4	6	8	10	12	14
Index circle to use............	39	39	39	39	39	49
Before cutting...	Move index crank the number of holes below more or less than for regular spacing.					
2d flute.....	8 less	4 less	3 less	2 less	4 less	3 less
3d flute.....	4 more	5 more	5 more	3 more	4 more	2 more
4th flute....	6 less	7 less	2 less	5 less	1 less	2 less
5th flute....	....	6 more	4 more	2 more	3 more	4 more
6th flute....	....	5 less	6 less	2 less	4 less	1 less
7th flute....	....	....	2 more	3 more	4 more	3 more
8th flute....	....	....	3 less	2 less	3 less	2 less
9th flute....	....	....	....	5 more	2 more	1 more
10th flute....	....	....	....	1 less	2 less	3 less
11th flute....	....	....	....	....	3 more	3 more
12th flute....	....	....	....	....	4 less	2 less
13th flute....	....	....	....	....	......	2 more
14th flute....	....	....	....	....	......	3 less

800 holes represent a complete revolution or 360 degrees, a movement of one hole = 360 ÷ 800, or nearly ½ degree; therefore, the number of holes required for a movement of 2 degrees equals 2 ÷ 0.5 = 4 holes. Now if the divisions were all to be equal we would move the index-crank 100 holes or 5 turns, but by varying the movement 4 holes one way or the other as near as this can be arranged, an irregularity of approximately 2 degrees is obtained. Thus the successive movements could be 96, 100, 103 and 101 holes, or 4 turns 16 holes, 5 turns, 5 turns 3 holes, and 5 turns 1 hole, respectively. Referring to diagram A, Fig. 8, flutes a and a_1 would be milled first diametrically opposite; then by indexing 96 holes the work would be located for milling flute b; after milling b_1 on opposite side, another movement of 100 holes would locate flute c; then after milling c_1 a movement of 103 holes would locate d; finally after fluting d_1 the cutter could be aligned with flute a_1 by a movement of 101 holes. The maximum amount of spacing between adjacent flutes is that represented by the spacing of flutes a_1 and b and equals 101 — 96 or 5 holes, which is equal to 2¼ degrees. When selecting the numbers of holes by which the indexing movements are to be varied, one should remember that the total sum of the numbers must equal one-half the number

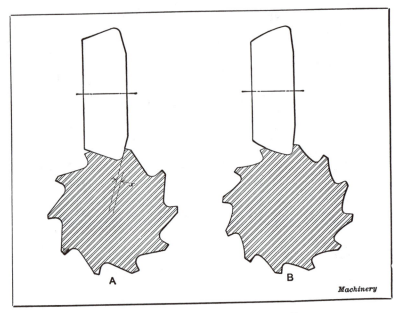

Fig. 9. Cross-sectional Views of Fluted Reamers

of holes representing a complete revolution, when each half of the reamer is spaced alike and indexed as just described; thus,

$$96 + 100 + 103 + 101 = 400.$$

When all the flutes are to have irregular spacing, the indexing movements may be obtained from the accompanying table "Irregular Spacing of Teeth in Reamers." To illustrate its application, suppose a reamer is to have 8 flutes. If the spacing were equal, 5 turns of the index-crank would be required ($40 \div 8 = 5$), but, as the table shows, the indexing movement for the second flute should be 5 turns minus 3 holes; for the third flute, 5 turns plus 5 holes, etc. In other words, the indexing movement for the second flute is $4\frac{36}{39}$ turns; for the third flute $5\frac{5}{39}$ turns; for the fourth flute $4\frac{37}{39}$ turns, etc., as shown by diagram B, Fig. 8. After milling the eighth flute, the cutter could be aligned with flute No. 1 by indexing $5\frac{3}{39}$ turns, as the illustration shows.

This last movement is equal to 40 (the total number of turns required for one revolution of the dividing head) minus the total number of turns for the 7 flutes. Of course, it will be understood that the irregularity in spacing can be obtained by variations in indexing, other than those given in the foregoing.

In general, reamers for steel and cast iron have the faces of the teeth either radial or slightly ahead of the center, but reamers for brass or bronze work better if the face of the tooth makes an angle of 5 or 6 degrees with the radial line, being that amount ahead of the center. The diagram A, Fig. 9, illustrates how negative rake is obtained by setting the cutter slightly ahead of the center. The amount of this offset x may be obtained by dividing the reamer diameter by the constant 23; thus for a diameter of 1.5 inch the offset would equal $1.5 \div 23 = 0.065$ inch. Hand reamers are ordinarily given negative rake, as they cut more smoothly than when the teeth are radial.

Generating Helical Grooves with Index-Head

Milling machines of the universal type are used sometimes for milling helical or "spiral" grooves in a cylindrical surface. For example, a milling machine may be used to mill the teeth of a helical gear, especially if a regular gear-cutting machine is not available. In order to do work of this kind, it is necessary to provide means of forming the helical grooves. Ordinarily, in helical milling operations, it is necessary to form a given number of equally spaced grooves around a circumference; consequently, provision must also be made for indexing. To illustrate, if a helical gear is being cut, it is necessary to index the gear after milling each helical tooth space in order to form a given number of equally-spaced teeth.

Index- or dividing-heads designed for helical milling are so arranged that they will rotate the work very slowly as it feeds past the cutter, thus forming a helical groove. After each groove is milled, this index-head is also used for turning the work whatever fractional part of a revolution is required for obtaining a given number of divisions. Index-heads of the class referred to are sometimes called "spiral heads" or "spiral index centers," assuming that they are adapted to helical milling.

What is a helical curve?

The top of a sharp V-thread, of uniform diameter and lead, represents a helical curve. Such a curve may be defined in simple language as one which winds around a cylinder and advances along this cylinder uniformly and in the direction of the axis. If a helical curve could be unwound so as to coincide with the flat surface, the helix would form the hypotenuse of a right-angle triangle as

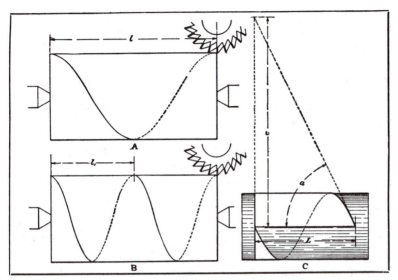

Fig. 1. Diagrams Illustrating the Principle of Helical Milling

shown by the diagram *C*, Fig. 1. Furthermore, side *c* would equal the circumference of the cylinder around which the helix extended and side *L* the lead of the helix or the distance it advances along the cylinder in a complete turn. If this procedure is reversed and a right-angle triangle is wrapped around a cylinder, the hypotenuse will form a helical curve.

In the case of a screw thread, the helical curve makes a number of complete turns, thus forming a continuous thread. The teeth of a helical gear include only a small part of a complete turn; nevertheless, each tooth, throughout its length, has helical curvature the same as a screw thread.

Helical curves are sometimes referred to as "spirals." Helical gears, for example, are often called "spiral gears." While this may be justified on the ground of common usage, a spiral is a curve located in one plane and it has a constantly increasing radius of curvature. Such a curve is illustrated by an ordinary watch spring, assuming that the successive coils extend outward along a true spiral.

The amount that a helical curve advances in one turn, or the distance between successive turns measured parallel to the axis, is known as the *lead*. Dimension l equals the lead of the helix shown at A, Fig. 1, and l_1 is the lead of the helix shown at B.

How is a groove of helical curvature generated?

When a helical groove is being milled in an ordinary universal milling machine, the work is turned slowly by the dividing-head as the table of the machine feeds lengthwise. As the result of these combined movements, a helical groove is generated by the milling cutter. The principle of helical milling is illustrated by the diagrams shown in Fig. 1. If a cylindrical part mounted between centers, as at A, is rotated, and, at the same time, moved longitudinally at a constant rate past a revolving cutter, a helical groove will be milled as indicated by the curved line. Evidently, the lead l, or distance that this helix advances in one revolution, will depend upon the ratio between the speed of rotation and the longitudinal feeding movement. If the speed of rotation is increased, the lead will be diminished, and vice versa, provided the rate of the lengthwise travel remains the same. If the cylinder moves a distance equal to its length while making one revolution, the dimension l as shown at A would equal the lead of the spiral generated, but, if the speed of rotation were doubled, the lead l as shown at B would be reduced one-half (assuming that the rate of lengthwise movement is the same in each case), because the cylinder would then make two revolutions while traversing a distance equal to its length. (In actual practice, the table of the milling machine would be set at an angle so as to locate the cutter in alignment with the helical groove, thus forming a groove of the same cross-sectional shape as the cutter. A practical example will be referred to later.)

What is the general method of generating a helix of given lead?

In helical milling, change-gears are used to rotate the index-head spindle at whatever rate is required to obtain a given lead. The method is similar in principle to that employed on lathes for cutting screw threads.

Fig. 2 shows an end and side view of an index-head mounted on the table of the machine and arranged for helical milling. The rotary movement of the spindle S and the work is obtained from the feed-screw L, which also moves the table longitudinally. This feed-screw is connected to shaft W by a compound train of gears a, b, c, and d, and the movement is transmitted from shaft W to the worm-shaft (which carries the indexing crank) through the spiral gears e, f, and spur gearing (not shown), which drives the index plate, crank- and worm-shaft. When a helical groove is to be milled, the work usually is placed between the centers of the dividing-head and the footstock, and change-gears a, b, c, and d are selected to rotate the work at whatever speed is needed to produce a helix of the required lead. When the change-gears are in place, evidently any longitudinal movement of the table caused by the rotation of feed-screw L will be accompanied by a rotary movement of the dividing-head spindle. As connection is made with the worm-shaft W_1 through the index plate and crank, the stop-pin G at the rear must be withdrawn for helical milling, so that the index plate will be free to turn.

What is the "lead" of a milling machine?

If change-gears with an equal number of teeth are placed on the table feed-screw and the worm-gear stud, then the *lead of the milling machine* is the distance the table will travel while the index spindle makes one complete revolution. This distance is a constant used in figuring the change-gears, as explained later. Assume that the combination of change-gears is such that 20 turns of screw L are required for one revolution of spindle S, and the screw has four threads per inch; then the table will advance a distance equal to $20 \div 4 = 5$ inches, which is the lead of the helix obtained with that particular gearing. The proper gears to use for producing a helix of any given lead can easily be determined if it is known what lead will be obtained when change-gears of equal diameter are used. Suppose gears of the same size are employed, so that feed-screw L (Fig. 2) and shaft W rotate at the same speed; then the feed-screw and worm-shaft W_1 will also rotate at the same speed, if the gearing which forms a part of the index-head and

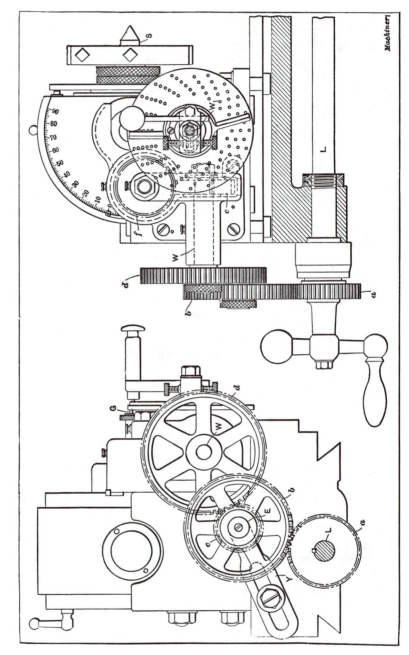

Fig. 2. The Indexing or Spiral Head of a Universal Milling Machine Geared for Helical or "Spiral" Milling

connects shafts W and W_1 is in the ratio of one to one, which is the usual construction. As a general rule, 40 turns of the worm-shaft are required for each revolution of spindle S; therefore, with change-gears of the same diameter, the feed-screw will also make 40 turns, and, assuming that it has four threads per inch, the table movement will equal $40 \div 4 = 10$ inches. This movement, then, of 10 inches, equals the lead of the helix that would be obtained by using change-gears of the same size, and 10 is also the *lead of the milling machine*. The lead of a milling machine may, therefore, be defined as the lead of the helix that will be cut when gears with an equal number of teeth are placed on the feed-screw and the worm-gear stud.

Rule.—To find the lead of a milling machine, place equal gears on the worm-gear stud and on the feed-screw, and multiply the number of revolutions made by the feed-screw to produce one revolution of the index head spindle, by the lead of the thread on the feed-screw. Expressing the rule as a formula:

$$\text{Lead of milling machine} = \frac{\text{Rev. of feed-screw for one}}{\text{revolution of index spindle}} \times \frac{\text{Lead of}}{\text{feed-screw}}$$
$$\text{with equal gears}$$

How is the lead of a milling machine used in calculating change-gears for helical milling?

The ratio represented by the lead of the helix and lead of the machine equals the ratio of the driven and driving gears. Expressed as a formula:

$$\frac{\text{Lead of helix to be cut}}{\text{Lead of machine}} = \frac{\text{Product of driven gears}}{\text{Product of driving gears}}.$$

The numerators equal the number of teeth in the *driven* gears, and the denominators, the number of teeth in the *driving* gears.

Rule for Finding Change-gears.—To find the change-gears to be used in the compound train of gears for helical milling, place the lead of the helix to be cut in the numerator and the lead of the milling machine in the denominator of a fraction; divide numerator and denominator into two factors each; and multiply each "pair" of factors by the *same* number until suitable numbers of teeth for the change-gears are obtained. (One factor in the numerator and one in the denominator are considered as one "pair" in this calculation.)

Example 1.—Assume that the lead of a machine is 10 inches, and that a helix having a 48-inch lead is to be cut. Following the method explained:

$$\frac{48}{10} = \frac{6 \times 8}{2 \times 5} = \frac{(6 \times 12) \times (8 \times 8)}{(2 \times 12) \times (5 \times 8)} = \frac{72 \times 64}{24 \times 40}$$

The gear having 72 teeth is placed on the worm-gear stud and meshes with the 24-tooth gear on the intermediate stud. On the same intermediate stud is then placed the gear having 64 teeth, which is driven by the gear having 40 teeth placed on the feed-screw. This makes the gears having 72 and 64 teeth the *driven* gears, and the gears having 24 and 40 teeth the *driving* gears. Either the driving or driven gears can be transposed without changing the lead of the helix. For example, the driven gear with 64 teeth could be placed on shaft W (Fig. 2) and the one having 72 teeth could be used as a second gear on the stud, if such an arrangement were more convenient. A reverse or idler gear is inserted in the train when it is desired to reverse the motion of the work for cutting a helix of the opposite hand, but this idler does not affect the ratio of the gearing.

Example 2.—The lead of a helical groove to be milled is 8.4 inches. What combination of change-gears will give this lead?

$$\frac{8.4}{10} = \frac{84}{10 \times 10} = \frac{12 \times 7}{4 \times 25} = \frac{72 \times 28}{24 \times 100}$$

Various other combinations of the same ratio might be used. The change-gear solution must, of course, include only those tooth numbers which are included in the set of gears supplied with the machine.

How is the accuracy of change-gear calculations proved?

The following rule may be used either for checking the accuracy of change-gear calculations or for determining what lead of helix will be obtained with any combination of gears that may be available.

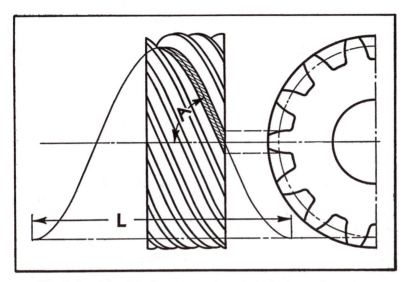

Fig. 3. Lead L of Helix and Angle A Relative to Axis of Gear

Rule.—Multiply the product of the *driven* gears by a number representing the lead of the machine, and divide the result by the product of the *driving* gears.

Example.—To illustrate the application of this rule, suppose the driven gears selected have 48 and 72 teeth, respectively, and the driving gears, 32 and 40 teeth, respectively. Assuming that the lead of the machine is 10, then

$$\text{Lead of helix} = \frac{72 \times 48 \times 10}{32 \times 40} = 27 \text{ inches.}$$

As the common lead of universal milling machines is 10 inches, the rule given can be simplified as follows: Divide 10 times the product of the driven gears by the product of the drivers, and the quotient will be the lead of the helix.

What is the angle of a helix and is it always designated in the same way?

There are two ways of designating the angle of a helical curve. One way is to give the angle relative to the axis. The other way is to give the angle relative to a plane perpendicular to the axis. The helix angles of helical gears are relative to the axis like angle *a*, Fig. 1, and also as represented by angle *A* in the diagram, Fig. 3. In the case of screw threads, worm threads, and hobs for gear cutting, the helix angle is measured from a plane perpendicular to the axis as at *A* in Fig. 4, and it is often called the *lead angle*. Dimension *L*, Fig. 3, is the lead. In designing and cutting helical gears, the angle relative to the axis is required; but for worm threads, etc., the angle from a plane perpendicular to the axis or the lead angle is more useful.

How is the helix angle found?

The angle of a helix depends both upon the lead and the diameter of the cylinder (real or imaginary) with which the helix coincides. For example, length *c*, Fig. 1, represents the circumference of the cylinder around which the helix extends. It is evident that if this circumference is increased, angle *a* will also be increased, assuming that the lead *L* is not changed. If the circumference *c* and the lead *L* are known, it is a simple matter to find the angle; but we must first know whether the angle required is measured from the axis or from a plane perpendicular to the axis.

The angle of a helix can be determined with a fair degree of accuracy by laying it out on a drawing-board. This is

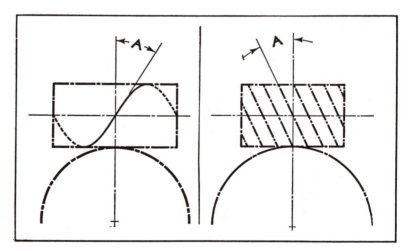

Fig. 4. Lead Angle A or Helix Angle Measured from Plane
Perpendicular to Axis

done by drawing a right-angle triangle. The length c of one side equals the circumference of the cylinder about which the helical curve winds and the base L equals the lead. The angle a between the hypotenuse representing the helical curve, and the axis, may be measured by using an ordinary protractor. While this method might be useful in some cases, helix angles usually are determined by calculation, and this is the accurate method.

If a helix angle is measured from the axis, how is it found?

Rule 1.—If the helix angle is measured from the axis (as shown at A, Fig. 3), divide the lead of the helix by 3.1416 times the diameter at which the angle is required. The quotient equals the cotangent of the helix angle. The angle equivalent to this cotangent then is found in the table of trigonometric functions.

In designing and cutting helical gears, the helix angle at the pitch diameter ordinarily is required. If the lead remains constant, the helix angle relative to the axis decreases as the diameter decreases; hence, a helical gear, for example, has many helix angles and the angle at the top of the tooth is greater than at the bottom. The angle at approximately half depth or at the pitch diameter is used ordinarily.

Example.—A helical gear has a pitch diameter of 4.005 inches and a lead of 30.374 inches. Find the helix angle at the pitch diameter.

$$\text{Cot helix angle} = \frac{30.374}{3.1416 \times 4.005} = 2.414$$

The corresponding angle is 22½ degrees.

If the helix angle is measured from a plane perpendicular to the axis, how is it found?

This method of measuring helix angles, as previously explained, is applied to worm threads, hobs for cutting wormwheels or spur gears, and to all forms of screw threads.

Rule 2.—If the helix or lead angle is measured from a plane perpendicular to the axis (as shown at A, Fig. 4), multiply the diameter at which the lead angle is required by 3.1416 and divide the product by the lead, thus obtaining the cotangent of the angle.

Example.—A worm thread has a pitch diameter of $3\frac{1}{2}$ inches and a lead of 2 inches. Find the lead angle at the pitch diameter or at half the thread depth.

$$\text{Cot lead angle} = \frac{3.5 \times 3.1416}{2} = 5.4978$$

Hence, the lead angle at the pitch diameter of this worm thread is 10 degrees 19 minutes (to the nearest minute). The foregoing rules might be arranged to find the *tangent* instead of the cotangent of the lead angle. The tangents are of smaller value than the cotangents for all angles less than 45 degrees, and the reverse is true for angles larger than 45 degrees. If the tangent is desired, the foregoing rules are reversed; thus, in Rule 1, the diameter times 3.1416 is divided by the lead to get the tangent. In Rule 2, the lead is divided by 3.1416 times the diameter to get the tangent.

In milling the teeth of a helical cutter, what is the general procedure?

As an example of helical milling, suppose we have a cylindrical milling cutter blank $3\frac{1}{4}$ inches in diameter in which right-hand helical teeth are to be milled. The blank is first mounted on an arbor which is placed between the centers with a driving dog attached. The arbor should fit tightly into the hole of the blank so that both will rotate as one piece, and it is also necessary to take up all play between the driving dog and faceplate.

The index-head is next geared to the feed-screw. If a table of change-gears is available, it will show what gears are needed, provided the lead of the helix is known. A small section of one of these tables is reproduced in Fig. 5 to illustrate the arrangement. (Complete tables will be found in engineering handbooks.) Suppose the lead given on the drawing is 48 inches; then this figure (or the nearest one to it) is found in the column headed, "Lead in Inches," and the four numbers to the right of and in line with 48 indicate the number of teeth in the four gears to be used. The numbers opposite 48 are 72, 24, 64 and 40, respectively, and the position for each of these gears is shown by the headings above the columns. As 72 is in the column headed "Gear on Worm," a gear d (see also Fig. 2) of this size is placed on shaft W. The latter is referred to as the "worm-shaft," although, strictly speaking, the worm-shaft W_1 is the one which carries the indexing crank and worm. The first gear c placed on the stud has 24 teeth, as shown by the table, and the second gear b on the same stud has 64 teeth, whereas gear a on the screw has 40 teeth.

	DRIVEN	DRIVER	DRIVEN	DRIVER		DRIVEN	DRIVER	DRIVEN	DRIVER		DRIVEN	DRIVER	DRIVEN	DRIVER
LEAD IN INCHES	GEAR ON WORM	1ST GEAR ON STUD	2ND GEAR ON STUD	GEAR ON SCREW	LEAD IN INCHES	GEAR ON WORM	1ST GEAR ON STUD	2ND GEAR ON STUD	GEAR ON SCREW	LEAD IN INCHES	GEAR ON WORM	1ST GEAR ON STUD	2ND GEAR ON STUD	GEAR ON SCREW
42 00	72	24	56	40	48.00	72	24	64	40	56.31	86	24	44	28
					48.38	86	32	72	40	57.14	100	28	64	40
42.23	86	28	44	32	48.61	100	24	56	48	57.30	100	24	44	32
42.66	100	28	86	72	48.61	100	24	28	24	57.33	86	24	64	40
42.78	56	24	44	24	48.86	100	40	86	44	58.33	100	24	56	40
42.86	100	28	48	40	48.89	64	24	44	24	58.44	100	28	72	44
42.86	72	24	40	28	49.11	100	28	44	32	58.64	86	24	72	44
43.00	86	32	64	40	49.14	86	28	64	40	59.53	100	24	40	28
43.00	86	28	56	40	49.27	86	24	44	32	59.72	86	24	40	24
43.00	86	24	48	40	49.77	100	24	86	72	60.00	72	24	64	32
43.64	72	24	64	44	50.00	100	24	56	40	60.00	72	24	56	28
43.75	100	32	56	40	50.00	100	24	48	40	60.00	72	24	48	24
43.98	86	32	72	44	50.00	72	24	40	24	60.61	100	24	64	44
44.44	64	24	40	24	50.00	100	32	64	40	61.08	100	32	86	44
44.64	100	28	40	32	50.17	86	24	56	40	61.43	86	28	64	32
44.68	86	28	64	44	50.26	86	28	72	44	61.43	86	24	48	28

Fig. 5. Section of Table Showing Change-gears Required in Milling Helix of Given Lead

After these gears are placed in their respective positions, the first and second gears c and b are adjusted to mesh properly with gears a and d by changing the position of the supporting yoke Y. As a right-hand helix is to be milled, which means that the milled teeth will advance by twisting or turning to the right as seen from one end of the cutter, an idler gear is not used with the design of index-head shown. In milling a left-hand helix, it is necessary to insert an idler gear in the train of gears, to rotate the work in a reverse direction; this idler has no effect, however, on the ratio of the gearing.

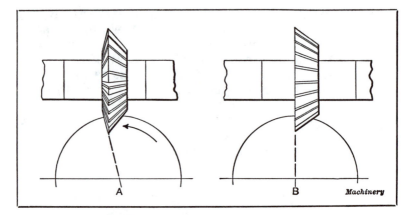

Fig. 6. Double- and Single-angle Cutters

What is the position of the cutter for helical milling?

If we assume that the grooves or flutes of the cutter are to be milled to an angle of 60 degrees, evidently the cutter must have teeth which conform to this angle. A type adapted for forming teeth of helical mills is shown at *A* in Fig. 6. The teeth have an inclination of 48 degrees on one side and 12 degrees on the other, thus giving an included angle of 60 degrees for the tooth spaces. This "double-angle" form of cutter is used in preference to the "single-angle" type shown at *B*, for milling helical teeth, because the 12-degree side will clear the radial faces of the teeth, and produce a smooth surface. (The single-angle cutter *B* is used for milling grooves that are parallel with the axis.) The cutter is mounted on an arbor, and in this case it is set in such a position that when the groove is cut to the required depth, the 12-degree side will be on a radial line, as shown by the sketch; in other words, it is set so that the front faces of the teeth to be milled will be radial.

A simple method of setting a double-angle cutter, for milling the teeth in helical mills, is as follows: The pointer of a surface gage is first set to the height of the index-head center and then the work is placed in the machine. The cutter is next centered with the blank, laterally, which can be done with a fair degree of accuracy by setting the knee

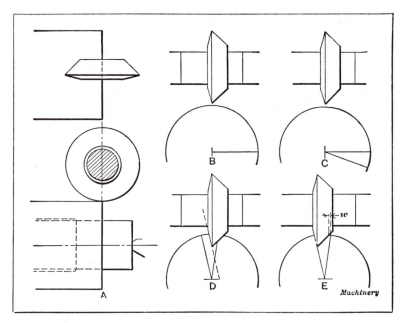

Fig. 7. Setting a Double-angle Cutter for Milling Teeth
of a Spiral Milling Cutter

to the lowest position at which the cutter will just graze the blank. The blank is then adjusted endwise until the axis of the cutter is in line with the end of the work, as shown by the side and plan views at A, Fig. 7. One method of locating the cutter in this position (after it has been set approximately) is to scribe a line on the blank a distance from the end equal to the radius of the cutter. The blade of a square is then set to this line, and the table is adjusted lengthwise until the cutter just touches the edge of the blade. The cutter can also be centered relative to the end (after it is set laterally) by first moving the blank endwise from beneath the cutter, and then feeding it back slowly until a tissue paper "feeler" shows that it just touches the corner of the blank. The relation between the cutter and blank will then be as shown at A.

The table is next set to the angle of the helix (as explained later) but its lengthwise position should not be changed. The surface gage, set as previously described, is then used to scribe lines which represent one of the tooth spaces on the end of the blank where the cut is to start. This is done by first drawing a horizontal line as at *B*. This line is then indexed downward an amount equal to one of the tooth spaces, and another horizontal line is drawn as at *C*. The last line scribed is then indexed 90 + 12 degrees, which locates it parallel with the 12-degree side of the cutter, as at *D*. The work is then adjusted laterally, and vertically by elevating the knee, until the cutter is so located that the 12-degree side cuts close to the scribed line, and, at the same time, the required width of land *w* (see sketch *E*) is left between the top edge of the groove and the line representing the front face of the next tooth. After the cutter is centered, as at *A*, the longitudinal position of the blank should not be changed until the cutter is set as at *E*, because any lengthwise adjustment of the work would be accompanied by a rotary movement (as the index-head is geared to the table feed-screw) and the position of the lines on the end would be changed.

How is the milling machine table set to the angle of the helix?

The table of the machine is set to the same angle that the helical grooves will make with the axis of the work unless the angle is beyond the range of the table's angular adjustment. The angular position of the table is shown by degree graduations on the base of the saddle. The reason for setting the work to the helix angle is to locate the cutter in line with the helical grooves which are to be milled by it. If the cutter were not in line, the shape of the grooves would not correspond with the shape of the cutter. The direction in which the table is turned depends upon whether the helix is right- or left-hand. For a right-hand helix the right-hand end of the table should be moved toward the rear, whereas if the helix is left-hand, the left-hand end of the table is moved toward the rear.

With most universal milling machines, it is inconvenient, if not impossible, to swivel the table to a greater angle than 45 degrees; hence, for greater angles, it is the general practice to leave the table in its normal position at right angles to the spindle of the machine, and use a special swiveling attachment for holding the cutter at the proper angle. An attachment of this kind is also used when milling helical grooves on a "plain" machine, as the table of this type cannot be swiveled.

How are helical grooves or flutes milled?

After the table of the machine is set to the required angle and is clamped in position, the work is ready to be milled. The actual milling of helical grooves is practically the same as though they were straight or parallel to the axis. When a groove is milled, it is well either to lower the table slightly or turn the cutter to such a position that the teeth will not drag over the work, when returning for another cut, to prevent scoring or marring the finished groove. If the work-table is lowered, it is returned to its original position by referring to the dial on the elevating screw.

After each successive groove is cut, the work is indexed by turning the indexing crank in the regular way. This operation of milling a groove and indexing is repeated until all the teeth are finished. The differential method of indexing cannot be employed in connection with helical milling, because with this system of indexing the worm-shaft of the index-head is geared to the spindle. When milling helical grooves, the position of the cutter with relation to the work should be such that the rotary movement for producing the spiral will be toward that side of the cutter which has the greater angle. To illustrate, the blank A, Fig. 6, should turn (as shown by the arrow) toward the 48-degree side of the cutter, as this tends to produce a smoother groove.

If a helical groove has parallel sides, how is it milled?

When a helical groove having parallel sides is required it should be cut with an end mill. If an attempt were made to mill a groove of this kind by using a side mill mounted on an arbor, the groove would not have parallel sides, because the side teeth of the mill would not clear the groove; in other words, they would cut away the sides owing to the rotary movement of the work and form a groove having a greater width at the top than at the bottom. This can be overcome, however, by using an end mill.

The machine is geared for the required lead of helix, as previously explained, and the work is adjusted vertically until its axis is in the same horizontal plane as the center of the end mill. This vertical adjustment can be obtained by moving the knee up until its top surface coincides with a line on the column marked *center;* the index-head centers will then be at the same height as the axis of the machine spindle.

In short-lead milling how are the change-gears arranged?

In milling helical grooves having an exceptionally short lead, the change gears of the dividing head may be arranged to transmit motion directly from the table feed-screw to the dividing head spindle, the worm being disengaged from the worm-wheel. If indexing is required, the change gear on the spindle should be some multiple of the number of divisions, to permit indexing by disengaging and turning the gear. With this direct drive, which may be utilized in milling worms, cylindrical cams of short lead, etc., the necessary rotation of the work is obtained with a much smaller driving gear on the feed-screw; consequently, the turning moment or twisting load is reduced proportionately and the danger of excessive feed-screw distortion or gear-tooth breakage is eliminated. If the table feed-screw has four threads per inch, which is standard for practically all milling machines, the change-gear ratio may be determined by the following formula:

$$\text{Change-gear ratio} = \frac{\text{Lead to be milled}}{0.25} = \frac{\text{Driven gears}}{\text{Driving gears}}$$

If the required lead is 0.5 inch and the lead of the machine is 10, then when the gearing is arranged in the usual manner, as shown in Fig. 2, the ratio of the driven to the driving gear equals 0.5 ÷ 10 = 0.05. With a direct drive, the ratio equals 0.5 ÷ 0.25 = 2, or 40 × 0.05, because the worm gearing is not included in the transmission.

Operations Requiring Angular Adjustment of Index-Head

When an index-head is used in connection with milling operations, the work may or may not be held between centers. Index-heads of the universal type, or those designed for handling a variety of work, are so constructed that the main spindle can be set to any angular position likely to be required. For example, the main spindle of some index-heads may be set to any angle from 5 or 10 degrees below the horizontal position to 10 or 15 degrees beyond the vertical position. In other words, there is a total angular adjustment of over 100 degrees. Such angular adjustment is needed when a part to be milled must not only be held at an angle, but must also be indexed or rotated as shown by examples to follow. The angular position of the index-head spindle is shown by degree graduations. The part to be milled may be held in a chuck mounted on the index-head spindle.

To what angle is the index-head set for milling a saw-tooth clutch?

In milling a saw-tooth clutch, the axis of the clutch blank should be inclined a certain angle a from the vertical, as shown at A in Fig. 1. If the teeth were milled with the blank vertical, the tops of the teeth would incline towards the center as at B, whereas, if the blank were set to such an angle that the tops of the teeth were square with the axis, the bottoms would incline upwards as at C. In either case, the two clutch members would not mesh completely, because the outer points of teeth cut as at B would bear at the bottom of the grooves in the opposite member, and the inner ends of teeth cut as at C would strike first, thus leaving spaces between the teeth around the outside of the

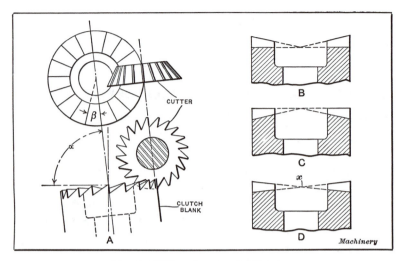

Fig. 1. Milling Saw-tooth Clutches

clutch. In order to secure better contact between the clutch teeth, they should be cut as indicated at *D*, or so that the bottoms and tops of the teeth have the same inclination, converging at a central point *x*. The teeth of both members will then engage across the entire width. The angle *a* required for cutting a clutch as at *B* can be determined by the following rule and formula in which *a* equals the required angle, and *N* the number of teeth:

Rule.—To determine the cosine of angle *a* (see diagram *A*, Fig. 1) find the tangent of the angle obtained by dividing 180 degrees by the number of teeth, and multiply this tangent by the cotangent of the cutter angle.

$$\cos a = \tan \frac{180 \text{ deg.}}{N} \times \cot \text{ cutter angle.}$$

Example.—A saw-tooth clutch is to have 8 teeth milled with a 60-degree cutter. To what angle *a* must the dividing-head be set?

$$\cos a = \tan \frac{180}{8} \times 0.57735 = 0.2391$$

From a table of trigonometric functions we find that the angle equivalent to this cosine 0.2391 is 76 degrees 10 minutes; hence, the index-head is set to this angle.

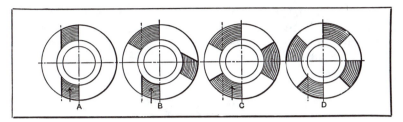

Fig. 2. Diagrammatical Views Showing Simple Method of Cutting Clutch Teeth

How are straight-tooth clutches milled?

A simple method of milling a straight-tooth clutch is indicated by the diagrams *A*, *B* and *C*, Fig. 2, which show the first, second, and third cuts required for forming the three teeth. The work is held either in the chuck of a dividing-head (the latter being set at right angles to the table), or in a plain vertical indexing attachment especially designed for this class of work. A plain milling cutter may be used (unless the corners of the clutch teeth are rounded), the side of the cutter being set to coincide with the centerline of the clutch. When the number of teeth in the clutch is odd, the cut can be taken clear across the blank as shown, thus finishing the sides of two teeth with one passage of the cutter. When the number of teeth is even, as at *D*, it is necessary to mill all the teeth on one side and then set the cutter for finishing the opposite side; therefore, clutches of this type commonly have an odd number of teeth. The maximum width of the cutter depends upon the width of the space at the narrow ends of the teeth. If the cutter must be quite narrow in order to pass the narrow ends, some stock may be left in the tooth spaces, which must be removed by a separate cut.

In milling end mills and side mills, what angular position is required for index-head?

When the end teeth of an end mill or the side teeth of a side mill are being cut, it is necessary to set the dividing-head at an angle as shown at *A* and *B*, in Fig. 3, in order to mill the lands or tops of the teeth to a uniform width.

Rule.—To determine this angle a, multiply the tangent of the angle between adjacent teeth (equals 360 ÷ number of teeth required) by the cotangent of the cutter angle B; the result is the cosine of angle a at which the dividing-head must be set.

$$\cos \text{ angle } a = \tan \frac{360°}{N} \times \cot \text{ cutter angle.}$$

Example.—An end mill is to have 10 teeth. At what angle should the dividing-head be set, assuming that a 70-degree fluting cutter is to be used?

The angle between adjacent teeth equals 360 ÷ 10 = 36 degrees. The tangent of 36 degrees is 0.72654 and the cotangent of 70 degrees is 0.36397; therefore, cosine of angle a to which the dividing-head should be set equals 0.72654 × 0.36397 = 0.26443. The angle whose cosine is 0.26443 is 74 degrees 40 minutes; hence, the dividing-head would be set to this angle.

The angle of elevation for cutting the side teeth of a side mill would be determined in the same way. Sketch B shows

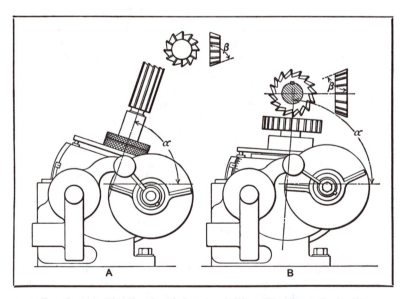

**Fig. 3. (A) Dividing-head Set for Milling Teeth of End-mill.
(B) Milling Side Teeth of Side-mill**

a dividing-head set for milling the side teeth of a side mill using a 70-degree cutter, the angle a being approximately 85½ degrees.

In milling angular cutters, how is angular position of dividing-head determined?

The angle to which the dividing-head must be set to mill teeth of uniform width at the top, depends upon two factors: The number of the teeth in the mill to be cut and the angle of the cutter with which the teeth are cut or fluted. The following formulas are for calculating the angle for setting the index-head:

a = setting-angle for dividing-head (Fig. 4) ;
B = angle of blank to be milled;
C = angle of fluting cutter;
T = tooth angle = 360 ÷ number of teeth to be milled;
D and E are angles indicated in Fig. 5.

Then:

$$\tan D = \frac{\cos T}{\tan B} \; ; \sin E = \tan T \cot C \sin D.$$

$$\text{angle } a = D - E$$

For end-mills or side milling cutters, where $B = 0$, the formulas can be reduced to:

$$\cos a = \tan T \cot C.$$

This formula, it will be seen, gives the same results as the one already given for side milling cutters.

Example.—To illustrate the application of the rule given, suppose that 18 teeth are to be milled in a 70-degree milling cutter blank, with a 60-degree single-angle cutter. To what angle a (see Fig. 4) must the dividing-head be set to obtain "lands" of uniform width? In this example, the angle T between adjacent teeth = 360 ÷ 18 = 20 degrees; blank angle B = 70 degrees; cutter angle C = 60 degrees.

$$\tan D = \frac{0.9396}{2.7474} = 0.342.$$

The angle the tangent of which is 0.342 is 18 degrees 53 minutes.

$$\sin E = 0.3639 \times 0.5773 \times 0.3236 = 0.0679.$$

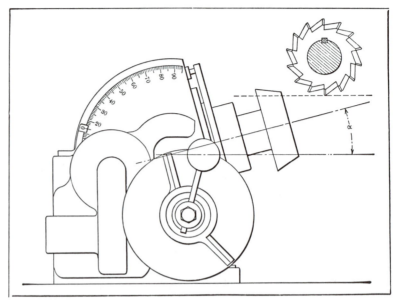

Fig. 4. Index-head Set at Angle of 15 Degrees for Cutting Teeth
in 70-degree Blank with 60-degree Cutter

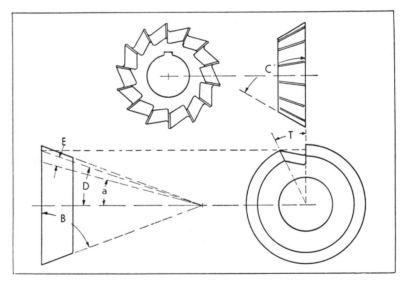

Fig. 5. Angles Involved in Calculation for Determining Position
of Index-head when Milling Teeth in Angular Cutter

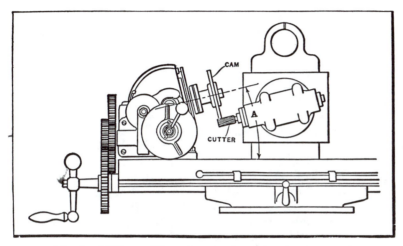

Fig. 6. Milling a Plate Cam by Use of Index-head and
Vertical Milling Attachment

The angle the sine of which is 0.0679 is 3 degrees 53 minutes, approximately. Therefore, angle a, at which the dividing-head should be set, equals the difference between angles D and E, or 18° 53′ — 3° 53′ = 15 degrees.

How is the index-head used for plate cam milling?

When cams having a constant rise are cut in the milling machine, the cam blank may be held on an arbor inserted in the spindle of the index- or dividing-head, and a vertical-spindle milling attachment used. The general arrangement of the index-head and the vertical milling attachment is shown by the diagram, Fig. 6. The index-head is geared to the table feed-screw, and suitable change-gears are selected to give the approximate lead to the cam (the lead represents the total rise for one complete revolution, assuming that it continued at a uniform rate). The index-head is elevated to obtain the exact lead, and the spindle of the vertical attachment is set in alignment with the index-head spindle so that the milling cutter will be parallel with the axis of the cam. The cutting is done by the teeth on the periphery of the end-mill, and the edge of the cam is milled at right angles to the sides.

Why is the index-head set at an angle for plate cam milling?

The reason why the spindle of the index-head is inclined, in order to obtain the required amount of lead, will be apparent by considering the result obtained when the spindle is in a vertical position and also the effect when it is in a horizontal position. When a cam is milled with the index-head set at 90 degrees or in a vertical position, the lead of the cam or its rise for one complete revolution will be the same as the lead for which the machine is geared. For instance, if a combination of gearing is used that advances the table 0.67 inch during one revolution of the spiral-head spindle, the lead of the cam for one complete revolution would equal 0.67 inch, assuming that the index-head was set in the 90-degree position. If the index-head is next set at zero or parallel to the surface of the table, and the spindle of the milling attachment is also set in the horizontal position, the cam will be milled concentric and the lead will be zero, because, as the table advances, the blank is rotated, but the distance between the axes of the index-head spindle and the milling cutter remains the same. By inclining the index-head, however, to any position between zero and 90 degrees, the cam can be given any lead, provided it is less than the lead for which the machine is geared; that is, less than the forward movement of the table for one revolution of the spiral-head spindle.

How is angular position of index-head calculated for plate cam milling?

The following formulas may be used for determining this angle of inclination. For a given rise of cam and with the machine geared for a certain lead, assume that:

A = angle to which index head and milling attachment are set (see Fig. 6);

r = rise of cam in given part of circumference;

R = lead of cam, or rise, if latter were continued at given rate for one complete revolution;

L = lead of spiral for which milling machine is geared;

N = part of circumference in which rise is required, expressed as a decimal in hundredths of cam circumference.

$$\sin A = \frac{R}{L}, \text{ and } R = \frac{r}{N}; \text{ hence, } \sin A = \frac{r}{N \times L}.$$

Example.—Suppose a cam is to be milled having a rise of 0.125 inch in 300 degrees or in 0.83 of the circumference, and that the machine is geared for the smallest possible lead, or 0.67 inch; then:

$$\sin A = \frac{r}{N \times L} = \frac{0.125}{0.83 \times 0.67} = 0.2247$$

This is approximately the sine of 13 degrees. Therefore, to secure a rise of 0.125 inch with the machine geared for 0.67 inch lead, the index-head is elevated to an angle of 13 degrees and the vertical milling attachment is also swiveled around to locate the cutter in line with the index-head spindle, so that the edge of the finished cam will be parallel to its axis of rotation.

When there are several lobes on a cam, having different leads, the machine can be geared for a lead somewhat in excess of the greatest lead on the cam, and then all the lobes can be milled without changing the index-head gearing, by simply varying the angle of the index-head and the cutter to suit the different cam leads. Whenever possible, it is advisable to mill on the under side of the cam, as there is less interference from chips; moreover, it is easier to see any lines that may be laid out on the cam face. To set the cam for a new cut, it is first turned back by operating the handle of the table feed-screw, after which the index-crank is disengaged from the plate and turned the required amount.

Milling Irregular Contours by Reproducing Shape of Model

There are two general methods of milling irregular forms. If the irregular shape extends along a part (or series of duplicate parts) in a straight line and does not vary in shape throughout its length, a formed milling cutter can be used as illustrated previously. In such cases, the cutter conforms to whatever cross-sectional shape is required on the work, and the milling process consists in merely reproducing this cutter shape as the work feeds along in a straight line past the revolving cutter.

A second general method produces an irregular contour or form by reproducing the shape of a master templet, form, or model. The machines employed in connection with the second method are so designed that the cutter follows a path conforming to the shape of the master or model, and, consequently, reproduces the shape of the master. Some of these machines are used simply for milling irregular contours, edges, or grooves. For this general class of work, the feeding movements are in one plane only, but they must be in any required direction in that plane. Other machines are designed to mill either raised forms or cavities which require three-dimensional movements. In other words, the universal movements in one plane (required for irregular contours) must be supplemented by movements perpendicular to this plane in order to mill either raised forms or cavities. Examples of both contour milling and three-dimensional form milling will be given in this section.

What general classes of work are milled on a profiling type of machine?

The profiling machine or "profiler" is one type of milling machine which is adapted for milling duplicate pieces having an irregular shape or contour, especially in connection with the interchangeable manufacture of parts for typewriters, sewing machines, fire-arms, and many other products. The milling cutter of a profiling machine, instead of revolving in a fixed position, is guided by a special former plate, the outline of which corresponds to the shape required on the work. Most "profilers" are hand-operated, so far as the feeding movements are concerned, although machines which are semi-automatic or entirely automatic, after the work is placed in position, are employed in some shops. Hand-operated machines usually have either one or two cutter spindles, the two-spindle type being commonly used. One spindle is often used for taking roughing cuts and the other for finishing cuts.

How are the cutter movements controlled in milling duplicate profiles?

Each cutter of a "profiler" has a former or guide pin which is located a fixed distance from the cutter and is guided around the former plate. This may be done by feeding the cutter-slide and pin laterally and the work-table in a longitudinal direction. Another method is to use a compound type of work-holding table or one which can be moved either lengthwise or crosswise. As the result of these movements, duplicate parts of irregular shape can be produced. As an example of profiling, an operation on the outside of a rifle, the finger lever, is shown in Fig. 1. One continuous cut is taken as indicated by arrow a to form the edges of that end of the lever which enters the rifle receiver and connects with the breech-bolt. The surfaces at the extreme end must be accurately machined as this end enters between the breech-bolt and a lug on the receiver (when the action is closed) to form a solid support for the bolt. The lever is located by pin A which enters the pivot-pin hole, and also by a small boss or pin at B which is solid with the lever and enters a hole in the base of the fixture. (This pin engages a cam slot in the breech-plug when the rifle is assembled.) The lever is held by a clamp C which is forced downward by a screw and lever D. The circular edge of the lever which is beneath this clamp is machined in a separate operation by a formed cutter, a milling machine being used. The former F is a plain type, shaped to correspond with the surface to be milled, and the path

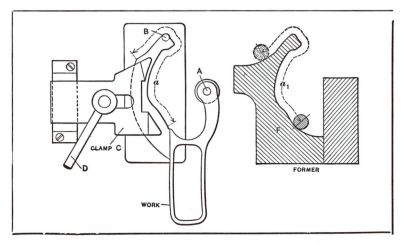

Fig. 1. Profiling Operation on Outside of Finger Lever

of the guide-pin is shown by the arrow a_1. The operation of a profiler is comparatively simple after it is properly set and adjusted, but many of the work-holding fixtures and former plates required for certain classes of work are ingenious.

How are profiling formers adapted to more than one operation?

Formers frequently have auxiliary plates which change the guiding surfaces for different operations. This general type of combination former is illustrated by the profiling operation shown in Fig. 2. The work is the action slide for a rifle. The operation consists in taking first a shallow cut, as indicated by the shaded area at A, and then a second cut, as indicated at B. These surfaces must be accurately machined both as to location and depth. In taking the first cut A, plate F on the former is in the working position. This plate swivels about screw G and is located by a pin H which is inserted through the plate and into the former base. The path of the guide-pin is indicated by the arrow a. When the first cut is completed, plate F is swung back out of the way, and plate F_1 is placed in the working position, as indicated by the dotted lines. The same pin H is also used for locating this plate. The second cut *B* is now taken, the guide-pin following path *b*. The hook-shaped end *J* on the former prevents the cut from running out at *K*. There is a long guard-plate near the right-hand side of the former (not shown in the illustration), which prevents the operator from moving the cutter over into contact with the left side of the former base. By having guards of this kind, the operator can shift the cutter-slide rapidly and without fear of feeding the cutter against some part of the work or fixture. The slide is located in the fixture by pin *L* at the end and the three half-pins along the right-hand edge, and is held by a single clamp which is cut away and beveled to clear the cutter.

Is the profile controlling templet ever mounted directly upon the part to be milled?

For some classes of work, the path followed by the cutter may be controlled by a more direct method than is employed on regular profiling machines. An example is shown in Fig. 3. The crankpin ends of master connecting-rods are milled to the required contour on a rotary milling machine. In this operation, a cutter 4½ inches in diameter by 3 inches

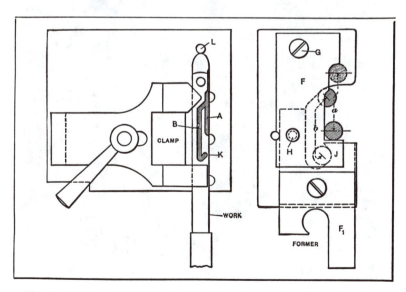

Fig. 2. Diagram Showing Combination Former Used for Profiling

wide is fed hydraulically toward the work fixture until a roller above the cutter comes into contact with a profile plate or templet mounted above the work; then, as the work revolves, the cutter-head moves according to the contour of the profile plate. To illustrate, the cutter-head is pushed back by the profile plate at the corners and permitted to advance under the hydraulic pressure between the corners. Stock up to a maximum of ¾ inch deep is removed. The master rod seen on the machine is approximately 20 inches long. It is bored and faced prior to the milling operation and is located from these finished surfaces for milling.

Fig. 3. Profile-controlling Templet Mounted upon the Work

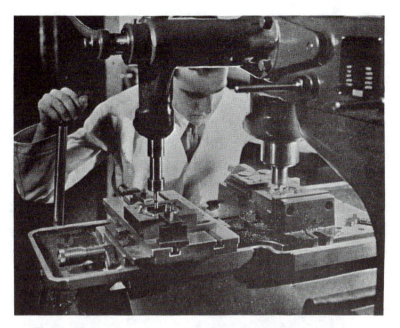

Fig. 4. Duplicating Type of Machine for Milling Dies, Molds, Etc., by Direct Reproduction of Model

What is the operating principle of a Gorton duplicator for die and mold milling?

The duplicator (see close-up rear view, Fig. 4) is equipped with a vertical milling spindle, a vertical tracer spindle, and a table capable of moving horizontally in any direction. These horizontal movements of the table, in conjunction with vertical movements of the cutter-spindle and tracer spindle, make it possible to reproduce or duplicate various classes of dies, molds, patterns, etc. The duplicator consists of a regular high-speed vertical-spindle milling machine equipped with a duplicator table and tracer head. In the illustration, the work is held in a machine vise mounted upon the duplicator table. At the operator's right, there is an auxiliary table for holding the master or model to be reproduced. This table may be adjusted longitudinally and laterally, and it is equipped with micrometers for setting the model in accurate relation to the work.

The vertical movements of the cutter-spindle and tracer spindle, which operate in unison, are controlled either by a hand-lever or handwheel manipulated by the workman's left hand. The right hand manipulates a hand-lever, as shown, for moving the duplicator table in whatever direction may be required. These combined movements make it possible to mill various classes of dies, plastic molds, glass molds, etc. The reproduction is always full size, there being no enlargement or reduction from the original model. The duplicator table, which is mounted on ball-bearing slides,

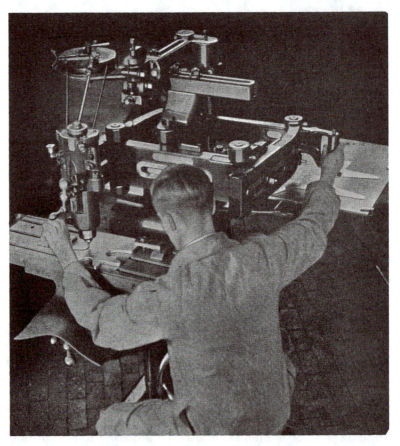

Fig. 5. Machine of Pantograph Type for Reproducing Master Templet or Model on Reduced Scale

may be clamped in position either when setting up work or for roughing out cavities by feeding with the regular milling machine table screws. These roughing cuts can be taken with standard end-mills; but for finishing operations, single-flute cutters are recommended. These cutters may have square, rounded, pointed, or other shapes, and they are easily resharpened or reground to whatever shape may be required.

Upon what principle does the pantograph type of profiler operate?

This type of machine is so designed that the required shape or contour can be reproduced from an *enlarged* model. For example, if a part must be milled to a certain shape or contour, an enlarged master templet is made to the same contour. Fig. 5 shows a typical application of a Gorton machine. The operator guides the tracing point around the master templet at the right and the cutter at the left reproduces the templet shape on a reduced scale. The master templet, which may be of brass or steel, frequently is from three to five times larger than the work, so that any slight errors which may exist in the templet are reduced proportionally on the milled part. The reduction is obtained by a pantograph mechanism. A pantograph is a combination of links which are so connected and proportioned as to length that any motion of one point in a plane parallel to that of the link mechanism will cause another point to follow a similar path either on an enlarged or a reduced scale. Such a mechanism may be used as a reducing motion. For instance, most engraving machines have a pantograph mechanism interposed between the tool and a tracing point which is guided along lines or grooves of a model or pattern. As the tracing point moves, the cutting tool follows a similar path, but to a reduced scale, and cuts the required pattern or design on the work. A pantograph mechanism, as applied to a profiling machine, may be designed to give reductions between the templet and work ranging from no reduction or 1 to 1 down to 6 to 1. The pantograph is provided with scales or graduations showing the adjustment required for obtaining any reduction within the minimum and maximum range. Machines of this kind are used in making blanking dies, templets, or for milling various irregular slots, contours, etc., on parts of machines or instruments.

How is the pantograph type of machine used for milling raised or sunken forms?

Machines are available for milling practically any shape required, by reproducing the shape of a model or pattern. These machines of the "three-dimension" type are used in milling molds for plastics or glass, molds for die-castings,

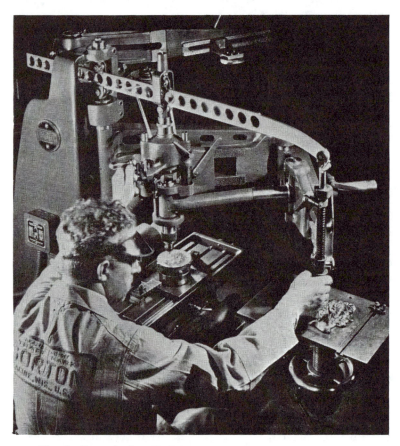

Fig. 6. Reproducing Model (seen at Right) on Machine of Three-dimensional Pantograph Type

dies for silverware, or other forming operations, and similar classes of work. A Gorton three-dimensional pantograph type of machine is shown in Fig. 6. The pattern or model to be reproduced (which may be seen at the right) is larger than the work, to insure accurate reproduction of all of the finer details. As a tracer point is moved over the surface of the model, the cutter follows a similar path but on a reduced scale, as determined by the adjustment of the pantograph mechanism. The pantograph bars may be graduated for reductions ranging from 2 to 1 to 8 to 1, or more. As a general rule, reductions of 2, 3 and 4 ordinarily are employed. When this machine is in use, both cutter-spindle and tracer spindle remain vertical to the work and to the model or pattern. In other words, any vertical movement of the tracer, as it is traversed across the model, is imparted to the cutter-spindle in a reduced ratio as determined by the pantograph adjustment.

What type of machine for irregular form milling automatically reproduces the master or model?

The Keller machine (see Fig. 7) may be operated either manually or with automatic control of the tracer and cutter. Machines of this general type are used for milling various sunken or raised forms on molds, dies, etc., and for milling certain machine parts. The master form or templet to be reproduced is the same size as the work and is attached to a vertical work-holder above the work itself. When the operation is simply that of contour milling, the cutter is set to the required depth and remains at this depth while the tracer and master templet control the horizontal and vertical movements of the machine. For these contouring operations, a thin sheet-metal templet ordinarily is used. Fig. 7 shows how four sprockets are milled simultaneously to the required form from steel disks. The desired contour of the sprocket teeth is produced by a tracer which automatically follows the sheet-steel templet seen above the work. A side-cutting end-mill as wide as the four sprockets is mounted on the tool-head. These sprockets are approximately 28 inches in outside diameter.

A second type of tracer control is used for three-dimensional work such as milling sunken or raised forms.

Tracer control in this case is automatic. The tracer continuously seeks contact with the surface of a solid master model, as the automatic longitudinal and transverse feeding movements cause the cutter to cover the entire surface, thus finally reproducing the complete shape of the model. This method is employed for milling various classes of molds, forming dies, forging dies and die-casting dies, as well as other kinds of work. Another type of tracer control regulates the horizontal and vertical movements, as in contour milling, by following around the edge of a master while, at the same time, controlling whatever depth variations may be required. This control may be utilized for such work as cutting a small groove or bead which extends irregularly over a surface of varying depth.

Cutters Used.—Several different forms of cutters have been designed especially for use on Keller machines. For

Fig. 7. Milling the Teeth on Four Tank Sprockets at a Time on a Keller Type of Machine

example, the end of the cutter may be square, partially rounded, or fully rounded to a ball shape. For ordinary profiling, a mill having a right-hand cut and left-hand spiral flutes is recommended. In milling molds or dies requiring a slight taper or draft around the sides, a cutter having a taper of 5 to 7 degrees may be used. For very fine delicate engraving work, where fine details must be duplicated accurately, a sharp pointed type of cutter is used. For some classes of work, special cutters are required.

Tracer Points.—Tracer points may be cylindrical or tapering. If cylindrical, the ends may be square, partly rounded, or fully rounded to a ball form to suit the type of cutter. These tracer points are also made to various sizes or diameters in order to adapt them to different classes of work. For example, a tracer of a certain size or shape might make proper contact with one model but not with another of different form.

Cutting Spur Gears by Milling with Formed Cutters

The gear-cutting processes utilized for producing different types of gears may be divided into three very general classes. One includes the use of tools or cutters which form gear teeth by reproducing the shape of the cutter itself; in another class are the generating processes whereby the proper tooth curves are formed through relative motions of the cutter and gear, as when a straight-sided cutting tool generates the required tooth curves due to the generating movements. The third general classification includes the use of templets or master formers, which control the path followed by the cutting tool, and consequently the curvature of the gear tooth. This method is applied chiefly to the cutting of very large gears. These various processes are utilized in manufacturing plants of different kinds, because the method of cutting gears, in common with other manufacturing processes, depends upon such factors as the type of gear, its size, the degree of accuracy necessary, and the required rate of production. For instance, the gear-cutting practice in a general repair shop would ordinarily differ from the methods used in the average plant manufacturing, say, machine tools; and still other variations are found in automobile plants where production is on a very large scale.

Under what conditions are milling machines used for gear cutting?

Standard types of milling machines are often used for gear cutting (as illustrated in Fig. 1), particularly in repair and jobbing shops, or wherever this work is done on such a small scale that the installation of a regular gear-cutting machine is not warranted. Milling machines are used especially for gears of small or medium size or for odd sizes, and they may be arranged for cutting spur gears, helical (or spiral) gears, bevel gears, and worm gears. The plain column-and-knee type of milling machine equipped with a dividing-head, may be employed either for cutting spur or bevel gears, but the universal type is required for helical gears unless a milling attachment is used having the angular adjustment required for holding the cutter in alignment with the teeth. The cutting of spur gears will be dealt with first.

What is the "pitch diameter" of a spur gear and why is it such an important dimension?

The pitch diameter D of a spur gear (see Fig. 2) is very important because it represents the *effective size* of the gear. Since the pitch diameter is the diameter of the *pitch circle*, we shall explain first the meaning of the latter term.

If plain disks were mounted upon parallel shafts with the edges or peripheries of these disks held firmly in contact, one shaft would drive or rotate the other provided the resistance of the driven shaft were not great enough to cause the driving disk to slip upon the driven disk. This

Fig. 1. Cutting a Spur Gear in a Milling Machine

simple arrangement, which is sometimes used, enables motion to be transmitted smoothly and uniformly, and it would be ideal were it not for the fact that most of the machines or devices in which gearing is used require a *positive* drive or one that is not dependent merely upon the frictional resistance of two disks or surfaces rolling together. To obtain a positive drive, gearing having intermeshing teeth is very generally used, and in cutting the teeth of spur

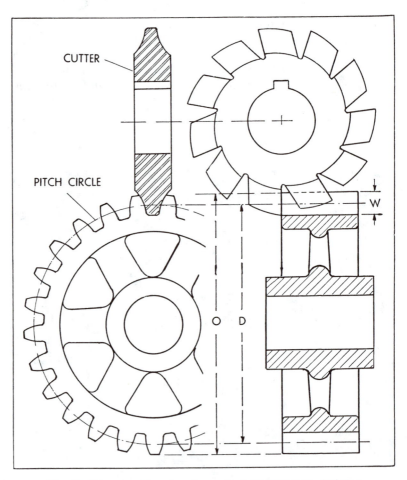

Fig. 2. Diagram Showing Important Dimensions Related to Spur Gear Cutting

gearing, the idea is to obtain teeth of such form or profile that the action of the gears will be like plain disks rolling together, the motion being transmitted smoothly and at a uniform rate. (Similarly, bevel gearing is intended to reproduce the action of two frustums of cones rolling in contact with each other.) There are various curves which might be applied to gear teeth in order to secure rotation between two gears having intermeshing teeth, but the involute curve (with more or less modification) is used almost universally because it has certain practical advantages. The pitch circles of two gears operating together correspond to the plain disks previously referred to, and they are tangent to each other at the point of intersection with the common center line. The ratio of the gearing depends upon the diameters of the pitch circles, and examples to follow will show how the pitch diameter is determined.

How are gear tooth sizes indicated on a drawing?

The gear tooth size is shown by a figure representing the pitch of the teeth. The common method of designating the size of the teeth of cut gearing is by giving the *diametral pitch*, which is a number representing the number of gear teeth per inch of pitch diameter. For example, a gear of 6 diametral pitch has 6 teeth around its circumference for each inch of pitch diameter. Therefore, diametral pitch is a ratio and not an actual dimension.

The *circular pitch*, which is sometimes used for designating the sizes of very large gear teeth and especially for cast gears, is equal to the actual dimension from the center of one tooth to the center of the next one measured along the pitch circle. If 3.1416 is divided by the circular pitch, the equivalent diametral pitch will be found, and inversely, if 3.1416 is divided by the diametral pitch, the result will equal the circular pitch.

The pitch of the teeth depends upon the power-transmitting capacity of the gearing and it is established by the gear designer. Allowable loads, especially for high-speed power-transmission gears, may be based either upon tooth strength or upon tooth wear, rather than strength, to avoid excessive wear.

If the pitch of a spur gear is known, how is the pitch diameter determined?

Ordinarily a pair of gears is designed for a given center-to-center distance, but involute gears might be separated somewhat and it would still be possible for one to transmit motion to the other at a uniform rate, although this would cause a certain amount of backlash or play between the intermeshing teeth. However, if two gears were assembled so that the center distance was somewhat greater than standard (as is sometimes done to meet a special condition) the pitch circles would be enlarged. The reason why the pitch circles and pitch diameters would become larger will be apparent when it is remembered that the radius of each pitch circle is equal to the distance from the gear center to the *pitch point* or the point where the line of action intersects the common center line. Notwithstanding the fact that the pitch diameters might be varied in this way, the term *pitch diameter* as ordinarily applied to gearing means the diameter obtained by dividing the number of teeth by the diametral pitch. The pitch diameter obtained in this way relates to the diameter corresponding to a standard center-to-center distance.

Rule 1.—If diametral pitch is known, divide number of teeth by the diametral pitch

$$D = \frac{N}{P}$$

Example.—A gear has 40 teeth of 4 diametral pitch; then

$$\text{Pitch diameter} = \frac{40}{4} = 10 \text{ inches}$$

Note: The term "diameter" as applied to a gear means the pitch diameter or diameter of the pitch circle.

Rule 2.—If circular pitch is known, multiply number of teeth by circular pitch and divide product by 3.1416.

Example.—A gear has 40 teeth of 0.7854 inch circular pitch; then

$$\text{Pitch diameter} = \frac{40 \times 0.7854}{3.1416} = 10 \text{ inches}$$

A circular pitch of 0.7854 is used in this example because it is the equivalent of 4 diametral pitch; hence, the same pitch diameter is obtained as in the preceding example.

How is the pitch and pitch diameter of a sample gear determined?

Suppose an old gear is to be replaced and the diametral pitch is required.

Rule.—To find diametral pitch, add 2 to the number of teeth and divide sum by outside diameter.

Example.—A sample spur gear has 40 teeth and the outside diameter, or diameter measured across the tops of the teeth, is 10.5 inches. Determine the diametral pitch so that a duplicate gear can be cut.

$$\text{Diametral pitch} = \frac{40 + 2}{10.5} = 4$$

Assume, in this case, that the outside diameter had been turned to $10\frac{15}{32}$ inches or about 1/32 inch under the standard size; then the diametral pitch, as obtained by the calculation, would be about 4.01, thus indicating that 4 is the pitch since the standard pitches are never odd fractional numbers.

Rule.—To find the pitch diameter, multiply number of teeth by the outside diameter and divide product by number of teeth plus 2.

Example.—A sample gear has 40 teeth and the outside diameter is 10½ inches; then

$$\text{Pitch diameter} = \frac{40 \times 10.5}{40 + 2} = 10 \text{ inches.}$$

How is a spur gear cut to a given pitch?

The pitch is obtained by using a cutter of whatever pitch is required, in conjunction with correct tooth spacing. This spacing is obtained with an index-head when a milling machine is used for gear cutting. Since the diametral pitch system is generally used, the cutter (which is obtained from a cutter manufacturer) usually conforms to some standard diametral pitch. A list of common pitches will be found in the table "Diametral Pitches and Equivalent Gear Tooth Dimensions." Gear designers try to avoid odd or special pitches as far as possible to permit using cutters which are standard and are carried in stock by cutter manufacturers. For example, if strength calculations or actual tests showed that teeth of 5 diametral pitch are not quite large enough, then 4 diametral pitch would be used because

Diametral Pitches and Equivalent Gear Tooth Dimensions*

Diametral Pitch	Circular Pitch	Thickness of Tooth on Pitch Line	Addendum	Working Depth of Tooth	Depth of Space below Pitch Line	Whole Depth of Tooth
½	6.2832	3.1416	2.0000	4.0000	2.3142	4.3142
¾	4.1888	2.0944	1.3333	2.6666	1.5428	2.8761
1	3.1416	1.5708	1.0000	2.0000	1.1571	2.1571
1¼	2.5133	1.2566	0.8000	1.6000	0.9257	1.7257
1½	2.0944	1.0472	0.6666	1.3333	0.7714	1.4381
1¾	1.7952	0.8976	0.5714	1.1429	0.6612	1.2326
2	1.5708	0.7854	0.5000	1.0000	0.5785	1.0785
2¼	1.3963	0.6981	0.4444	0.8888	0.5143	0.9587
2½	1.2566	0.6283	0.4000	0.8000	0.4628	0.8628
2¾	1.1424	0.5712	0.3636	0.7273	0.4208	0.7844
3	1.0472	0.5236	0.3333	0.6666	0.3857	0.7190
3½	0.8976	0.4488	0.2857	0.5714	0.3306	0.6163
4	0.7854	0.3927	0.2500	0.5000	0.2893	0.5393
5	0.6283	0.3142	0.2000	0.4000	0.2314	0.4314
6	0.5236	0.2618	0.1666	0.3333	0.1928	0.3595
7	0.4488	0.2244	0.1429	0.2857	0.1653	0.3081
8	0.3927	0.1963	0.1250	0.2500	0.1446	0.2696
9	0.3491	0.1745	0.1111	0.2222	0.1286	0.2397
10	0.3142	0.1571	0.1000	0.2000	0.1157	0.2157
11	0.2856	0.1428	0.0909	0.1818	0.1052	0.1961
12	0.2618	0.1309	0.0833	0.1666	0.0964	0.1798
13	0.2417	0.1208	0.0769	0.1538	0.0890	0.1659
14	0.2244	0.1122	0.0714	0.1429	0.0826	0.1541
15	0.2094	0.1047	0.0666	0.1333	0.0771	0.1438
16	0.1963	0.0982	0.0625	0.1250	0.0723	0.1348
17	0.1848	0.0924	0.0588	0.1176	0.0681	0.1269
18	0.1745	0.0873	0.0555	0.1111	0.0643	0.1198
19	0.1653	0.0827	0.0526	0.1053	0.0609	0.1135
20	0.1571	0.0785	0.0500	0.1000	0.0579	0.1079
22	0.1428	0.0714	0.0455	0.0909	0.0526	0.0980
24	0.1309	0.0654	0.0417	0.0833	0.0482	0.0898
26	0.1208	0.0604	0.0385	0.0769	0.0445	0.0829
28	0.1122	0.0561	0.0357	0.0714	0.0413	0.0770
30	0.1047	0.0524	0.0333	0.0666	0.0386	0.0719
32	0.0982	0.0491	0.0312	0.0625	0.0362	0.0674
34	0.0924	0.0462	0.0294	0.0588	0.0340	0.0634
36	0.0873	0.0436	0.0278	0.0555	0.0321	0.0599
38	0.0827	0.0413	0.0263	0.0526	0.0304	0.0568
40	0.0785	0.0393	0.0250	0.0500	0.0289	0.0539
42	0.0748	0.0374	0.0238	0.0476	0.0275	0.0514
44	0.0714	0.0357	0.0227	0.0455	0.0263	0.0490
46	0.0683	0.0341	0.0217	0.0435	0.0252	0.0469
48	0.0654	0.0327	0.0208	0.0417	0.0241	0.0449
50	0.0628	0.0314	0.0200	0.0400	0.0231	0.0431

*The addendum and depth dimensions apply to standard full-depth teeth.

Circular Pitches and Equivalent Gear Tooth Dimensions*

Circular Pitch	Diametral Pitch	Thickness of Tooth on Pitch Line	Addendum	Working Depth of Tooth	Depth of Space below Pitch Line	Whole Depth of Tooth
4	0.7854	2.0000	1.2732	2.5464	1.4732	2.7464
3½	0.8976	1.7500	1.1140	2.2281	1.2890	2.4031
3	1.0472	1.5000	0.9549	1.9098	1.1049	2.0598
2¾	1.1424	1.3750	0.8753	1.7506	1.0128	1.8881
2½	1.2566	1.2500	0.7957	1.5915	0.9207	1.7165
2¼	1.3963	1.1250	0.7162	1.4323	0.8287	1.5448
2	1.5708	1.0000	0.6366	1.2732	0.7366	1.3732
1⅞	1.6755	0.9375	0.5968	1.1937	0.6906	1.2874
1¾	1.7952	0.8750	0.5570	1.1141	0.6445	1.2016
1⅝	1.9333	0.8125	0.5173	1.0345	0.5985	1.1158
1½	2.0944	0.7500	0.4775	0.9549	0.5525	1.0299
1⁷⁄₁₆	2.1855	0.7187	0.4576	0.9151	0.5294	0.9870
1⅜	2.2848	0.6875	0.4377	0.8754	0.5064	0.9441
1⁵⁄₁₆	2.3936	0.6562	0.4178	0.8356	0.4834	0.9012
1¼	2.5133	0.6250	0.3979	0.7958	0.4604	0.8583
1³⁄₁₆	2.6456	0.5937	0.3780	0.7560	0.4374	0.8154
1⅛	2.7925	0.5625	0.3581	0.7162	0.4143	0.7724
1¹⁄₁₆	2.9568	0.5312	0.3382	0.6764	0.3913	0.7295
1	3.1416	0.5000	0.3183	0.6366	0.3683	0.6866
¹⁵⁄₁₆	3.3510	0.4687	0.2984	0.5968	0.3453	0.6437
⅞	3.5904	0.4375	0.2785	0.5570	0.3223	0.6007
¹³⁄₁₆	3.8666	0.4062	0.2586	0.5173	0.2993	0.5579
¾	4.1888	0.3750	0.2387	0.4775	0.2762	0.5150
¹¹⁄₁₆	4.5696	0.3437	0.2189	0.4377	0.2532	0.4720
⅔	4.7124	0.3333	0.2122	0.4244	0.2455	0.4577
⅝	5.0265	0.3125	0.1989	0.3979	0.2301	0.4291
⁹⁄₁₆	5.5851	0.2812	0.1790	0.3581	0.2071	0.3862
½	6.2832	0.2500	0.1592	0.3183	0.1842	0.3433
⁷⁄₁₆	7.1808	0.2187	0.1393	0.2785	0.1611	0.3003
2/5	7.8540	0.2000	0.1273	0.2546	0.1473	0.2746
⅜	8.3776	0.1875	0.1194	0.2387	0.1381	0.2575
1/3	9.4248	0.1666	0.1061	0.2122	0.1228	0.2289
⁵⁄₁₆	10.0531	0.1562	0.0995	0.1989	0.1151	0.2146
2/7	10.9956	0.1429	0.0909	0.1819	0.1052	0.1962
¼	12.5664	0.1250	0.0796	0.1591	0.0921	0.1716
2/9	14.1372	0.1111	0.0707	0.1415	0.0818	0.1526
1/5	15.7080	0.1000	0.0637	0.1273	0.0737	0.1373
³⁄₁₆	16.7552	0.0937	0.0597	0.1194	0.0690	0.1287
1/6	18.8496	0.0833	0.0531	0.1061	0.0614	0.1144
1/7	21.9911	0.0714	0.0455	0.0910	0.0526	0.0981
⅛	25.1327	0.0625	0.0398	0.0796	0.0460	0.0858
1/9	28.2743	0.0555	0.0354	0.0707	0.0409	0.0763
¹⁄₁₀	31.4159	0.0500	0.0318	0.0637	0.0368	0.0687
¹⁄₁₆	50.2655	0.0312	0.0199	0.0398	0.0230	0.0429

* The addendum and depth dimensions apply to standard full-depth teeth.

this is the next larger size and is a standard or commercial pitch. Note that tooth sizes increase as the numbers indicating diametral pitch decrease. For example, a tooth of 1 diametral pitch has a thickness on the pitch circle of 1.5708 inches and a total depth of 2.1571 inches. The thickness for 5 diametral pitch is only 0.3142 inch and the depth 0.4314 inch.

Can the pitch be regulated to obtain any gear diameters and center distances required?

The diametral pitch system is so arranged as to provide a series of tooth sizes, just as the pitches of screw threads are standardized. Inasmuch as there must be a whole number of teeth in each gear, it is apparent that gears of a given pitch vary in diameter according to the number of teeth. Suppose for example, that a series of gears are of 4 diametral pitch. Then the pitch diameter of a gear having, say, 20 teeth will be 5 inches; 21 teeth, $5\frac{1}{4}$ inches; 22 teeth, $5\frac{1}{2}$ inches, and so on. It will be seen that the increase in diameter for each additional tooth is equal to $\frac{1}{4}$ inch for 4 diametral pitch. Similarly, for 2 diametral pitch the variations for successive numbers of teeth would equal $\frac{1}{2}$ inch, and for 10 diametral pitch the variations would equal 1/10 inch, etc.

The center-to-center distance between two gears is equal to one-half the total number of teeth in the gears divided by the diametral pitch. While it may be desirable at times to have a center distance which cannot be obtained exactly by any combination of gearing of given diametral pitch, this is an unusual condition and ordinarily the designer of a machine can alter the center distance whatever slight amount may be required for gearing of the desired ratio and pitch. By using a standard system of pitches all calculations are simplified, and it is also possible to obtain the benefits of standardization in the manufacturing of gears and gear-cutters.

To what diameter is a spur gear blank turned before the teeth are cut?

When driving and driven gears are placed together, or in the running position, there is a clearance space between the tops of the teeth and the bottoms of adjacent tooth spaces. This clearance is to prevent the tops of the teeth from bearing upon the bottoms of mating tooth spaces. If the outside diameter O, Fig. 2, of a gear is correct, the clearance space equals 0.157 divided by the diametral pitch; for example, if the diametral pitch is 4, the clearance equals 0.039 inch or a little over 1/32 inch. If a slight error in the outside diameter only affected the clearance space, it would be of no practical importance. There is an advantage, however, in having this outside diameter correct, so that it can be used in adjusting the machine for the right depth of cut as explained later.

Rule 1.—If diametral pitch is known, add 2 to the number of teeth and divide sum by diametral pitch to obtain the outside diameter.

Example.—A gear is to have 40 teeth of 4 diametral pitch. Determine outside diameter or diameter to which gear blank is turned.

$$\text{Outside diameter} = \frac{40+2}{4} = 10\ 1/2 \text{ inches}$$

Rule 2.—If circular pitch is known, add 2 to number of teeth and multiply sum by circular pitch; then divide product by 3.1416.

Example.—A gear has 40 teeth of 0.7854 inch circular pitch; then

$$\text{Outside diameter} = \frac{(40+2) \times 0.7854}{3.1416} = 10.5 \text{ inches}$$

How is the spacing of the gear teeth accurately controlled?

When a milling machine is used, the required tooth spacing is obtained usually by using a dividing- or index-head. A common method of holding a gear blank in position for cutting is by mounting it upon a mandrel (see Fig. 1). This mandrel should preferably have a taper shank that fits into the dividing-head spindle, but if a mandrel having centers in each end is employed, it is important to take up all play between the driving dog and faceplate to insure holding the gear securely and obtain accurate indexing. If the gear is too large to swing between the dividing-head centers, the swing may be increased by inserting parallel blocks beneath the dividing-head and tailstock. Another method is to place

Fig. 3. Index Centers for Holding Gears that are too Large
for the Regular Centers

the dividing-head spindle in a vertical position and hold
the gear on a short arbor and as close as possible to the
dividing-head. On the cutting side, the gear rim should also
be supported by one or two vertical studs or rods to take
the downward thrust of the cut. For milling comparatively
large spur gears, a gear-cutting attachment is preferable to
the regular dividing-head. This attachment in its usual
form is similar to a dividing-head, but is larger and heavier
in construction (see Brown & Sharpe attachment, Fig. 3).
Still another method of cutting fairly large gears on the
milling machine is by using a circular milling attachment
(see Fig. 4). The gear blank is clamped onto the table of
the attachment, although parallels are usually placed beneath it to provide clearance for the cutter which is held
on the regular arbor in the usual manner. The gear is
located in a central or concentric position by the finished
bore of the hub, which fits over a plug inserted in the central

hole of the attachment table. The tooth spaces are milled by using the vertical feed of the machine. It will be noted that this attachment is equipped with an index plate and crank for obtaining the required number of teeth. By holding a large gear on the circular attachment, the rim is more rigidly supported than when the gear is held on an arbor mounted between the centers of a dividing head.

Why is a milling cutter for spur gears selected to suit both the pitch and the number of gear teeth?

When a spur gear is cut by milling, the shape of the milled space (or curvature of the teeth) is a direct reproduction of the cutter shape. A cutter of given pitch, how-

Fig. 4. Cutting Spur Gear which is Held on
Circular Milling Attachment

ever, cannot be employed for milling *any* gear of that pitch because the shapes of the spaces between the teeth vary according to the number of teeth in a gear. For instance, the tooth spaces of a small gear or pinion are not of the same shape as the spaces of a large gear of equal pitch. Theoretically, there should be a different cutter for every tooth number, but such refinement is unnecessary in practice.

Series of Cutters for Each Pitch.—The involute cutters commonly used are made in series of eight cutters for each diametral pitch. Cutter No. 1 is intended for gears ranging from 135 teeth to a rack; No. 2 from 55 to 134 teeth; No. 3 from 35 to 54 teeth; No. 4 from 26 to 34 teeth; No. 5 from 21 to 25 teeth; No. 6 from 17 to 20 teeth; No. 7 from 14 to 16 teeth; and No. 8 for 12 and 13 teeth. The shape of each cutter in this series is correct for a certain number of teeth only, but it can be used for other numbers within the limits given. For instance, a No. 6 cutter may be used for gears having from 17 to 20 teeth, but the tooth outline is correct only for 17 teeth or the lowest number in the range, which is also true of the other cutters listed. When this cutter is used for a gear having, say, 19 teeth, too much material is removed from the upper surfaces of the teeth, although the gear meets ordinary requirements.

Intermediate Series or Half Numbers.—When greater accuracy of tooth shape is desired to insure smoother or quieter operation, an intermediate series of cutters having half numbers is used. The half numbered cutters made by the Brown & Sharpe Mfg. Co. are for the following ranges of tooth numbers: Cutter No. $1\frac{1}{2}$, 80 to 134 teeth; No. $2\frac{1}{2}$, 42 to 54; No. $3\frac{1}{2}$, 30 to 34; No. $4\frac{1}{2}$, 23 to 25; $5\frac{1}{2}$, 19 and 20; $6\frac{1}{2}$, 15 and 16; $7\frac{1}{2}$, 13 teeth. There are seven cutters in this series, No. $8\frac{1}{2}$ being omitted since this would be for a pinion with less than 12 teeth.

Why should the cutter be centered accurately relative to gear-holding mandrel?

It is very important to have the cutter accurately centered in relation to the work-holding mandrel so that the center lines of the teeth will be radial (see sectional view of cutter, Fig. 2). One method of centering the cutter requires a true mandrel which is placed between the dividing-head centers. The machine table is adjusted horizontally and vertically until one corner of the cutter just grazes the mandrel on one side, and as far down as possible; the dials on both horizontal and vertical feed screws are then set at zero. The machine table is next adjusted until the mandrel is in a similar position on the opposite side of the cutter, the latter just grazing it again when the vertical feed screw dial is in the zero position. The backlash of both feed screws should also be taken up in the same direction as with the first setting of the mandrel relative to the cutter. The distance between the first and second positions of the mandrel is now determined by referring to the dial of the horizontal feed screw and this dimension is divided by 2 to obtain the central location for the mandrel. In adjusting the mandrel to this position beneath the cutter, the backlash between the feed screw and nut should again be taken up in the same direction as previously.

How is the trial-cut method applied in testing the position of the cutter?

An accurate method of testing the position of the cutter relative to the mandrel is by first milling a tooth space in a trial blank having the same diameter as the gear blank. The position of this trial blank is then reversed on the mandrel, and the blank, which should have a sliding fit, is pushed along until the cutter enters in the tooth space previously milled. The cutter is next revolved slowly by hand, and if it is accurately centered, the second cut will follow the first. On the contrary, if it is not exactly central, some metal will be removed from the upper part of one tooth and the lower part of the other. In order to center the gear blank relative to the cutter, the cross-slide should be moved toward that side of the tooth space which was milled away at the bottom. Another trial cut is then taken, and the test is repeated. The saddle should be clamped in position after the cutter has been set.

To what depth should the gear-tooth spaces be milled?

The next step is to set the cutter for milling tooth spaces of the proper depth, W, Fig. 2. If the outside diameter of the gear blank is accurate, this can be done by first adjusting the blank upward until the revolving cutter just grazes its surface. The dial of the elevating screw is then set at zero, after which the blank is moved horizontally to clear the cutter, and then vertically the required amount, as shown by the micrometer dial. This vertical adjustment should equal the total depth of the tooth space. The rules which follow apply to full-depth teeth and not to "stub teeth."

Rule 1.—To find total depth of tooth space, divide 2.157 by the diametral pitch.

Example.—A gear is to have teeth of 4 diametral pitch; determine total depth or depth of cut.

$$\text{Whole depth} = \frac{2.157}{4} = 0.5393 \text{ inch}$$

Rule 2.—To find total depth when circular pitch is given, multiply circular pitch by 0.6866.

Are roughing and finishing cuts required for milling gear teeth?

After making the adjustments previously described, a gear tooth is formed by first milling a tooth space across the gear rim and then indexing whatever fractional part of a revolution is required for the number of teeth to be cut. This operation of milling a tooth space and indexing is repeated until all the teeth are formed. If the gear teeth are of fairly coarse pitch, it may be desirable to take roughing and finishing cuts. This is often done in cutting gears coarser than 6 or 7 diametral pitch, although a definite dividing line cannot be drawn owing to variations both in regard to the cutting capacity of the machine itself and the cutting qualities of the stock. When roughing cuts are considered desirable, special stocking cutters are often used for the roughing operations, as they are more efficient for removing stock rapidly than the regular formed cutters. When a second cut is taken for finishing, it is important to have the roughed-out tooth spaces central with the cutter to avoid removing unequal amounts of metal on the sides, since this tends to wear the cutting edges unevenly and produce inaccurate teeth. If roughing cuts are taken, the "stocking" cutter should be set to mill the teeth the full depth, the allowance for finishing being on the sides.

In checking the size of a gear tooth, how is chordal thickness found?

It is advisable to check the accuracy of the adjustments by measuring the thickness of the first tooth milled. The cutter is allowed to feed in at one side of the blank far enough to form a complete tooth surface for a short distance and then the gear is indexed, after which another trial cut is taken part way across the gear blank, thus forming a short section of a tooth. A vernier gear tooth caliper is then used to measure the thickness (see Fig. 5). Such a caliper does not, of course, measure the thickness along the pitch circle, but the *chordal* thickness T; consequently, this chordal thickness, which is slightly less than

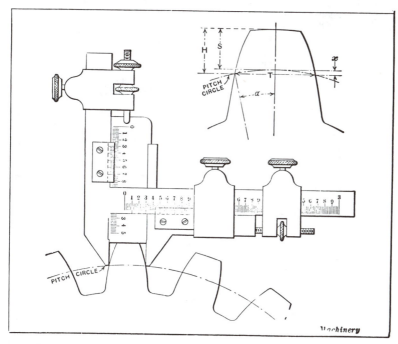

Fig. 5. Vernier Caliper for Measuring the Thickness of Gear Tooth at the Pitch Circle

Chordal Thicknesses and Corrected Addenda of Gear Teeth*

No. of Teeth	Chordal Thickness	Corrected Addenda	No. of Teeth	Chordal Thickness	Corrected Addenda	No. of Teeth	Chordal Thickness	Corrected Addenda
10	1.56435	1.06156	59	1.57061	1.01046	108	1.57074	1.00570
11	1.56546	1.05598	60	1.57062	1.01029	109	1.57075	1.00565
12	1.56631	1.05133	61	1.57062	1.01011	110	1.57075	1.00560
13	1.56698	1.04739	62	1.57063	1.00994	111	1.57075	1.00556
14	1.56752	1.04401	63	1.57063	1.00978	112	1.57075	1.00551
15	1.56794	1.04109	64	1.57064	1.00963	113	1.57075	1.00546
16	1.56827	1.03852	65	1.57064	1.00947	114	1.57075	1.00541
17	1.56856	1.03625	66	1.57065	1.00933	115	1.57075	1.00537
18	1.56880	1.03425	67	1.57065	1.00920	116	1.57075	1.00533
19	1.56899	1.03244	68	1.57066	1.00907	117	1.57075	1.00529
20	1.56918	1.03083	69	1.57066	1.00893	118	1.57075	1.00524
21	1.56933	1.02936	70	1.57067	1.00880	119	1.57075	1.00519
22	1.56948	1.02803	71	1.57067	1.00867	120	1.57075	1.00515
23	1.56956	1.02681	72	1.57067	1.00855	121	1.57075	1.00511
24	1.56967	1.02569	73	1.57068	1.00843	122	1.57075	1.00507
25	1.56977	1.02466	74	1.57068	1.00832	123	1.57076	1.00503
26	1.56986	1.02371	75	1.57068	1.00821	124	1.57076	1.00499
27	1.56991	1.02284	76	1.57069	1.00810	125	1.57076	1.00495
28	1.56998	1.02202	77	1.57069	1.00799	126	1.57076	1.00491
29	1.57003	1.02127	78	1.57069	1.00789	127	1.57076	1.00487
30	1.57008	1.02055	79	1.57069	1.00780	128	1.57076	1.00483
31	1.57012	1.01990	80	1.57070	1.00772	129	1.57076	1.00479
32	1.57016	1.01926	81	1.57070	1.00762	130	1.57076	1.00475
33	1.57019	1.01869	82	1.57070	1.00752	131	1.57076	1.00472
34	1.57021	1.01813	83	1.57070	1.00743	132	1.57076	1.00469
35	1.57025	1.01762	84	1.57071	1.00734	133	1.57076	1.00466
36	1.57028	1.01714	85	1.57071	1.00725	134	1.57076	1.00462
37	1.57032	1.01667	86	1.57071	1.00716	135	1.57076	1.00457
38	1.57035	1.01623	87	1.57071	1.00708	136	1.57076	1.00454
39	1.57037	1.01582	88	1.57071	1.00700	137	1.57076	1.00451
40	1.57039	1.01542	89	1.57072	1.00693	138	1.57076	1.00447
41	1.57041	1.01504	90	1.57072	1.00686	139	1.57076	1.00444
42	1.57043	1.01471	91	1.57072	1.00679	140	1.57076	1.00441
43	1.57045	1.01434	92	1.57072	1.00672	141	1.57076	1.00439
44	1.57047	1.01404	93	1.57072	1.00665	142	1.57076	1.00435
45	1.57048	1.01370	94	1.57072	1.00658	143	1.57076	1.00432
46	1.57050	1.01341	95	1.57073	1.00651	144	1.57076	1.00429
47	1.57051	1.01311	96	1.57073	1.00644	145	1.57077	1.00425
48	1.57052	1.01285	97	1.57073	1.00637	146	1.57077	1.00422
49	1.57053	1.01258	98	1.57073	1.00630	147	1.57077	1.00419
50	1.57054	1.01233	99	1.57073	1.00623	148	1.57077	1.00416
51	1.57055	1.01209	100	1.57073	1.00617	149	1.57077	1.00413
52	1.57056	1.01187	101	1.57074	1.00611	150	1.57077	1.00411
53	1.57057	1.01165	102	1.57074	1.00605	151	1.57077	1.00409
54	1.57058	1.01143	103	1.57074	1.00599	152	1.57077	1.00407
55	1.57058	1.01121	104	1.57074	1.00593	153	1.57077	1.00405
56	1.57059	1.01102	105	1.57074	1.00587	154	1.57077	1.00402
57	1.57060	1.01083	106	1.57074	1.00581	155	1.57077	1.00400
58	1.57061	1.01064	107	1.57074	1.00575	156	1.57077	1.00397

*The values in this table are for spur gears of one diametral pitch. For any other diametral pitch, divide the given value by the r quired pitch. This table may also be applied to helical gears. See instructions on pages 258 and 259.

one-half the circular pitch, should be determined. This may be done as follows:

Rule.—First divide 90 degrees by the number of teeth in the gear, and then find the sine of the angle thus obtained. Next, multiply this sine by the pitch diameter; the product equals the chordal thickness.

Example 1.—A spur gear has 40 teeth of 4 diametral pitch. Find the chordal thickness, assuming that there is no allowance for play or "backlash."

$$\text{Chordal thickness} = 10 \times \sin \frac{90°}{40} = 10 \times \sin 2° \ 15' = 0.3926 \text{ inch.}$$

The chordal thickness, in this case, is practically the same as the arc thickness along the pitch circle which equals 1.5708 ÷ diametral pitch = 1.5708 ÷ 4 = 0.3927 inch. The difference between the chordal and arc thicknesses increases for smaller gears and pinions.

Example 2.—A pinion has 15 teeth of 3 diametral pitch; the pitch diameter = 15 ÷ 3 = 5 inches; hence

$$\text{Chordal thickness} = 5 \times \sin \frac{90}{15} = 5 \times \sin 6° = 0.5226 \text{ inch}$$

In this case, the thickness on pitch circle is 0.5236 inch.

At what depth should the chordal thickness be measured?

Before measuring the chordal thickness of a gear tooth, it is necessary to set the vertical scale of the vernier gear tooth caliper so that the caliper jaws come into contact with the sides of the tooth at the pitch circle.

Rule.—To obtain chordal or "corrected" addendum, H, Fig. 5, subtract from 1 the cosine of an angle obtained by dividing 90 by number of teeth; multiply the difference by pitch *radius* of gear and add product to addendum S.

Example.—A pinion has 15 teeth of 3 diametral pitch; then pitch radius = 2.5 inches; addendum = 1 ÷ diametral pitch = 0.3333 inch; therefore

Chordal addendum = 2.5 × (1 − cos 6°) + 0.3333 = 0.347″

Note: The chordal thickness and corrected addendum may be determined quickly by finding the values in the accompanying table and dividing by the diametral pitch of the gear to be measured.

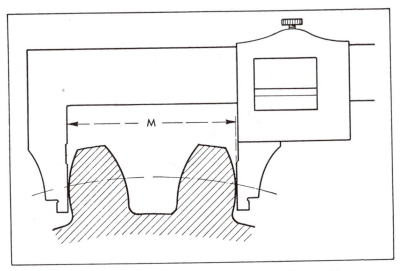

Fig. 6. Testing Gear Size by Measuring over Two or More
Teeth with Vernier Caliper

How are spur gear sizes checked by taking a chordal measurement over two or more teeth?

Gear-tooth sizes may be checked by measuring as at M (see Fig. 6) over two or more teeth instead of measuring the thickness of one tooth. This measurement across two or more teeth is not affected by errors in the outside diameter of the gear blank and it is not necessary to calculate a chordal addendum as when measuring the thickness of one tooth. Chordal dimension M depends upon the number of teeth included in the measurement or the number between the caliper jaws, upon their pitch and the pressure angle of the gear. (The meaning of the term "pressure angle" is explained in section on Generating Methods of Forming Gear Teeth.) Two rules will be given for determining what measurement M should be when the tooth size is correct. The first one applies to a pressure angle of $14\frac{1}{2}$ degrees, and the second one to a pressure angle of 20 degrees—two angles that are applied extensively in gear design.

Rule 1—Pressure Angle 14 1/2 Degrees.—To find measurement M over spur gear teeth, multiply the total number of gear teeth by 0.00537; then add an amount depending upon the number of teeth measured over (as shown in the following table) and divide the sum thus obtained by the diametral pitch of the gear to be measured. Finally, subtract from this quotient one-half the total amount of backlash, assuming that backlash is to be obtained by reducing the teeth of both mating gears.

For 12 to 18 teeth, incl., add   4.56229 and meas. over 2 teeth.
For 19 to 37 teeth, incl., add   7.60380 and meas. over 3 teeth.
For 38 to 50 teeth, incl., add 10.64533 and meas. over 4 teeth.
For 51 to 62 teeth, incl., add 13.68685 and meas. over 5 teeth.
For 63 to 75 teeth, incl., add 16.72838 and meas. over 6 teeth.

Example.—Find chordal measurement M for a spur gear having 40 teeth of 4 diametral pitch and a pressure angle of 14½ degrees. The total backlash or play between the teeth is to equal 0.006 inch.

$$M = \frac{0.00537 \times 40 + 10.64533}{4} - 0.003 = 2.712 \text{ inches}$$

Note that this is the measurement over 4 teeth because, as the table shows, this is the number for all gears having from 38 to 50 teeth, inclusive.

Rule 2.—Pressure Angle 20 Degrees.—To find measurement M, multiply the total number of gear teeth by 0.014; then add an amount depending upon the number of teeth measured over (as shown in the following table) and divide the sum by the diametral pitch of the gear to be measured. Finally, subtract from the quotient one-half the total amount f backlash.

For 12 to 18 teeth, incl., add   4.42819 and meas. over 2 teeth.
For 19 to 27 teeth, incl., add   7.38033 and meas. over 3 teeth.
For 28 to 36 teeth, incl., add 10.33225 and meas. over 4 teeth.
For 37 to 45 teeth, incl., add 13.28459 and meas. over 5 teeth.
For 46 to 54 teeth, incl., add 16.23672 and meas. over 6 teeth.

Why does number of teeth included in the chordal measurement vary for different tooth numbers?

Measurements over the teeth are along a chordal line that is tangent to the base circle of the gear. (Radius of base circle = pitch circle radius × cosine pressure angle.) The contact points between the caliper jaws and gear teeth should be as close as possible to the pitch circle because that part of the profile on and adjacent to the pitch circle is the most important. This part ordinarily is a true involute, but the tips and roots of the teeth often are modified. As the number of gear teeth increases, the distance between the pitch circle and base circle also increases, and, consequently, it is necessary to measure over a larger number of teeth in order to keep the contact points between the caliper jaws and tooth profiles as near as possible to the pitch circle.

How are gear sizes checked by measuring over pins or rolls?

When spur-gear sizes are checked by the pin or roll method, cylindrical pins of known diameter are placed in diametrically opposite tooth spaces; or, if the gear has an odd number of teeth, the pins are located as nearly opposite as possible. The measurement over these pins is then checked by using any sufficiently accurate method of measurement. The general formulas for determining what the measurement over the pins should be, when the gear size is correct, are not included in this treatise because they are cumbersome to apply and involve the use of a table of involute functions. The measurement for a given number of teeth and standard pin size may be determined readily from tables in MACHINERY'S HANDBOOK. This method of checking gear sizes by measurement over pins is especially useful in shops having a limited gear-inspection equipment.

Are formed cutters used on machines designed expressly for gear cutting?

One general type of gear cutting machine uses a formed milling cutter. These machines are designed primarily for cutting spur gears, although some gear-cutting machines of the formed-cutter type are designed for cutting either spur or bevel gears. Such machines are so arranged that the cutter-slide can be set parallel to the work-spindle for spur gears, and at an angle for bevel gears. Regular gear-cutting machines of the formed-cutter type are commonly known as "automatic gear-cutting machines" because, after the gear blank or blanks are in the cutting position and the machine is properly adjusted, all the gear teeth are cut automatically without further attention on the part of the operator. Other types of gear-cutting machines also operate automatically to the same extent, although the term "automatic" generally is not used in designating them.

What are the principal features of an automatic gear-cutting machine?

Automatic gear-cutting machines of the formed-cutter type include a main spindle for holding and driving the work-holding arbor, a cutter-slide arranged to move parallel with the axis of the work-spindle, a mechanism for feeding the cutter-slide at a suitable rate and returning it to the starting point, and a mechanism for indexing the gear blank after each tooth space is milled. The arrangement of different makes of these machines varies somewhat. For instance, the cutter-slide of some machines has a horizontal feeding movement along the main bed and the work-spindle is carried by a slide having vertical adjustment along a column, to suit the gear diameter. The cutter-slides of other machines feed vertically along the face of the column and the work-spindle is held in a vertical position on a slide or table mounted on horizontal ways formed on the main bed. The exact method of setting up or adjusting automatic gear-cutting machines varies somewhat for different makes. After a cutter of the right pitch and number has been selected, this cutter with its arbor is placed in the working position. The cutter-spindle bearing is then adjusted to locate the cutter central with the work-spindle so that the center line of each tooth space cut in the gear blank will be radial. This is usually done by means of a centering gage or indicator supplied with the machine.

The indexing mechanism is geared to agree with the number of teeth to be cut, the proper combination of gearing for a given number of teeth being shown by the chart accompanying the machine. The cutter speed and rate of feed are regulated by means of change-gears, according to the pitch of the gear to be cut and the kind of material the gear is made of.

Fig. 7 shows a close-up view of an automatic gear-cutting machine cutting a fairly large spur gear. Incidentally, large

cutters are sometimes made with high-speed steel blades
attached to the main body of the cutter which is made of
a cheaper steel. An example of this construction is shown.
The lower end of each blade fits into a notch formed in the
cutter body, and it is prevented from shifting laterally by
projections on each side extending below the notched part.
A single cap-screw holds each blade in position. The inserted-blade type of cutter is especially adapted for very
coarse pitches because it is cheaper to make.

When are roughing or stocking cutters used?

When the pitch of a gear is large enough to warrant
taking both roughing and finishing cuts, the "stocking cut-

Fig. 7. Cutter of Inserted-blade Type

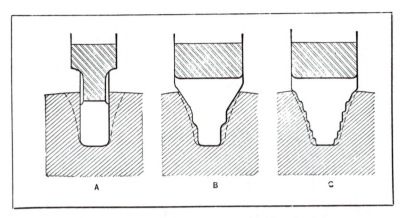

Fig. 8. Shapes of Roughing or Stocking Cutters

ter" used for roughing removes the bulk of the metal, leaving a small amount on the sides of the teeth for finishing. Several different types of stocking or roughing cutters are in use. Some of these simply form straight-sided slots in the gear blank, as illustrated by diagram A, Fig. 8. The dotted line on this and the other diagrams represents the shape of the finished tooth space. When a stocking cutter of this kind is used, it is sometimes followed by a second roughing cutter having a shape similar to a finishing cutter, except that it is about 1/32 inch smaller all around the edge, to allow for a finishing cut. The straight or parallel-slot type of stocking cutter has the disadvantage of leaving too much metal. It is desirable to finish gears with not over two cuts, and several designs of stocking cutters have been made for roughing out tooth spaces so that they conform quite closely to the finished shape. An improved type of stocking cutter has stepped teeth in order to break up the chips and secure a better cutting action. These stepped stocking cutters differ in regard to the number and arrangement of the steps. For instance, some have more steps than others (see diagrams B and C) and all the teeth of some cutters are stepped, whereas others have stepped teeth alternating with plain teeth, in which case most of the cutting is done by the plain teeth and the stepped edges project beyond the outline of the plain teeth only enough to break up the chips.

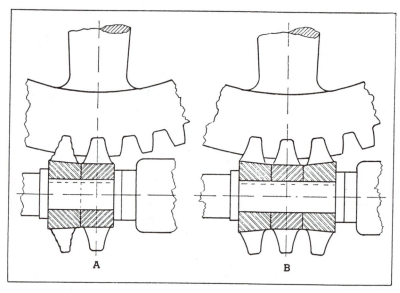

Fig. 9. (A) Gang Cutter for Roughing and Finishing. (B) Gang
Cutter for Finishing Two Teeth

When is a gang cutter used for milling
two or more tooth spaces simultaneously?

A gang of two or more cutters is sometimes mounted on
an arbor, so that all the cutters operate simultaneously.
One method is to place a stocking cutter in advance of the
finishing cutter, as illustrated at *A*, in Fig. 9, the finishing
cutter being in the central position. This plan is also followed in cutting rack teeth, a common method being to
place a stocking cutter and a finishing cutter side by side.

Another arrangement of gang cutters is illustrated at *B*,
which shows how two teeth may be finished at each indexing by using a gang of three cutters. It will be noted that
the end cutters are of special shape, which is one disadvantage of this method. When accurate gears are required,
the gang-cutter method is not advisable; when used, it is
generally applied to cutting gears of small pitch, especially
when required in large quantities. In general, the use of
gang cutters is not recommended for the best classes of
work. Ordinarily when the finishing cutter is preceded by

a stocking cutter, the teeth are not milled as accurately as when these cutters are used separately; moreover, gang cutters for finishing two or more teeth simultaneously are inferior to the single-cutter method as regards accuracy.

Why is the rate of feed reduced for taking the first cut with a gang cutter?

In taking the first cut, as illustrated at A in Fig. 9, both cutters must cut a tooth space through solid stock, whereas for the following cuts the finishing cutter simply removes what has been left by the roughing or stocking cutter. Because of this heavy cutting at the start, a slower feeding movement should be used, and to avoid feeding by hand or changing the gears, some gear-cutting machines are equipped with a feed-changing mechanism which automatically changes the slow feed used for the first cut to a feeding movement twice as fast for the remaining cuts. The feed is again reduced for taking the first cut on the next blank, by means of a control lever or knob.

In milling rack teeth, how is the pitch determined?

A rack may be compared to a spur gear which has been straightened out so that all of the teeth are in one plane. In the design of various classes of machines, a pinion or gear meshes with the rack in order to transmit a linear or straight-line motion to a machine table slide or other part. If a rack is to mesh properly with the pinion, the pitch of the rack teeth, or the center-to-center distance between adjacent teeth, must equal the circular pitch of the mating pinion. Assume that a pinion has teeth of 4 diametral pitch. In that case, the pitch of the rack teeth must equal the circular pitch equivalent to 4 diametral pitch, or 0.7854 inch. The rack teeth are also similar to the mating pinion teeth in regard to depth. For example, if the pinion has full-depth teeth of 4 diametral pitch, then the depth of both pinion and rack teeth should equal 2.157 divided by this diametral pitch, or 0.6866 times the equivalent circular pitch. The teeth of a true involute rack have straight sides; however, for rack milling, it is common practice to use a

No. 1 spur-gear cutter of whatever pitch may be required, since this number is intended for spur gears varying from 135 teeth up to a rack.

How is the rack indexed to obtain a given pitch?

The method of cutting rack teeth in a milling machine may depend upon the length of the rack. If a rack is comparatively short, it may be clamped in the machine vise in a position parallel to the cutter arbor or at right angles to the work table. The teeth are then milled by feeding the table longitudinally and indexing for milling successive tooth spaces, by using the graduated dial on the cross-feed screw. As most racks are too long to permit holding them in this cross-wise position, special rack-cutting attachments are frequently used. One of these attachments is shown in Fig. 10. The cutter-spindle is parallel with the work-table, thus permitting the rack to be held in a lengthwise position. This cutter-spindle is driven from the main spindle

Fig. 10. Cutting Rack Teeth on Milling Machine

of the machine through bevel and spur gearing. Instead of using the feed-screw dial for indexing, the feed-screw is connected through change-gears with a locking disk A, having a notch that is engaged by a locking pin B. The gear on the index disk and the first gear on the stud are changed to secure different indexing movements, but it is not necessary to change the gear on the screw or the second gear on the stud. The ratio of the change-gears used in any case is such that the work-table and the rack will move a distance equal to the pitch of the rack teeth, while disk A makes one revolution excepting for diametral pitches of 3, $3\frac{1}{2}$ and $4\frac{1}{2}$, which require two revolutions. For instance, in order to index, pin B is withdrawn and feed-screw handle C is turned until pin B again engages the notch in disk A after the latter has made a revolution, or two revolutions in the case of the three pitches referred to. The change-gears to use for any pitch are determined by referring to a table which accompanies the rack-cutting attachment.

Indexing with Feed-Screw Dial. To find the feed-screw rotation, divide linear pitch of rack by lead of feed-screw (lead usually is 0.25 inch). Then multiply the *decimal* part of the feed-screw rotation, by 250 (number of graduations on dial) to obtain the dial reading in thousandths for the fractional part of the indexing movement.

Example: If pitch of rack is 0.7854 inch, 0.7854 ÷ 0.25 = 3.1416 feed-screw turns or 3 turns plus 0.1416 × 250 = 35.4 thousandths on dial.

When are gang cutters used for rack milling?

Racks may be milled either by using a single cutter or by using gang cutters consisting of two or more cutters on the same arbor. Diagram A, Fig. 11, illustrates the single-cutter method, which has been described in connection with Fig. 10. Diagram B indicates how roughing and finishing cutters are often used together especially on machines designed expressly for rack cutting. The roughing cutter, which should preferably be of the stepped form to break up the chips, is mounted on the same arbor as the finishing cutter and precedes the latter. Roughing cutters which simply mill straight-sided slots are sometimes used instead of the stepped form.

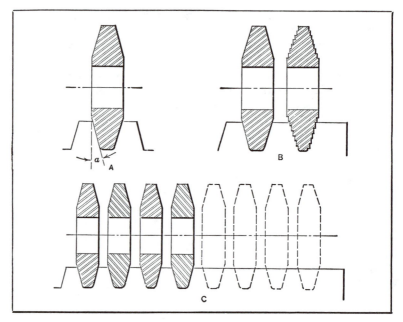

Fig. 11. Different Methods of Milling Rack Teeth

The lower diagram *C* illustrates the application of a gang of finishing cutters, which is used to mill four tooth spaces simultaneously. The cutters are spaced on the arbor to suit the pitch of the rack to be cut, and the indexing movement between successive cuts equals the number of cutters in the gang multiplied by the linear pitch of the rack. Where racks are cut by the gang method, it is applied, as a rule, to the smaller pitches.

Special rack-cutting machines have been designed along the general lines of milling machines, with certain features introduced to make them more efficient and better adapted for cutting racks on a quantity basis. Rack teeth may also be generated on a gear shaper equipped with a rack-cutting attachment.

Milling Bevel, Helical and Worm-Gears

A correctly formed bevel gear tooth has the same sectional shape throughout its length but on a diminishing scale from the large to the small end. This accounts for the fact that a bevel gear cannot be cut properly by using a formed milling cutter which simply reproduces its shape, because if this cutter were of the exact curvature required at the large end of the tooth, it would not be correct for any other part, and the error would be considerable at the small end of the tooth. Consequently, accurate bevel gears are cut by a generating process in order to form teeth having the proper curvature on a diminishing scale, the tooth tapering toward the apex of its pitch cone. Bevel gears often are cut by means of formed milling cutters in connection with repair or miscellaneous work. These milled gears are sufficiently accurate for many purposes, especially when the speeds are relatively low, but they are unsatisfactory for high speeds and where accurate tooth curves are of importance.

What attachment is used for holding a bevel gear while milling the teeth?

When a bevel gear is cut in a milling machine, the gear blank is mounted on an arbor inserted in the dividing-head spindle, and the latter is set to the cutting angle as shown by Fig. 1 and by the diagrams in Fig. 2. A formed cutter is used, and it is necessary to take two cuts through each tooth space, with the gear blank slightly off center, first on one side and then on the other, to obtain a tooth of approximately the correct form. The gear blank is also rotated proportionately to obtain the proper tooth thickness at the large and small ends. The different steps or operations connected with cutting a bevel gear are as follows: Setting

241

the blank to the cutting angle; selecting the cutter; taking a trial cut; determining the amount of offset; and milling the teeth. These different operations will be referred to in the order given.

At what angle is a bevel gear blank set for milling the teeth?

The cutting angle a (see Fig. 2) at which the dividing-head spindle should be set for milling the teeth equals the pitch-cone angle β, minus the *addendum* angle θ of the tooth. Ordinarily, the cutting angle for bevel gears equals the pitch-cone angle minus the *dedendum* angle; when cutting the teeth by milling with a formed cutter, the first rule is preferable as it gives a uniform clearance at the bottom

Fig. 1. Milling Machine Arranged for Cutting a Bevel Gear

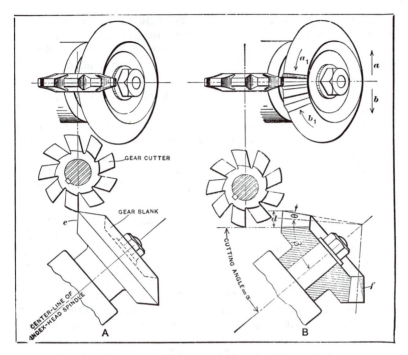

Fig. 2. Diagrams Showing Relative Positions of Cutter
and Bevel Gear Blank

of the tooth spaces and a somewhat closer approximation
to the theoretically-correct tooth shape.

For bevel gears with shafts at right angles, the tangent
of the pitch-cone angle of the pinion is found by dividing
the number of teeth in the pinion by the number of teeth
in the gear; the tangent of the pitch-cone angle of the gear
equals the number of teeth in the gear divided by the number of teeth in the pinion; the tangent of the addendum
angle equals the addendum divided by the pitch-cone radius;
the addendum equals 1.0 divided by the diametral pitch.

In milling bevel gears, can the same cutter be used for both gear and pinion?

A cutter for milling bevel gears is made thinner than a
spur gear cutter because it must pass through the narrow
tooth spaces at the inner ends of the teeth. For 14½-degree

involute teeth, there are eight cutters numbered from one to eight for each pitch and suitable for cutting bevel gears from a twelve-tooth pinion to a crown gear. The cutter to use, in any case, must not only be of the required diametral pitch but the right number in the series. The number of the cutter depends upon the number of teeth in the gear or pinion. When cutting miter gears, only one cutter is needed, but, if one gear is larger than the other, two cutters of the same pitch but of different numbers may be required. The number of teeth for which to select the cutter is not the actual number of teeth in the gear, but is found as follows:

Rule.—Divide the actual number of teeth in the gear by the cosine of the pitch-cone angle.

Example.—As an example illustrating the method of selecting a cutter, assume that the pitch-cone angle of a bevel gear is 60 degrees and that it is to have 40 teeth of 6 diametral pitch. What cutter number should be used? The cosine of 60 degrees is 0.5, and $40 \div 0.5 = 80$. Hence, a 6 diametral pitch cutter of No. 2 shape should be used because this shape is intended for all tooth numbers from 55 to 134, as indicated by the list previously given for spur gear cutters.

The reason why a bevel gear cutter is not selected with reference to the actual number of teeth, as in the case of spur gear cutters, is because the shape of the tooth space of a bevel gear varies not only for different numbers of teeth but for different pitch cone angles. Thus, for a bevel gear, the idea is to use a cutter having a shape suitable for a spur gear, the radius of which is equal to the back cone distance of the bevel gear. This explains why the expression "number of teeth in an equivalent spur gear" is some-times applied to the number used in selecting bevel gear cutters.

What is the meaning of "diameter" as applied to bevel gears?

The term "diameter" as applied to bevel gearing means the pitch diameter at the outer ends of the teeth. Smooth cones, mounted as shown by the left diagram, Fig. 3, so as to transmit motion from one shaft to another by frictional resistance, represent the pitch cones of bevel gears. The

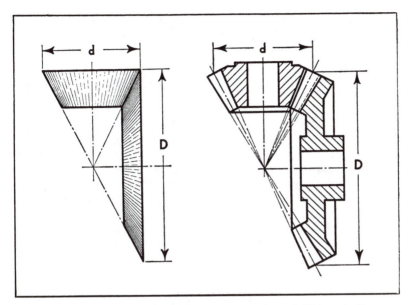

Fig. 3. Diagrams Illustrating the Meaning of "Diameter"
as Applied to Bevel Gearing

intermeshing teeth of bevel gears have the faces so formed
or curved as to transmit a uniform motion, the same as
though smooth cones were actually used. The diameters D
and d represent the pitch diameters which are used in
figuring speeds or ratios.

Rule 1.—Pitch diameter D of gear equals number of gear
teeth divided by diametral pitch, the same as for a spur
gear.

Rule 2.—Pitch diameter d of pinion equals number of
pinion teeth divided by diametral pitch.

Example.—Gear has 60 teeth; pinion, 15 teeth; and
diametral pitch is 3.

$$D = \frac{60}{3} = 20 \text{ inches} \qquad d = \frac{15}{3} = 5 \text{ inches}$$

Does the pitch and depth of bevel gear teeth apply to the large ends?

The term "pitch," as applied to a bevel gear, relates to
the proportions at the large end of the tooth. For instance,

the circular pitch is the distance from the center of one tooth to the center of the next tooth, as measured along the pitch circle which coincides with the large ends of the teeth. Similarly, the diametral pitch equals the number of teeth in the gear divided by the diameter of this same pitch circle. Ordinarily the diametral-pitch system is applied to bevel gears as well as to spur gears and other types, unless the teeth are very large when circular pitch is generally used.

The gear blank mounted upon the dividing head is first set in a central position relative to the cutter, as at A, Fig. 2. This may be done by placing a true center in the dividing-head spindle and setting the point in line with the center line on the gear cutter. After this centering operation, the machine table is adjusted vertically and horizontally, until the cutter just grazes the edge of the gear blank at the large end, preparatory to making the adjustment for the depth of the cut. The depth of a bevel gear tooth at the large end may be calculated by dividing 2.157 by the diametral pitch, or a gear tooth depth gage may be used to scribe a depth line on the back cone surface. The rule just given for determining the whole depth of the tooth space is the same as the one used for spur gears. In the case of a bevel gear, however, the depth measurement is perpendicular to the pitch line or pitch cone, and not perpendicular to the axis of the gear, as in the case of spur gearing.

When milling bevel gears, how much offset adjustment is required to improve the tooth form?

If the tooth spaces of a bevel gear were milled with the cutter in a central position, the teeth would not have the tapering form required. Teeth of coarse pitch may be roughed out while the cutter is centered, but invariably the teeth are finished by taking two series of cuts, the blank being set slightly off center first in one direction and then in the other. If the amount of offset is not known, it may be determined by the "cut-and-try" method, although it can be calculated quite accurately by means of the following rule:

Rule.—First find the factor, in the accompanying table, which corresponds to the number of the cutter used and to the ratio of the pitch-cone radius to the face width; divide this factor by the diametral pitch, and subtract the result from one-half of the cutter thickness at the pitch line.

Example.—A bevel gear of 24 teeth, 6 diametral pitch, 30 degrees pitch cone angle, and $1\frac{1}{4}$ inch face width is to be cut. By applying the rule, it will be found that a No. 4 cutter of 6 diametral pitch will be required. The pitch-cone radius equals 4 inches, since this radius equals the pitch diameter divided by twice the sine of the pitch cone angle. Thus the pitch diameter equals $24 \div 6 = 4$, and the sine of 30 degrees equals 0.5. Hence, the pitch-cone radius equals $\dfrac{4}{2 \times 0.5} = 4$ inches. In order to obtain the factor from the table, the ratio of pitch-cone radius to width of face must be known. This ratio is $\dfrac{4}{1.25} = \dfrac{3.2}{1}$, or about $\dfrac{3\frac{1}{4}}{1}$. The factor in the table for this ratio with a No. 4 cutter is 0.280.

Table for Obtaining Offset for Cutting Bevel Gears

No. of Cutter	RATIO OF PITCH-CONE RADIUS TO WIDTH OF FACE					
	$\frac{3}{1}$	$\frac{3\frac{1}{4}}{1}$	$\frac{3\frac{1}{2}}{1}$	$\frac{3\frac{3}{4}}{1}$	$\frac{4}{1}$	$\frac{4\frac{1}{4}}{1}$
1	0.254	0.254	0.255	0.256	0.257	0.257
2	0.266	0.268	0.271	0.272	0.273	0.274
3	0.266	0.268	0.271	0.273	0.275	0.278
4	0.275	0.280	0.285	0.287	0.291	0.293
5	0.280	0.285	0.290	0.293	0.295	0.296
6	0.311	0.318	0.323	0.328	0.330	0.334
7	0.289	0.298	0.308	0.316	0.324	0.329
8	0.275	0.286	0.296	0.309	0.319	0.331

No. of Cutter	$\frac{4\frac{1}{2}}{1}$	$\frac{4\frac{3}{4}}{1}$	$\frac{5}{1}$	$\frac{5\frac{1}{2}}{1}$	$\frac{6}{1}$	$\frac{7}{1}$
1	0.257	0.258	0.258	0.259	0.260	0.262
2	0.274	0.275	0.277	0.279	0.280	0.283
3	0.280	0.282	0.283	0.286	0.287	0.290
4	0.296	0.298	0.298	0.302	0.305	0.308
5	0.298	0.300	0.302	0.307	0.309	0.313
6	0.337	0.340	0.343	0.348	0.352	0.356
7	0.334	0.338	0.343	0.350	0.360	0.370
8	0.338	0.344	0.352	0.361	0.368	0.380

Next measure the cutter at the proper depth for 6 pitch, which is found by dividing 1.157 by the diametral pitch. This depth equals 0.1928 inch. Suppose that the thickness of the cutter at this depth is 0.1745 inch. The dimension will vary with different cutters, and will vary in the same cutter as it is ground away, since formed bevel gear cutters are commonly provided with side relief. Using these values, the offset is determined as follows:

$$\text{Offset} = \frac{0.1745}{2} - \frac{0.280}{6} = 0.0406 \text{ inch.}$$

How are the offset and rotary adjustments applied in milling the bevel gear teeth?

If the pitch of the gear is large enough to warrant taking roughing cuts, this is done with the cutter set in a central position as previously mentioned. If roughing cuts are not considered necessary, two tooth spaces are usually milled to the full depth with the cutter in a central position, thus forming a roughly shaped tooth upon which trial cuts are taken to check the accuracy of the offset adjustment. The cross-feed dial should be set to zero while the cutter is in a central position; then the cross-slide is moved an amount equal to the offset, either as determined by calculation as just described, or as estimated. This first offset adjustment is indicated by arrow b, Fig. 2. The dividing head is then used to rotate the gear blank (as indicated by arrow b_1) so that one side of the cutter will mill the entire surface of the trial tooth on one side. Having thus formed one side of the tooth so that it is a reproduction of the cutter's curvature, the work must be offset, as shown by arrow a, an equal amount on the opposite side of the central position, taking the usual precautions to avoid errors from backlash. The blank is then rotated in an opposite direction (see arrow a_1) to again align the cutter with the tooth space, after which another trial cut is taken, thus milling the tooth on both sides. This trial tooth is milled to the proper thickness by rotating the blank toward the cutter, moving the crank around the dial for the rough adjustment, and bringing it to an accurate thickness by such means as may be provided in the head. On one type of dividing head, this

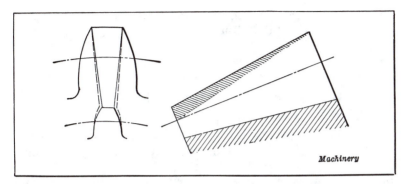

Fig. 4. Diagram Showing Part of Bevel Gear Tooth
that is Corrected by Filing

fine adjustment is effected by two thumbscrews near the hub of the index-crank, which turn the index worm with relation to the crank.

The thickness of the tooth is tested by measuring with a vernier caliper, assuming that fixed gages are not available. The chordal thickness at the large end of the tooth and depth at which thickness should be measured, can be obtained by using the rules previously given for spur gears.

Why are milled bevel gear teeth corrected by filing?

Since the shape of a formed cutter for bevel gears can be correct for only one section of the tooth, the cutter outline is based on the tooth shape at the large end; consequently, the small end of the tooth is too thick above the pitch line and it does not have enough curvature, as indicated by the dotted lines in the end view of a tooth in Fig. 4. If the cutter outline conformed to the shape of the teeth at the small ends, too much metal would be removed from the large ends above the pitch line. Therefore the cutter is shaped to suit the large end, because then it is possible to correct the teeth by removing the surplus metal at the small end. This error in thickness above the pitch line, which progressively increases from the large to the small end, is corrected ordinarily by filing the teeth. The triangular area indicated by the fine shade lines shows what part of the tooth surface should be corrected by filing. No metal should be removed at or below the pitch line, because if the gear has been milled properly the pitch line thicknesses at both large and small ends will be correct.

Is the milling process adapted to helical gear cutting?

Most helical or "spiral" gears are cut in regular gear-cutting machines of the hobbing type, although milling machines are sometimes used either in connection with jobbing work or in shops where such gear-cutting operations are done on a small scale. The cutting of a helical gear consists in milling the required number of equally spaced

Fig. 5. Cutting a Helical or Spiral Gear in a Milling Machine

helical (often called spiral) teeth around the gear blank. To do this, it is necessary to use a dividing-head in order to rotate the work for milling the helical tooth spaces and also to provide means of indexing for milling successive tooth spaces. A formed cutter is used, and this must be in alignment with the groove being milled in order to reproduce, as far as possible, the shape of the cutter. Ordinarily a universal type of milling machine is used, so that the required angular adjustment can be obtained, by swiveling the table (see Fig. 5). It is possible, however, to mill a helical gear with the machine table set at right angles to the spindle, provided a vertical-spindle milling attachment is used that has the required angular adjustment to permit holding the cutter in alignment with the helical tooth spaces.

Are helical gears cut to a given diametral pitch like a spur gear?

In the case of spur gears, the diametral pitch of the gear (and of the cutter used in forming the teeth) equals the number of teeth divided by the pitch diameter. The pitch of the cutter for a helical gear depends not only upon the number of teeth and pitch diameter, but also upon the helix angle of the teeth. Helical gears, like spur gears, usually are designed for *diametral* pitch rather than *circular* pitch; but the diametral pitch of a helical gear differs from that of a spur gear because of the effect of the helix angle of a helical gear on its circular pitch. As the diagram, Fig. 6, shows, the circular pitch P_c of a helical gear, in a plane perpendicular to its axis, is greater than the *normal circular pitch* which is measured in a plane perpendicular to the teeth; furthermore, the normal circular pitch will be reduced as the helix angle A is increased. This shows that the pitch of the cutter for a helical gear, whether based upon this circular pitch or upon the diametral pitch system, is affected by the helix angle. If a cutter were proportioned to suit circular pitch, this would be the *normal* circular pitch and not the pitch P_c in the plane of rotation. When a cutter is proportioned to suit diametral pitch, which is a general practice, this *normal diametral pitch*, as it is called, is equal to 3.1416 divided by the normal circular pitch. It may also be determined as follows:

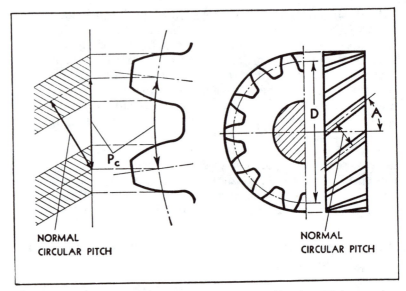

Fig. 6. Diagrams Illustrating Normal Circular Pitch

Rule.—To find the normal diametral pitch of a helical gear, divide the number of teeth by the product of the pitch diameter and the cosine of the helix angle.

Example.—A helical gear has 38 teeth, a helix angle of 45 degrees, a pitch diameter of 6.718 inches. What is the normal diametral pitch of the cutter required for this gear?

$$\text{Pitch of cutter} = \frac{38}{6.718 \times 0.70711} = 8$$

This cutter is the same as would be used for a spur gear of 8 diametral pitch; however, when the cutter is applied to a helical gear, it represents *normal* diametral pitch.

How is the pitch of the cutter established for helical gears?

In cutting helical gears (or any other type), it is desirable, as far as possible, to use a cutter conforming to some standard or commercial diametral pitch. This is made possible, in the case of a helical gear, by using whatever number of teeth, pitch diameter, or helix angle conform to a standard pitch, as illustrated later. The pitch of the cutter, and of the teeth formed by it, must be such as to give the necessary tooth strength, and this pitch may be determined either by tests or on the basis of past experience with gears applied to a similar class of service. When the pitch or required tooth size has been determined, then the number of teeth and pitch diameter for a given helix angle can be established to suit this pitch.

Example.—Assume that two helical gears of the same size (1 to 1 ratio) are to drive shafts located at right angles to each other. The helix angle of the teeth on each gear is to be 45 degrees and a normal diametral pitch of 8 is required. The pitch diameter of each gear should be about 6.7 inches, but the design permits some variation in the center-to-center distance. How much change in pitch diameter will be required to permit using a cutter of 8 normal diametral pitch?

Rule.—To find the number of helical gear teeth, multiply the normal diametral pitch by the pitch diameter and then multiply this product by the cosine of the helix angle.

This rule will be applied to determine, first, the approximate number of teeth.

Approximate No. teeth = 8 × 6.7 × 0.70711 = 37.9
Since 38 is the nearest whole number, the next step is to find the actual pitch diameter for 38 teeth.

Rule.—To find the pitch diameter of a helical gear, divide the number of teeth by the product of the normal diametral pitch and the cosine of the helix angle.

$$\text{Pitch diameter} = \frac{38}{8 \times 0.70711} = 6.718 \text{ inches}$$

The actual pitch diameter in this case is 6.718 inches or 0.018 inch larger than the diameter desired, thus permitting the use of a cutter of 8 pitch and a helix angle of 45 degrees.

The foregoing example has been used chiefly to show the relationship between the pitch of a helical gear and other related factors. The actual design of helical gears includes unequal ratios, shafts located at various angles, and, in some cases, designing to suit an exact center distance. The solution of these various problems relates primarily to gear design and is beyond the scope of this treatise.

In milling helical gears with a formed cutter, how is the cutter number determined?

In connection with the preceding example it was found that a cutter of 8 normal diametral pitch should be used, this cutter being the same as would be employed for a spur gear of 8 diametral pitch. The next step is to determine the required number of the cutter, since there are eight cutters of different shape for each pitch. Now, in the case of helical gears, it is necessary to consider the helix angle as well as the number of teeth in determining the cutter number, because the shape of the tooth spaces depends both upon the inclination of the teeth and their number.

Rule.—The required cutter number may be determined by dividing the number of teeth by the cube of the cosine of the helix angle.

Example.—What cutter number is required for a helical gear having 38 teeth and an angle of 45 degrees?

The cosine of 45 degrees is 0.7071 and $38 \div 0.7071^3 =$ 107, which represents the number of teeth for which the cutter should be selected. Hence, the No. 2 shape is used, as this is intended for all tooth numbers between 55 and 134, according to the system for spur gears (see list of cutter numbers previously given for spur gears). The cutter number just referred to is not related to the pitch of the cutter, the latter being 8 diametral pitch for this particular gear, as previously determined. As the foregoing example illustrates, the number of teeth for which a cutter for helical gears is selected may be much larger than the actual number of teeth in the gear. For high helix angles and small numbers of teeth, the number of teeth from which to select the cutter number may need to be modified somewhat as explained in MACHINERY'S HANDBOOK.

Is the normal diametral pitch used in determining the outside diameter of a helical gear?

Yes, the pitch diameter and the pitch of the cutter, or normal diametral pitch of the gear, are required in finding the outside diameter or the diameter to which the gear blank is turned before cutting the teeth.

Rule 1.—To obtain the outside diameter, divide 2 by the normal diametral pitch and add quotient to pitch diameter.

Note: The addendum equals 1 ÷ normal diametral pitch; hence, 2 ÷ normal diametral pitch is equal to twice the addendum.

Example.—If pitch diameter = 6.718 inches and the normal diametral pitch is 8, then

$$\text{Outside diameter} = 6.718 + \frac{2}{8} = 6.968 \text{ inches}$$

Why is the lead of helical gear teeth required in milling them?

The milling machine must be geared to suit the lead of the teeth in order to obtain teeth of helical form. The dividing-head must be connected to the lead-screw by change-gears selected according to the lead of the gear to be cut, the lead being the distance that any one tooth would advance if it made a complete turn around the gear. In designing a helical gear, the pitch diameter of the gear and the helix angle of the teeth are usually made to suit conditions; consequently, the lead may be an odd dimension that cannot be obtained exactly with any combination of change-gears available. Ordinarily, however, the gears supplied with a milling machine will give a lead that is accurate enough for practical purposes. The change-gears connect the index-head and table feed-screw (as shown in Fig. 5) and rotate the gear blank as the table feeds lengthwise.

The combination of gearing for obtaining a given lead is usually determined by referring to a table of change-gears. Such tables will be found in MACHINERY'S HANDBOOK in connection with the selection of change-gears for helical milling. The method of calculating change-gears for helical milling is explained in the section on Generating Helical Grooves with Index-head (page 174).

Rule.—The lead of a helical gear may be determined by multiplying the pitch circumference by the cotangent of the helix angle.

What angular position of the machine table is required for helical gear milling?

Before the teeth of a helical gear can be milled the table of the machine must be set to the helix angle. This is done so that the cutter will produce grooves and teeth of the proper shape. The angle of a helix depends upon the lead and the circumference of the cylindrical surface around which the helix is formed. The smaller the circumference, the smaller the angle, assuming that the lead remains the same. The angle, then, that the teeth of a helical gear makes with the axis gradually diminishes from the tops to the bottoms of the teeth, and if it were possible to cut a groove right down to the center or axis, its angle would become zero. Hence, if the table of the machine is set to the angle at the top of a tooth, the cutter will not be in line with the bottom of the groove, and, consequently, the teeth will not be milled to the correct shape. It is a common practice to set the table to the angle at the pitch line (which is nearly halfway between the top and bottom of the tool), although some contend that if the angle near the bottom of the groove is taken, teeth of better shape will be obtained.

Rule.—The cotangent of the helix angle, at the pitch diameter, equals the lead of the helix divided by the pitch circumference.

Example.—The gear has a pitch diameter of 4.46 inches and a lead of 20 inches. Then

$$\text{Cot helix angle} = \frac{20}{4.46 \times 3.1416} = 1.427$$

In this case the helix angle is 35 degrees; hence, the machine table is adjusted from its normal position at right angles to the spindle, to an angle of 35 degrees. Before adjusting the machine table, the cutter should be centered relative to the gear, this centering operation being the same as is required before cutting a spur gear.

Is the angular adjustment of the machine table always equal to the helix angle of the gear?

When the milling cutter is held on an arbor that is inserted in the main spindle of the machine, as shown in Fig. 5, the angular adjustment of the table equals the helix angle of the gear. If the helix angle is large, it may not be possible to swing the machine table around far enough when the cutter is held as shown in Fig. 5. In cases of this kind, the cutter may be held at right angles to its usual position. Fig. 7 shows how a vertical milling attachment is used to change the position of the cutter. This attachment may be set in any position from the vertical to the horizontal, and, in this case, it is horizontal, thus locating the cutter at an angle of 90 degrees from its normal position. With this arrangement, the angular position of the table equals 90 degrees minus the helix angle of the gear. For example, if the helix angle of the gear is 60 degrees, then the table should be set at an angle of 30 degrees in order to align the cutter with the helical teeth having the 60-degree angle.

Is the depth of a helical gear tooth based upon the normal pitch?

The tooth proportions of a helical gear are based upon the *normal* pitch. The whole depth of the tooth may be obtained by dividing 2.157 by the normal diametral pitch of the gear. The normal tooth thickness at a pitch line equals 1.571 divided by the normal diametral pitch.

After a milling machine has been properly geared and adjusted and a cutter of the right pitch and number has been selected, the cutting of a helical gear consists of milling the required number of equally spaced tooth grooves of correct depth. The tooth spacing is simply a matter of indexing. If the teeth of a helical gear are coarser than 10 or 12 diametral pitch, it is well to take roughing and finishing cuts. In taking the final cut, the cutter should be prevented from dragging over the tooth surfaces while it is being returned to the starting point preparatory to cutting the next tooth space. The method usually recommended is to stop the machine and turn the cutter to such a position that its edges will not scrape the finished tooth surfaces. Another method is to lower the knee slightly, but since this involves resetting for each tooth, the first method referred to is ordinarily preferred.

In checking the size of a helical gear tooth, how is the chordal thickness found?

The normal chordal thickness is required in measuring the tooth thickness with a vernier gear tooth caliper.

Rule.—Divide number of teeth by *normal* diametral pitch and multiply quotient by the sine of an angle equal to 90 divided by number of teeth.

Example.—A helical gear has 45 teeth of 5 normal diametral pitch. Then

$$\text{Normal chordal thickness} = \frac{45}{5} \times \sin\frac{90}{45} = 0.31409 \text{ inch}$$

In this example the chordal thickness is only slightly less than the normal circular or arc thickness. The chordal thickness is especially important when the number of teeth is small.

The section on spur gears includes a table (page 228) giving corrected addenda and chordal thicknesses for various numbers of spur gear teeth of 1 diametral pitch. This table may also be used for helical gears. When such a table is

Fig. 7. Use of Vertical Milling Attachment for Milling Helical Gear of Large Helix Angle

used for *spur* gears, values for pitches other than 1 diametral are obtained simply by dividing the given values by the required diametral pitch.

In applying the table to helical gears, especially when the number of teeth is small and the helix angle large, divide the actual number of helical gear teeth by the cube of the cosine of the helix angle; then use the quotient as the number of teeth in entering the table, instead of the actual number; finally, divide the table values by the *normal* diametral pitch to get the normal chordal thickness and the normal chordal addendum. Example: Helical gear has 18 teeth of 1 normal diametral pitch and helix angle of 45 degrees. Then, $18 \div \cos^3 45° = 51 =$ number of teeth used in entering table.

In cutting worm-gears on a milling machine, what operations are required?

Worm-gears may be cut on a milling machine instead of using a gear-hobbing machine or a special type designed expressly for cutting worm-gears. If a milling machine is employed, two operations are required: First, gashes or grooves are milled around the worm-gear blank to form the teeth roughly, and then a hob is used for finishing. Gashing is done preferably by the use of an involute spur gear cutter of a number and pitch corresponding to the number and pitch of the teeth in the worm-gear. If it is necessary to use a plain milling cutter, the width of the latter should not exceed three-tenths of the circular pitch of the worm-gear, and the cutter should have rounded corners to prevent removing the fillets from the bottoms of the teeth.

The cutter used for gashing is held on an arbor in the usual way, and the gear blank is also mounted on an arbor which is supported between the centers of a dividing head (see Fig. 8). The cutter should be centered in both crosswise and lengthwise directions relative to the gear. It is also necessary to locate the table at an angle so that the inclination of the gashes will correspond to the lead angle or helix angle of the worm thread.

Rule.—The tangent of the lead angle of the worm may be found by dividing the lead of worm thread by the circumference of its pitch circle.

Fig. 8. Gashing Teeth of a Worm-wheel in a Milling Machine

Fig. 9. Hobbing the Teeth of a Worm-wheel

If the diameter of the gashing cutter is larger than the diameter of the hob to be used later, the whole depth of the tooth should be marked on the side of the blank to avoid gashing too deeply. After the cutter has been sunk to the depth indicated by this line, the reading on the dial should be noted in order to cut the remaining gashes to the same depth.

Tooth Depth.—According to recommended practice of the American Gear Manufacturers Association, the tooth depth for single and double thread worms equals 0.686 x pitch; for triple and quadruple threads the tooth depth equals 0.623 x pitch.

How are worm-gear teeth formed by hobbing on a milling machine?

After the gashes have been cut, the machine table is set at right angles to the cutter spindle, and the gashing cutter is replaced with a hob (see Fig. 9). The dog is removed from the work arbor or the dividing-head gearing is disengaged to permit free rotation of the spindle, and the hob is placed in mesh with the gashes in the blank; consequently, as the hob is revolved, the worm-gear revolves with it. As this occurs, the gear blank is gradually elevated so that teeth are formed on it which mesh accurately with those of the hob. Since the hob is a duplicate of the worm except for those provisions necessary to convert it into a cutting tool, it is evident that the worm gear produced in this manner will also mesh with the worm. The necessary clearance between the worm and the worm-wheel is obtained by making the outside diameter of the hob and also its root diameter slightly greater than corresponding dimensions of the worm. It is advisable to cut the worm thread before hobbing the worm gear, since the worm can be used to advantage for testing the center-to-center distance between the worm and the gear. In making this test, the knee of the machine should be lowered sufficiently to make room for the worm beneath the hob, but the lengthwise position of the table should not be disturbed.

Cutting Very Large Spur Gears and Bevel Gears

Large gears of coarse pitch may be cut either by planing on a templet or form-copying type of machine, by milling with a formed cutter, or by hobbing. Most gear manufacturers use the templet planer for the very large gears which may range from two or three feet in diameter to ten or fifteen feet in diameter or larger in some cases. One advantage of this type of machine is that simple, inexpensive tools are used, and this is very important, as often only one of these large gears is required, and the cost of making a formed cutter or hob would be prohibitive. Gear-cutting machines of the templet type are also used for cutting large bevel and herringbone gears; in fact, gear planers of this class are used invariably for cutting very large bevel gears. Some gear planers are designed for cutting spur gears exclusively, but there are also combination types which may be applied to either spur or bevel gears.

How are gear teeth formed on a templet type of gear planer?

A characteristic feature of the templet type of gear planer is the templet or master former which serves to guide the planing tool, thus causing it to plane teeth having the correct shape or curvature. When the planer is at work, a slide or head which carries the tool is given a reciprocating motion, and as the tool feeds inward for each stroke, the path it follows is controlled by the templet. The traversing movement of the tool-slide is derived from a crank on some gear planers, whereas others have a reversing screw. Still another method of traversing the head is by means of a rack and pinion, the latter being arranged to rotate in opposite directions.

There are two general methods of machining the teeth on one of these planers. One is to rough out the teeth with a single-pointed tool and then finish with a formed tool which removes the feed marks and gives the teeth a smooth finish. The other method is to take both roughing and finishing cuts with single-pointed tools. The use of the formed tool for finishing is impracticable for the larger pitches which are finished by a single-pointed tool. The number of cuts required depends upon the size of the tooth, amount of stock to be removed, and the kind of material.

A templet type of planer designed for cutting external and internal spur gears is shown in Fig. 1. The tool A is held by a slide which can move vertically. This slide is con-

Fig. 1. Gear Planer of the Templet or Form-copying Type
with Finished Spur Gear in Position

nected with another slide that is given a horizontal feeding movement after each cutting stroke of the tool. A roller *B* attached to the vertical slide, rests upon a templet *C*, which is stationary. When the horizontal slide feeds inward for taking a finishing cut, the tool planes one side of the tooth to the required curvature, because the path it follows is controlled by the shape of templet *C*. Sets of these templets are supplied with a gear planer, one templet of each pair being for the upper sides and the other for the lower sides of the teeth. Each pair of templets in the set covers a certain range of diameters and can be used for planing any pitch from the smallest up to the pitch stamped on the templet, which requires the full length of curve of that templet. The entire tool head is mounted upon a large main slide that is traversed by a crank type of drive on this particular machine.

The indexing mechanism consists of a large dividing wheel which is engaged by a worm connecting with a mechanism for controlling the indexing movement. After one side of a tooth is planed, the index mechanism is tripped by the operator, and then the gear is automatically rotated an amount equal to the circular pitch of the gear being planed. The power for this indexing movement is supplied by a small auxiliary motor which forms part of the indexing mechanism. As a general rule, gears having a circular pitch of two inches or smaller are planed with tools which are used for both the sides of the teeth and the fillets at the bottom, but for coarser pitches, different tools are used for the sides and the fillets.

What is the most practical method of cutting very large internal gears?

The most practical method of cutting very large internal gears is on a planer of the templet or form-copying type. A regular spur gear planer is equipped with a special toolholder for locating the tool in the position required for cutting internal teeth. The holder is of a heavy, rigid design, which prevents excessive deflection of the tool. The templets used to control the path followed by the tool conform to the shape required for internal teeth, and differ in shape from those used for external gears of the same size.

Very satisfactory internal gearing may be cut on planers of this type, which are especially adapted for large work. The procedure in cutting internal gears is practically the same as for external spur gearing, after the machine has been equipped with the special tool-holder referred to.

Why are the teeth of very large gears cast to the approximate shape required?

Rough or unplaned blanks for large gears, such as are cut on the templet type of gear planer, often have rough-cast teeth, which are left thick enough to allow for planing. When the teeth are cast, it is not necessary to remove so much metal when machining them, although these cast teeth have a hard scale that offsets, more or less, the advantage mentioned. Because of this fact it is easier to cut the teeth "from the solid" in some cases; however, there is a good reason for casting the teeth to the approximate shape required. When the teeth of large gears are cast, this tends to insure soundness in the rim of the casting and tooth surfaces free from defects. On the contrary, if coarse teeth are cut from a solid rim, blow-holes or other interior defects might be encountered, especially in those rim sections adjoining the spokes or arms. According to one general rule, there is no advantage in casting the teeth if they are smaller than one diametral pitch.

How is a gear planer arranged to cut either spur or bevel gears?

Some planers are designed for cutting either spur or bevel gears. These planers differ from the type used for spur gears only, especially in regard to the arrangement of the main slide upon which the head is traversed. One end of this slide is provided with both vertical and horizontal bearings. The vertical bearing permits it to swing horizontally, and the horizontal bearing provides for a vertical movement. A feeding mechanism controls the horizontal motion, whereas the vertical movement is controlled by a former engaged by a roller attached to the outer end of the main slide. These combined horizontal and vertical movements cause the tool to follow the converging form of a bevel gear tooth. The tool-holder used for bevel gears does not have a cross-feed relative to the main slide, but for planing spur gears another tool-head having this lateral feeding movement is used and then the main slide is held stationary in a position parallel to the axis of the gear. On planers of the combination type, the head which carries the gear blank has not only a side adjustment to care for different diameters, but also an adjustment that makes it possible to set bevel gears of different angles so that the apex of the pitch cone of the gear will coincide with the "center of the machine."

Why are gear planers of the templet type always used for very large bevel gears?

Very large bevel gears, and many of medium size, are cut by using gear planers of the templet or form-copying

Fig. 2. Bevel Gear Planer of the Single-tool Form-copying Type

type. While such machines are often used for cutting the larger sizes of spur gears, their application to very large bevel gears is universal. More distinct types of machines have been designed for cutting spur gears of various sizes than for cutting bevel gears, since the former present a simpler gear-cutting problem. For instance, spur gears may be cut readily with formed cutters, and a great many are produced by this simple method; but as the tooth of a bevel gear has the same cross-sectional shape on a diminishing scale toward the apex of the pitch cone, it cannot be given the correct shape with a formed cutter, although this method has been used to a limited extent for bevel gears. The templet principle, however, provides a relatively simple method of cutting bevel gears of medium and large sizes, especially in gear shops or wherever a variety of pitches or sizes is encountered; this accounts for the general use of the templet or form-copying type of gear planer.

What are the main features of a single-tool, templet type of bevel gear planer?

Some bevel gear planers of the form-copying type use a single tool, and others are designed to use two tools. A single-tool planer of 24-inch size is shown in Fig. 2. When this planer is at work, the tool-slide is given a reciprocating motion by a crank adjusted for a stroke equal to the face width of the gear being cut, plus about ¾ inch over-travel, or possibly 1 inch on larger machines. The arm carrying the tool-slide is pivoted so it can swing about a horizontal axis. The entire tool-head (with its tool-slide, crank drive, and pivoted arm) is mounted on a turret or baseplate supported by the main bed.

This turret swings about a vertical axis to feed the tool inward, and at the same time the tool-slide arm swings about its horizontal axis, provided the supporting roller R is resting on one of the curved forms or templets B or C, which are used in taking finishing cuts. The inward feeding movement and the backward motion to withdraw the tool for indexing are obtained from a cam acting through an adjustable lever for varying the angle through which the turret swings. This angular movement is regulated according to the depth of the gear teeth, and is equal to the angle subtended by the whole depth of the tooth, plus 1½ degrees for clearance. The turret is set to suit the pitch cone angle of the gear being cut.

In cutting teeth from the solid, the tooth spaces are first roughed out, as described later, and then the teeth are finished first on one side all around the gear and then on the opposite side. The holder to which the forms are attached can be turned about a horizontal axis for locating the different forms in the working position. The indexing movement of the gear being cut is controlled automatically through change-gears located at E, which are selected by referring to an index table furnished with the machine. The planer illustrated in Fig. 2 will cut miter gears up to 24 inches pitch diameter, and bevel gears of a 2 to 1 ratio up to 28 inches pitch diameter. Machines of the form-copying type are made for cutting gears up to the largest sizes ordinarily encountered in commercial work, and such machines have the advantage of using tools of a simple inexpensive form.

How is the bulk of the metal removed in cutting large bevel gears?

When a bevel gear is cut from the solid (instead of planing teeth which have been roughly cast to shape, as is customary for very large gears) there are several methods of roughing or stocking out the teeth. For roughing steel or cast-iron gears up to about 2 diametral pitch, it is well to use a single V-shaped tool, as indicated by the diagram A, Fig. 3. While this tool may have plain cutting edges, the stepped or corrugated form indicated is preferable, as it breaks up the chips. The size of the roughing tool depends upon the width of the tooth space at the small end. The tool should be wide enough to leave about 1/32 inch of stock on each side of the teeth at the small end, and the length of the cutting end must be at least equal to the full depth of the teeth at the large end.

Sometimes in cutting "flat gears" (gears having a large pitch cone angle) of, say, 2 diametral pitch, especially if the section is light and the gear blank tends to spring, it might be necessary first to cut parallel slots in the center of each tooth space, as indicated by diagram B, and then

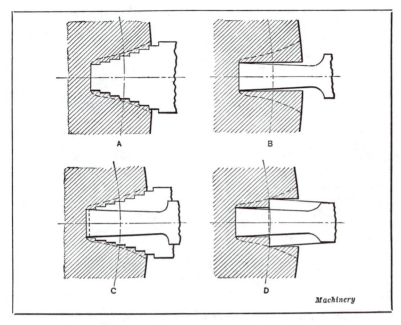

A   B

C   D

Machinery

**Fig. 3. Different Forms of Tools Used for Roughing
or Stocking Cuts**

follow with a V-shaped roughing tool. Gears coarser than
2 diametral pitch and with pitch-cone angles of less than
45 degrees may be roughed out by using a combination slotting and V-tool, as indicated by diagram *C*. The combination slotting and V-tool arrangement is not readily applicable to the 24-inch gear planer, so it finds use only on the
larger sizes. The slotting tool is always set to cut slightly
deeper than the V-tool, so that the latter never cuts on the
point. In cutting "flat gears," or those that have large
pitch-cone angles (above 45 degrees), there is sometimes a
tendency for the work to spring in toward the tool and stall
the machine when a heavy cut is being taken, which makes
it necessary to relieve the cutting action by not setting the
V-tool in to the full depth.

Another combination tool that is sometimes used is illustrated by diagram *D*. This is a narrow slotting tool extending to the full depth, followed by a wider slotting tool
which goes in as far as the pitch line. When the work is

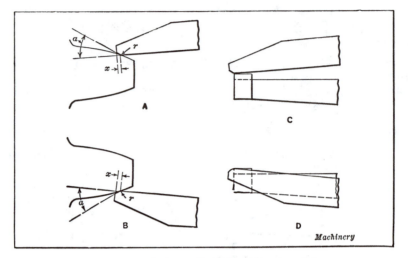

**Fig. 4.   Tools Used for Finishing Upper and Lower
Sides of Teeth**

such that combination tools cannot be used, and the pitch
of the teeth is too large to permit roughing from the solid
with a V-shaped tool, it is necessary to cut plain slots first
and then follow with a V-shaped tool.

What form of single-point tool is used for the finishing cuts?

Diagrams *A* and *B*, Fig. 4, show the type of tool used for
finishing the upper and lower sides of the teeth, respectively.   The end of the finishing tool must be small enough to
pass through the space of the small end without touching
the tooth on the opposite side.   The tools are rounded at *r*
just enough to leave a smooth cut.   The distance *x* and the
angle *a* (which is about 35 degrees) should be the same for
both finishing tools.   The toolpost is so arranged as to allow
the tool to swing away from the work during the return
stroke, the same as on an ordinary planer.   The tool is
always returned to the cutting position by means of a stop,
which should be properly adjusted before machine is used.

The tool-holders are so constructed that the tools may be
adjusted for height by means of a wedge, and for lengthwise position by means of a stop-nut.   A gage placed on

the tool-slide arm shows when the tools are properly set. Diagrams C and D show the ends of upper and lower finishing tools set in correct relation with the end of this gage. The roughing tool first is inserted in the toolpost, and the end is set even with the end of the gage. The tool is also adjusted for height until it is central with the end of the gage. The roughing tool is next replaced by the lower-side finishing tool, which is adjusted so that the cutting point r coincides with the outer corner of the gage, thus locating the tool vertically and lengthwise. The upper tool is next set in the same way and is left in the machine for further adjustments.

When a pair of planed gears is in mesh, it will be seen that the clearance between the end of one tooth and the bottom of the corresponding tooth space is the same at both the large and small ends; consequently, the clearance at the small end is larger in proportion to the size of the tooth at this end, and it might be inferred from this that the gears are not properly in mesh. The clearance should, however, be the same at both ends, because when the tools are set as illustrated by diagrams C and D, part of the tool projects beyond the cutting point an amount represented by dimension x (see diagrams A and B). When the tool is set properly, the cutting point r travels along a line coinciding with the apex A of the pitch cone, as indicated by the diagram Fig. 5. This cutting point r will always travel toward A, or the center of the machine, as the tool feeds in across the surface of the tooth. As the clearance space is formed by that part of the tool projecting beyond the cutting point, the width of the clearance space at the large end of the tooth is the same as at the small end.

What adjustments are required in locating the tool-slide and gear in the cutting position?

The turret which supports the tool-slide is set to the pitch-cone angle by means of graduations on its base in conjunction with a fixed pointer. When this adjustment is made, an auxiliary centering form is attached to the form-holder in place of form C, Fig. 2, for the lower side. This centering form has a V-shaped notch, which engages roller R. The roller must be in contact with both sides of the V-slot

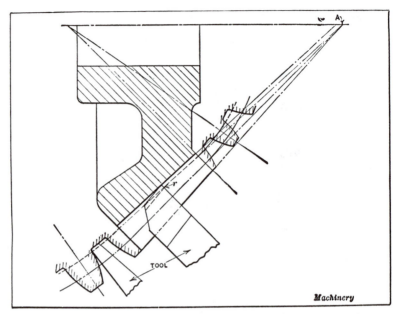

Fig. 5. Diagram Showing how Clearance Space is Formed by Planing Tool

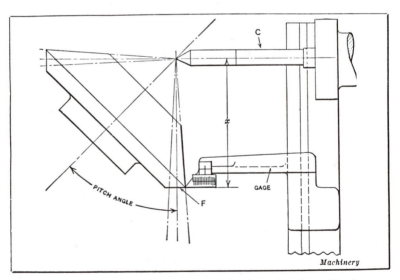

Fig. 6. Cone Distance Gage Used for Locating Gear Blank in Correct Position Axially

in the centering form when the turret has been adjusted to the required pitch-cone angle.

After the turret has been set with reference to the pitch angle, the head D (Fig. 2) carrying the work-spindle is adjusted along the bed to locate the gear blank in the proper cutting position. The gear blank will be in the right position when the apex of the pitch cone coincides with the "center of the machine," or with the point of intersection between the vertical axis about which the turret swivels and the horizontal axis about which the tool-slide swivels. The cone distance gage (Fig. 6) is used in making this adjustment. The vernier on the gage is first set to conform to the pitch depth or addendum of the gear to be cut. The gage is then placed on the tool-slide arm. The head carrying the work-spindle is next moved along the bed until the pointer of the gage stands at zero when the extreme outside edge of the gear blank is in contact with the vernier, as shown by the illustration. The gear blank is then in the correct position.

In locating a pinion having a pitch-cone angle of, say, 15 degrees or less, it is preferable to set the blank by measuring the cone distance x instead of using the cone distance gage. The cone distance or pitch-cone radius can be determined by dividing half the pitch diameter by the sine of the pitch-cone angle. This calculated dimension should correspond with distance x from the point of center-pin C to face F of the gear blank. The center-pin should be pushed all the way into its socket to locate the point at the center of the machine. A scale that is graduated in hundredths is preferable for checking this dimension. A line, representing the full depth of the teeth at the large end, is scratched on the back angle or face of the gear blank, and is used in setting the tools. This may be done accurately when the gear blank is turned. The index mechanism must be properly geared and adjusted, and the stroke, as well as the cutting position of the tool, must be so regulated that the tool has an over-travel of, say, $5/8$ inch at the large ends and $1/8$ inch at the small ends of the teeth. The rate of feed and number of strokes per minute must also be varied to suit the gear to be cut, which is done by using different combinations of speed and feed gears.

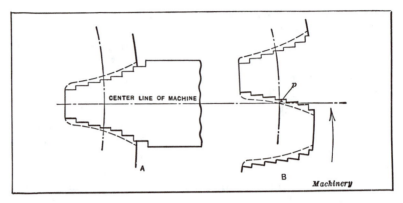

Fig. 7. (A) Position for Roughing. (B) Position of
Tooth for Finishing

What is the general procedure in taking stocking and finishing cuts?

After the type of stocking tool to use has been determined, the first tooth space is roughed out. As the tool feeds straight in or radially, a straight templet is used instead of a curved form, and this is shown in the operating position in Fig. 2. When a tool is approaching the bottom of the first tooth space, care should be taken to see that it does not cut below the full-depth circle.

The forms that control the path followed by the tools in taking finishing cuts are selected according to the pitch-cone angle, by referring to a chart accompanying the machine. The roughing tool is replaced by the upper-side finishing tool, and the form B (see Fig. 2) for the upper sides of the teeth is swung around to the operating position. A latch controlled by a handle engages a slot and locates the form-carrying plate. The plate should be tightened after the form to be used is in position.

When the gear is being stocked out, the roughing tool is in a central position relative to the center line of the machine or to a horizontal plane coinciding with the axis of the gear (see diagram A, Fig. 7). Before taking the finishing cuts, the gear must be rotated until pitch point p (diagram B) coincides with the center line of the machine. In order to change the position of the gear, a clutch at K

(Fig. 2) connecting with the indexing gear train, is disengaged and crank L on the indexing worm-shaft is used to turn one section of the clutch relative to the other section. If two cuts are to be taken for finishing each side (the second being a light finishing cut) crank L is turned to the right (assuming the upper sides of the teeth are to be planed) until graduation No. 1 on clutch K is one notch to the left of the zero line on the other half of the clutch. The pitch point p is now in the position indicated by diagram B, Fig. 7. After tightening the clutch lock-nut and engaging the feed clutch, a cut is taken over the upper sides of all the teeth. Then when the turret and tools are at the limit of their outer movement, the feed clutch is disengaged and graduation No. 1 on the index clutch is set opposite the zero position; finishing cuts are then taken on the upper sides of the teeth.

Before finishing the bottom sides of the teeth, it is necessary to replace the upper-cutting tool with the one used for under-cutting, and also place the form C (Fig. 2) for the lower sides in the working position under the roller. The gear must now be shifted, so that the pitch point on the side opposite p (diagram B, Fig. 7) coincides with the

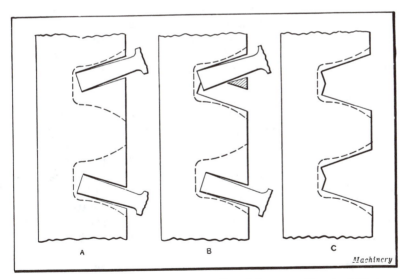

Fig. 8. One Method of Stocking out Bevel Gears on Two-tool Planer

center line of the machine. The clutch K (Fig. 2) is again disengaged, and the two parts are shifted in the opposite direction or until graduation No. 3 is one notch to the right of the zero position. Cuts taken with the clutch in this position will leave the teeth somewhat wide. When the final finishing cuts are taken, with graduation No. 3 in the zero position, the teeth should be about the right size, although as soon as the first tooth has been planed, its width should be tested either by using a solid gage of the right diametral pitch or an adjustable vernier caliper. If a slight adjustment is necessary, this can be made by shifting a clutch on a worm-shaft which has a large number of teeth, in order to provide a finer adjustment than could be obtained with the clutch at K. The amount of stock that the tool will remove from the tooth when this clutch is shifted one notch depends upon the size of the gear, increasing as the diameter of the gear increases.

Why are some bevel gear planers of the templet type equipped with two tools?

A bevel gear planer of the templet or form-copying type that is designed to use two tools simultaneously, saves from 45 to 50 per cent of the cutting time, as compared with a single-tool planer. In taking roughing cuts, one tool is operating in one tooth space while the other tool works in the space following (see diagrams, Fig. 8). In taking finishing cuts, both tools operate on opposite sides of the same tooth.

The head carrying the work-spindle is given a feeding movement about a vertical axis coinciding with the apex of the pitch cone or center of the machine. When the cutting tools have reached the proper depth, the work-head swings back to clear the tools for indexing. The feeding movement is disengaged automatically when the tools reach the bottom of a tooth space, and there is a dwell at this point before the gear swings away from the tools, so that the tools make one complete stroke at full depth, thereby finishing all of the teeth to the same depth. The extent of the feeding movement is regulated by adjustable dogs on a graduated dial connected with the reversing mechanism. The two tool-slides are inclined in relation to each other on account of the converging form of bevel gear teeth, the angle between the slides being varied to suit the gear. The axis of the pivot or bearing about which the tool-slides are adjusted coincides with the center of the machine.

The bevel gears cut on a two-tool type of machine usually are finished from the solid. While it is possible to rough out the tooth spaces by the methods previously described in connection with single-tool machines, an effective method, especially for relatively coarse pitches, is indicated in Fig. 8. Square-nosed end-cutting tools are used, and these are set at angles that are equal approximately to the inclination of the sides of the tooth. One tool is inclined upward and the other downward, since they are used for roughing opposite sides of the tooth spaces. A straight form or templet is used. Two slots are formed as indicated by diagram *A*. The gear then indexes an amount equal to the circular pitch, so that one tooth space is roughed out as shown by diagram *B*, the central part (indicated by the shaded lines) falling out after it has been separated by the tool. As this operation is repeated, a tooth space is roughed out at each indexing, to the approximate shape indicated by diagram *C*. This method of roughing would be employed for stocking out gears of the larger pitches. When it is practicable to use corrugated V-shaped tools, as in roughing cast-iron gears and especially those of the smaller pitches, two tooth spaces are stocked out simultaneously, and the indexing mechanism is set to index the gear an amount equal to twice the circular pitch.

Generating Methods of Forming Gear Teeth

The generating method of cutting gears is used extensively in the machine building industry. In cutting gears by using a generating type of machine, the gear teeth are formed as the result of certain relative motions between gear blank and cutter, instead of simply reproducing the shape of a formed cutter. When a series of formed cutters is used for cutting spur gears, for example, it is evident that the curvature of any cutter of a set can be absolutely correct only for a certain number of teeth. The error in shape for other tooth numbers, within a limited range, may be negligible for ordinary requirements, but for many gear applications it is essential to utilize a method that gives the curvature conforming to any given number of teeth. This may be accomplished by a generating method, although it should not be inferred that a generating type of machine always results in greater accuracy than is obtained by using well-made formed cutters, since much depends in either case upon the condition of both machine and cutter and other factors of a mechanical nature. A generating process, however has the inherent advantage of being theoretically correct and of enabling a cutter of a given pitch to cut gears having different numbers of teeth to the correct shape, except for purely mechanical errors such as occur in varying degrees with any method.

How are gear tooth curves generated by using a gear-shaped cutter?

In order to illustrate the principle of the generating process of gear-cutting, assume that a finished gear having teeth of correct form is revolved while in contact with a blank rotating at the same speed and, for purposes of illustration, is assumed to be made of some soft, plastic material. The result of this rolling action would be to form or

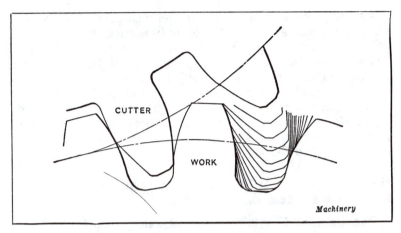

CUTTER

WORK

Machinery

Fig. 1. Diagram Illustrating How a Gear-shaped Cutter is Used to
Form Spur Gear Teeth by a Generating Action

generate teeth on the plastic blank. Thus, the teeth on the
finished gear, as they roll into contact with the blank, form
teeth having the curvature required for meshing properly
with the generating teeth. This is a simple illustration of
the principle of the generating process. Now, if this tooth-
forming or generating gear were hardened, and its teeth
given suitable clearance, the cutter thus formed could be
used to generate teeth in a cast-iron or steel blank, provided
the cutter hâd a reciprocating action parallel to the axis of
the blank, while both cutter and blank slowly revolved
together, the same as two gears in mesh.

This method of using a gear-shaped cutter is employed on
a Fellows gear shaper. The type of cutter used on this machine resembles a spur gear. This cutter is reciprocated
vertically, and in starting to cut a gear it is first fed in to
depth; then one gear tooth after another is formed as cutter and work slowly rotate together just as though two
finished gears were in mesh. The action is illustrated by
the diagram Fig. 1. Successive positions of one cutter tooth
are indicated by the succession of outlines. The feeding
movement per stroke of the cutter will, however, be less
than the amounts shown by these outlines. A gear-shaped
cutter may be applied to internal as well as external gears.
Fig. 2 shows a Fellows gear shaper cutting an internal gear.

As the tooth curves on the cutter are ground by a generating process after the cutter is hardened, very accurate gear teeth can be produced by this method. The involute curvature on the cutter teeth is obtained as the result of a rolling movement of the cutter past the flat surface of a grinding wheel. The rolling movement is similar to a gear rolling along a rack and the grinding surface corresponds to the side of a rack tooth; hence, this gear-cutting process is based upon the rack tooth principle even though a spur gear type of cutter is used.

How are gear tooth curves generated by using a rack-shaped cutter?

The diagram, Fig. 3, shows a rack-shaped cutter and how teeth are generated with this form. If a rack were moved

Fig. 2. Cutting the Teeth of a 12-inch Internal Gear on a Fellows Gear Shaper

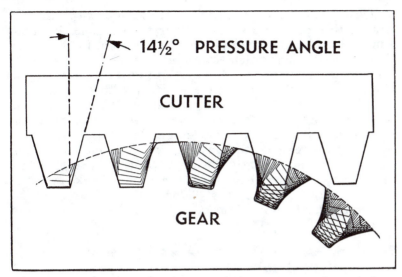

14½° PRESSURE ANGLE

CUTTER

GEAR

Fig. 3. Diagrams Showing How Spur Gear Teeth are Generated
by Rolling the Gear Blank Relative to a Cutter
Representing a Short Section of a Rack

horizontally and the gear rolled with it at the same rate,
then gear teeth conforming to the standard rack would be
formed provided, for purposes of illustration, the blank were
made of some soft plastic material. The same general result
might be obtained with a gear blank of cast iron or steel
if a rack-shaped cutter were used, assuming that the cutter
operates with a vertical motion, and after each cutting
stroke, the gear blank is rolled forward slightly just as
though it were a finished gear rolling on a rack. The cutting action occurs while the blank is held stationary; as the
cutter returns, the blank receives a slight rotary motion relative to the cutter, which is in effect a feeding movement.
The diagram indicates successive positions of the gear
blank relative to the cutter, and shows how the straight-
sided rack teeth would generate the involute tooth
curves.

The rack-shaped form of cutter is seldom used, but various forms of gear cutters, in general use on regular gear
cutting machines, are related to this basic rack form. For
example, the hobs which are very generally used, have teeth

which represent the rack. It is possible to employ either a gear-shaped or a rack-shaped cutter for the following reason: A rack can be designed, for any system of interchangeable gearing, which will mesh correctly with a range of gear sizes of the same pitch. Moreover, all gears that will mesh properly with the rack will also mesh with one another. Gear-cutting processes of the generating class are based on this interchangeable feature. The cutter represents either a rack or a gear of the interchangeable series, and it cuts or generates teeth as the uncut gear blank and cutter are given movements, relative to each other, similar to a finished gear running in mesh either with a rack, as illustrated by the diagram, Fig. 3, or with another gear, as shown by the diagram, Fig. 1.

How are rack teeth generated by using a gear-shaped cutter?

Rack teeth can be generated merely by reversing the process illustrated by the diagram, Fig. 3. When this method is employed, the cutter is a circular or gear-shaped form and the rack teeth are generated as the rack moves longitudinally past the cutter which rotates slowly the same as if it were in mesh with the rack. As the cutter rotates, it also has a reciprocating motion for traversing it across the rack. The design of the Fellows rack shaper is based upon this principle. The table of this rack shaper is traversed by a lead-screw and change-gears. The cutter-head is mounted on a base located at right angles to the table. The position of the head when the cutter is in operation is controlled by a feed-cam which, in conjunction with change-gears, controls the depth of cut and the depth feed per stroke of the cutter. The saddle carrying the cutter-spindle is of the "relieving" type. The machine has complete electrical control, comprising push-buttons and selector switches which can be set to start the cutter at either end of the rack or at its center. The switches can also be set for controlling the depth of feed and traverse feed motions, as well as for stopping the machine when the rack is completed. A rack can be finished automatically in either one or two cuts. This machine will cut racks for meshing either with spur or helical pinions.

Why is a generating cutter of given pitch applicable to any number of teeth?

A great advantage of the generating method of cutting gears is that a cutter of given pitch may be used for different numbers of gear teeth and they will all have the correct shape and be interchangeable. (The term "interchangeable" means that gears of various sizes will mesh with the basic rack and with one another, assuming that they are all of the same pitch.) The shape of spaces between the teeth of a gear having, say, 20 teeth is quite different from the shape of the spaces of a gear having, say, 40 teeth. The generating process, however, automatically produces the correct shape in both cases, but this is not true when gear teeth are cut by milling because the milled space is a direct reproduction of the profile of the cutting edges.

What is the pressure angle of gearing and how does it affect the actual cutting of a gear?

In all gear cutting, the shape of the cutter used and the resulting gear-tooth form is established with reference to some pressure angle. To illustrate, if the pressure angle is 14 1/2 degrees, this is obtained by using a cutter having a form or profile that will produce gears conforming to this angle. (The cutter shown in Fig. 3 would produce gears having a pressure angle of 14 1/2 degrees.) The one who actually cuts the gear may not understand the meaning of the term pressure angle; but, nevertheless, all cutters for forming gear teeth, whether by milling or in some other way, represent a certain pressure angle. Since this angle must be referred to at times in connection with certain branches of gear cutting, it will be explained briefly.

When one gear is driving another, the points of contact between successive pairs of involute teeth, as they roll into and out of mesh, is along a straight line. This line is known as the *line of action*. The angle between this line of action along which the pressure is exerted, and a line perpendicular to the center line of the two gears, is known as the *pressure angle* of the gearing. This line of action is tangent to the base circle of each gear. These base circles are so named because the involute curvature of the teeth is derived from them, as explained in treatises on gear design.

What are the common pressure angles?

In the design and manufacture of gearing, certain pitches, pressure angles and tooth proportions have been used so extensively that they are generally accepted as standards. While 14 1/2 degrees is the most common pressure angle, as applied to miscellaneous classes of gearing, 20 degrees may be regarded as the standard sanctioned by common usage for certain types of transmission gears. Bevel gears, as well as spur gears, are designed for both 14 1/2- and 20-degree pressure angles, and other angles are sometimes employed. Gearing is commonly designated by giving the pressure angle. For instance, the expression 14 1/2 degrees, 20 degrees, etc., as applied to gearing, relates to the pressure angle.

When the pressure angle is changed, what is the effect on the tooth form?

A 14 1/2-degree angle is unsatisfactory for small pinions (especially if the teeth are of standard or ordinary full-depth proportions) because of under-cutting which weakens the teeth and causes poor tooth action when two gears revolve together. Under such conditions, the tooth form may be improved by increasing the pressure angle. The usual increase is to 20 degrees. As the pressure angle is increased, the gear teeth have a greater increase in width from the point to the base; the involute curvature also extends farther below the pitch circle and forms a larger part of the tooth curve or profile. There are differences of opinion as to preferable applications of different gear tooth standards and pressure angles, but it is advisable to be guided by prevailing practice at least until the selection of a standard can be based upon actual experience with a given class of gearing and service.

In cutting gears, how is a given pressure angle obtained?

As previously pointed out, the pressure angle is obtained by using a cutter conforming to the required angle. The relationship between the angle represented by the cutter and the pressure angle of the gear it produces is illustrated by the diagram, Fig. 3. This diagram shows a rack having straight-sided teeth which is the form of rack tooth that would mesh properly with a rotating involute gear. When gears are generated by using a rack-shaped cutter, the nominal pressure angle of the gear teeth thus formed equals one-half the included angle of the rack teeth or the angle on one side, as the diagram shows. The actual pressure angle depends upon the inclination of the line of action as previously explained, and if the gears were separated somewhat, thus making the center-to-center distance greater than standard, this would increase the angle of the line of action, and, consequently, the actual pressure angle. However, the term pressure angle, as used in gear cutting practice, relates to that angle of the line of action which would occur between gears operating at the standard center-to-center distance.

What is the meaning of "pressure angle" as applied to a gear-shaped cutter?

When a gear is cut by rolling it in unison with a gear-shaped cutter, as illustrated by the diagram, Fig. 1, then the pressure angle is equivalent to the angle of the rack from which the cutter tooth curvature was derived when the cutter was made.

In producing formed cutters for milling gears, the cutter is given practically the same shape as the tooth space which would be generated by a rack type of cutter, and the milling cutter represents the same pressure angle as the rack. For example, if the rack tooth has an included angle of 29 degrees, or 14 1/2 degrees on each side, then a milling cutter shaped to conform with the tooth space generated by this basic rack would also represent a pressure angle of 14 1/2 degrees. In actual practice, a formed cutter of any one pitch is used for a range of tooth numbers as explained in the section on spur gear cutting, although the profile of the cutter is only correct for one of these numbers.

Why are gear-tooth standards represented by rack teeth?

When the generating process of forming gear teeth is understood, the reason why the rack tooth is the basis for

14½-Degree Full-depth Involute System

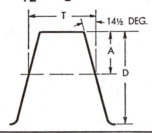

ADDENDUM A =
  1 ÷ Diametral pitch
  0.3183 × Circular pitch

TOTAL DEPTH D =
  2.157 ÷ Diametral pitch
  0.6866 × Circular pitch

BASIC THICKNESS T =
  1.5708 ÷ Diametral pitch
  0.5 × Circular pitch

20-Degree Full-depth Involute System

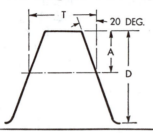

ADDENDUM A =
  1 ÷ Diametral pitch
  0.3183 × Circular pitch

TOTAL DEPTH D =
  2.157 ÷ Diametral pitch
  0.6866 × Circular pitch

BASIC THICKNESS T =
  1.5708 ÷ Diametral pitch
  0.5 × Circular pitch

20-Degree Stub Involute System

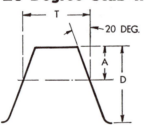

ADDENDUM A =
  0.8 ÷ Diametral pitch
  0.2546 × Circular pitch

TOTAL DEPTH D =
  1.8 ÷ Diametral pitch
  0.5729 × Circular pitch

BASIC THICKNESS T =
  1.5708 ÷ Diametral pitch
  0.5 × Circular pitch

14½-Degree Full-depth Composite System

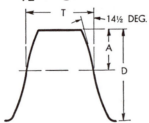

ADDENDUM A =
  1 ÷ Diametral pitch
  0.3183 × Circular pitch

TOTAL DEPTH D =
  2.157 ÷ Diametral pitch
  0.6866 × Circular pitch

BASIC THICKNESS T =
  1.5708 ÷ Diametral pitch
  0.5 × Circular pitch

Fig. 4. American Standard Basic Rack Tooth Forms

a system of interchangeable gearing becomes apparent. Thus, when gear teeth are formed by a generating process, the cutter (in the case of spur and helical gears) represents a rack, directly or indirectly. Moreover, this cutter forms the tooth curves as both cutter and gear blank move relative to each other, the same as gearing in actual operation, and a cutter of given pitch may be used for various numbers of gear teeth. Consequently, an entire system of standard gear teeth can be established by first proportioning a basic rack tooth of suitable proportions.

Standard tooth forms differ in regard to the depth of the tooth for a given pitch, and also in regard to the angle (pressure angle) or form of the rack tooth. The upper diagram, Fig. 4, shows the basic rack of the American Standard 14 1/2-degree full-depth involute tooth. A great many gears in use today have teeth conforming to this basic rack. The second diagram shows another American Standard basic rack. This differs from the upper rack tooth in regard to the angle which is 20 degrees on each side instead of 14 1/2 degrees. This angle, in all cases, represents the pressure angle of the gearing. The third diagram from the top shows the American Standard basic rack or the 20-degree stub involute system. This is known as a "stub tooth" because it is shorter than the full-depth teeth. For example, if the diametral pitch is 4 in each case, then depth D of a "full-depth" tooth $= 2.157 \div 4 = 0.539$ inch; depth D of an American Standard stub tooth $= 1.8 \div 4 = 0.450$ inch.

Cutters conforming to these straight-sided racks will generate gear teeth having involute curves. Just why such curves are generated is explained in treatises on gear design. While practically all gears in use today are thought of as involute gears, there are certain practical reasons why a great many do not have the true involute form. This is true, for example, of gears conforming to the basic rack shown by the fourth diagram. This is the American Standard 14 1/2-degree composite system. The straight-sided or involute form of rack is modified by introducing some curvature above and below the pitch line. The 14 1/2-degree composite tooth form was developed originally for use with the form milling process and gear teeth conforming to this

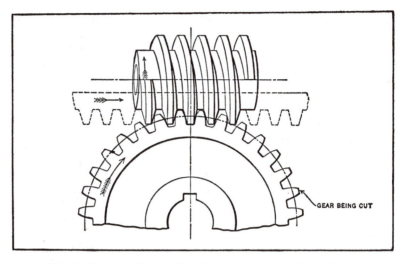

Fig. 5. Diagram Illustrating the Principle of the Hobbing
Process of Forming Spur Gears

standard generally are cut by form milling. They can, however, be produced readily on hobbing or other generating machines by making a hob or cutter of the basic rack form. If a hob is used, the relieving tool can be made to the form of the basic rack tooth. The line of tooth action is longer with the composite system than with the pure involute tooth form.

How are gear teeth generated by hobbing?

Gear teeth cut by the hobbing process are given the required shape or curvature by a generating action resulting from the rotation of the gear blank relative to a cutter of the hob form. A hob resembles a worm (see Fig. 5), excepting that cutting teeth are formed by milling gashes across the threads and relieving the teeth to provide clearance, as shown in Fig. 6. The hob type of cutter has cutting teeth of the same cross-sectional shape as teeth of a rack of corresponding pitch, except for minor variations such, for example, as increasing the length of the hob teeth to provide for clearance at the bottom of the tooth spaces. The hob teeth coincide with those of an imaginary rack as indicated in Fig. 5. As the hob teeth lie along a helical path (like a screw thread) the hob is set at an angle to align the teeth on the cutting side with the axis of the gear blanк. When the hob is inclined an amount depending upon its helix angle, the teeth on the cutting side represent a rack.

When a hobbing machine is in operation, the gear blank and hob revolve together, the ratio depending upon the number of teeth in the gear and the number of threads on the hob—that is, whether the hob has a single or a multiple thread. This rotation of the hob causes successive teeth to occupy positions corresponding to the teeth of a rack, assuming that the latter were in mesh with the revolving gear and moving tangentially. In conjunction with the rotary movement of the hob, the slide on which it is carried is given a feeding movement parallel to the axis of the gear blank. As this feeding movement continues across the gear blank (or blanks when several are cut together) all of the gear teeth are completely formed. In other words, hobbing is a continuous operation, since the teeth around the entire circumference of the gear are finished together (instead of one tooth being cut at a time) and ordinarily by one passage of the hob. Gear-hobbing machines are commonly applied to the cutting of spur, helical, and worm gearing, and hobbing is the most rapid method of cutting gears by a generating process.

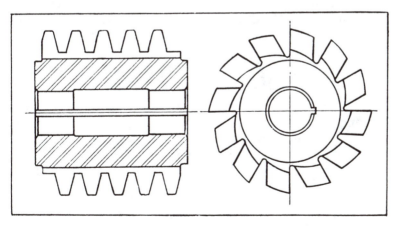

Fig. 6. Hob for Cutting Gears by Generating Process

Why is the normal pitch of a hob equal to the circular pitch of the gear to be cut?

Hobs for spur gears are made by cutter manufacturers for cutting gears of various diametral pitches, the pitch being stamped on the end of the hob. The normal pitch of the hob should be the same as the circular pitch of the gear. The "normal pitch" is the shortest center-to-center distance between adjacent teeth, the pitch being measured at right angles to the thread. To illustrate, if the diametral pitch of a hob is 4, this is equivalent to a circular pitch of 0.7854 inch; hence, the pitch of the hob, measured perpendicular to the teeth, should be 0.7854 inch. A spur gear cut with this hob will also have a circular pitch of 0.7854 or a diametral pitch of 4. If the pitch of the hob, as measured parallel to its axis, were made equal to the circular pitch of the gear, an error would be introduced owing to the fact that the hob must be inclined to locate it in the working position. However, the difference between the normal pitch and the pitch measured parallel with the axis will be very small when there is little inclination of the hob teeth relative to the axis.

Although a hob must conform to the diametral pitch and pressure angle of the gear to be cut, it is not necessary to consider the number of teeth in the gear, as when formed milling cutters are used. In other words, a hob of given pitch is applicable to gears having any number of teeth, and this is also true of cutters used on any machine of the generating type.

In arranging a machine for hobbing spur gears, what adjustments are required?

The exact method of adjusting or setting up a gear-hobbing machine varies somewhat with different designs and makes, but there are certain general adjustments common to different machines. The hob must be set to the correct angle, and an axial adjustment for centering a tooth relative to the gear blank may be considered desirable. The machine must be so geared that the hob and work revolve at the proper speed ratio. This is taken care of by means of change-gears. The rate at which the hob feeds across the gear blank must be varied to suit the quality of finish required on the teeth and the cutting properties of the material. Change-gears in some form are also used for this purpose. This feeding movement is parallel to the axis of the gear being cut, and the amount per revolution of the gear blank or work-table may vary from 0.015 to 0.030 inch on some jobs and from 0.100 to 0.200 on others.

The method of holding or chucking the gear blank (or blanks) is very important. First, the blanks must be supported rigidly to prevent any decided springing action when the hob is at work; second, the gear blanks must be concentric with the axis of the work-spindle or table and not be sprung out of shape by clamping. It is well to test the concentricity of the blanks after they have been placed on the machine and before they are cut, by the use of a dial indicator.

The next step is to adjust the blank relative to the cutter. or vice versa, so that gear teeth of the correct depth will be cut. This may be done on most machines by adjusting the cutter-slide until the hob is opposite the uncut blank; then the work-table is moved until the teeth of the hob, which should be revolving, just graze the blank. The hob is next moved far enough to clear the rim of the blank, and the work-slide is adjusted toward the hob an amount equal to the whole depth of the tooth. Gear hobbers have graduated dials on the feed-screws or special gaging devices to facilitate making these adjustments. When the hob is set for the depth of cut, as just described, the teeth should be correct provided the outside diameter of the gear blank is the right size.

A reliable method of checking the accuracy of the teeth is to measure the chordal thickness, using a vernier gear tooth caliper. This chordal thickness should be checked just as soon as the hob has been fed into the gear blank far enough to produce completely formed teeth. The stop which serves either to disengage the feed or stop the entire machine automatically after the gear teeth are cut, is set before starting the machine, so that it will come into action after the center of the hob clears the edge of the gear blank. Spur gears usually are finished at one passage of the hob, although sometimes a roughing cut is taken with the hob set slightly less than full depth, and this is followed by a finishing cut in order to obtain greater refinement. According to one rule, two cuts should be taken in cutting gears of 3 diametral pitch or coarser, if gears of the best grade are required.

At what angle is a hob set for cutting a spur gear?

The angle at which the hob-spindle or swivel slide is set depends upon the lead of the hob thread and its diameter, since the object of inclining the hob is to bring the teeth on the cutting side into alignment with the axis of the gear blank. This angle is equal to the helix or lead angle of the hob thread at the pitch line, measured from a plane perpendicular to the hob axis, and is often called the "end angle." To avoid the necessity of making calculations, this angle is usually stamped on the hob. If the angle is not known, its tangent may be determined simply by dividing the lead of the hob thread by the pitch circumference.

Gears may be cut with left-hand hobs, although hobs threaded right-hand are used ordinarily. The hob is inclined from the horizontal position in one direction when it is right-hand, and in the opposite direction when it is left-hand. The proper direction may be determined readily by simply considering which way it is necessary to turn the swivel slide in order to bring the teeth on the cutting side of the hob parallel with the work-spindle or gear tooth grooves.

How many revolutions are made by the hob per revolution of the gear?

In cutting spur gears with a single-threaded hob, which is the kind generally used, the number of revolutions made by the hob per revolution of the gear is equal to the number of teeth to be cut. For example, if a gear is to have forty teeth, the machine would be geared to revolve the hob forty times during one revolution of the gear or work-table. The combination of gears required for cutting any gear would ordinarily be determined simply by referring to a table or chart accompanying the machine.

In calculating these gears, it is necessary to consider the fact that in all gear-hobbing machines there are certain gears that form a permanent part of the machine and serve to transmit motion from the main driving shaft to the hob-spindle and to the work-table. Since the ratios of these permanent gears vary in machines of different makes, these ratios enter into change-gear calculations. In the equation given below, which may be used to calculate change-gears on machines of different makes, there is a constant or fixed value for each machine. These fixed numbers are based on the ratio of the permanent gearing in each machine, and serve to allow for this ratio. In the equation n equals the number of threads on hob (that is, n equals 1 for a single-threaded hob, 2 for a double-threaded hob, etc.), N equals number of teeth to be cut in gear. The constant for machines of different makes varies. For example, it may be some number such as 24, 30, 32, 60, etc.

$$\frac{\text{Product of No. of Teeth in Driving Gears}}{\text{Product of No. of Teeth in Driven Gears}} = \frac{\text{Constant} \times n}{N}$$

To illustrate how change-gears are calculated, assume as an example that the gears to be hobbed have 45 teeth, a single-threaded hob is to be used, and the constant of the machine is 60. Then

$$\frac{\text{Product of Driving Gears}}{\text{Product of Driven Gears}} = \frac{60 \times 1}{45} = \frac{5 \times 12}{5 \times 9}$$

The numbers in this expression are now raised to higher values by multiplying with trial numbers in the usual manner, thus obtaining larger numbers corresponding to the numbers of teeth in available change-gears. Thus,

$$\frac{5 \times 12}{5 \times 9} = \frac{(5 \times 16) \times (12 \times 5)}{(5 \times 16) \times (9 \times 5)} = \frac{80 \times 60}{80 \times 45}$$

The numbers above the line represent the driving gears which have 80 and 60 teeth, respectively, and the numbers below the line the driven gears, with 80 and 45 teeth, respectively. Of course there are other combinations that would give the same ratio, and it might be necessary to calculate them by selecting other trial numbers, assuming, for example, that it were not possible to find change-gears corresponding to one or more of the numbers obtained by the calculation.

Why are multiple-threaded hobs sometimes used for cutting spur gears?

Although hobs that have a single thread or row of teeth are generally used for cutting spur gears, multiple-threaded hobs are sometimes employed. When a multiple-threaded hob is used, it is necessary to change the ratio of the gearing controlling the relative speeds of the hob and the work, in proportion to the number of threads in the hob. For instance, if driving and driven change-gears of 3 to 4 ratio are used for a triple-threaded hob, for a single-threaded hob the ratio should be 1 to 4. Much greater production is obtained with multiple-threaded hobs, if the same rate of speed is maintained.

Example.—A single-threaded hob is to be used for cutting a gear having 128 teeth; the hob diameter is 3 inches, and the speed 80 revolutions per minute, thus giving a cutting speed of about 63 feet per minute. Find the cutting time and the reduction in time when a triple-threaded hob is used.

Assume that the feeding movement of the hob per revolution of the gear is 0.160 inch, and that the hob has to travel 2 inches for cutting the teeth; then the gear will make 12.5 revolutions while the teeth are being milled (2 ÷ 0.160 = 12.5), and the total number of hob revolutions will equal 12.5 × 128 = 1600, since the hob makes 128 revolutions to one of the gear. Hence, the actual cutting time will equal 1600 ÷ 80 = 20 minutes when a single-threaded hob is used.

Now suppose that the same gear is cut with a triple-threaded hob having the same diameter as in the preceding case and that the same cutting speed and feed per revolution of the gear are employed. Since the triple-threaded hob makes 42 2/3 revolutions to one of the gear, the total number of hob revolutions during 12.5 revolutions of the gear will equal 42 2/3 × 12 1/2 = 533 1/3. Therefore, the cutting time equals 533 1/3 ÷ 80 = 6 2/3 minutes, or one-third of the time required when a single-threaded hob is used for the same work. Notwithstanding the relatively high production obtained with multiple-threaded hobs, the single-threaded type is preferable for hobbing spur gears, as a general rule, because it generates more accurate teeth.

What is the most common method of cutting helical gears?

Helical gearing usually is cut by some generating method, although milling machines are sometimes used, especially when such gears are not required in quantity. The most common generating method employed is that of hobbing, but the shaping or planing processes are also used in many shops. The general method of cutting helical gears by hobbing is practically the same as cutting spur gears, after the machine is properly geared and adjusted. The angular position of the hob must be determined with reference to the helix angles of both the hob and the gear to be cut, and the machine must be geared to generate helical teeth having the required helix angle with relation to the axis. When the

Fig. 7. Cutting a Helical Gear by Hobbing Process

machine is at work, the hob has a feeding movement parallel to the axis of the gear, and, ordinarily, all the teeth are finished during one passage of the hob. The principle governing the generation of helical gears by hobbing is illustrated by the action of a helical pinion meshing with a rack. When a helical gear is to be cut with a hob, the latter is set at such an angle that its teeth, as they come around to the cutting position, coincide with the teeth of an imaginary rack in mesh with the revolving gear. Fig. 7 shows the cutting of a helical gear by hobbing. The hob, set at the proper angle, begins cutting at one side of the blank, and the horizontal feeding movement is parallel to the axis of the gear being cut. This gear has fifty teeth, of 5 normal diametral pitch and a pressure angle of 20 degrees. The face width is 3 inches.

Many hobbing machines are so designed that the hob feeding movement is vertical. One machine of this design is shown in Fig. 8. The operation in this particular case is cutting the teeth of a double-helical or herringbone gear. This gear is about 6 feet in diameter and it is a marine reduction gear.

How are helical teeth formed by a rotating hob?

In cutting a *spur* gear with a single-threaded hob, the number of revolutions of the hob for each revolution of the gear equals the number of teeth in the gear, but for helical gears, the ratio of the change-gears on the machine is affected not only by the number of teeth in the gear to be cut, but also by the lead of the teeth, the relation between the hand of the hob and the gear, and the rate of the hob feeding movement. The formation of helical teeth, as the hob feeds across the gear blank, is accomplished by accelerating or retarding the motion of the work-table. The principle will be made clear by comparing the hobbing of spur and helical gears.

Suppose, for example, that a single-threaded hob were used for cutting a spur gear having 48 teeth; in this case the work-table would make just one revolution to 48 revolutions of the hob. Now, assume that a left-hand spiral gear with 48 teeth is to be cut with a left-hand hob; then it would be necessary for the table to revolve somewhat faster than for a spur gear with the same number of teeth. It is this acceleration or increase of rotary motion of the table, as compared with the rotation for cutting a spur gear, that causes the hob, as it feeds across the blank, to develop helical teeth.

In some cases, however, the motion is retarded instead of being accelerated, since this depends upon the kind of hob used. In cutting a left-hand gear with a left-hand hob, or a right-hand gear with a right-hand hob, the table revolves at a faster rate than it would in cutting a spur gear having the same number of teeth. On the contrary, in cutting a gear of opposite hand, the rotation is retarded in order to secure the same effect in generating the helical

Fig. 8.   Hobbing a Large Turbine Reduction Gear of the
Double-helical or Herringbone Form

teeth. The plan is to gear the machine so that the table would either gain or lose one complete turn during the time required for the hob to feed a distance equal to the lead of the gear being cut. In actual practice, of course, the total feeding movement is usually only a small part of the lead. since the latter represents the distance that a tooth would advance if it made a complete turn about the gear.

Should a hob for helical gears be right-hand or left-hand?

Whether the hob should be right-hand or left-hand depends upon the gear. As a general rule, it is advisable to use right-hand hobs for right-hand gears, and left-hand hobs for left-hand gears, particularly when the gear is to have teeth of large helix angle. The reason that the hob and gear should be of the same hand for large helix angles, is that the hob has a better cutting action relative to the work, the direction of cut being against the rotation of the gear blank. When the helix angles are not large, a hob may be used interchangeably for either right- or left-hand gears. Hobs for helical gears are selected according to the normal diametral pitch.

At what angle is a hob set for cutting helical gears?

The angle at which the hob is set for cutting helical gears depends upon the helix angle of the gear and the angle of the hob itself. When a right-hand hob is used for a right-hand gear, or a left-hand hob for a left-hand gear, the hob spindle is inclined from the horizontal position an amount equal to the *difference* between the angles of the gear and hob. The helix angle of the gear is measured from the axis, and the helix angle of the hob (which should be stamped on it to avoid measurement and calculations) is measured from a plane perpendicular to the hob axis and is often referred to as the "end angle." If a right-hand hob is used for a left-hand gear or vice versa, the hob spindle is inclined an amount equal to the *sum* of the gear and hob angles. When a hob has the same angle as the gear to be cut, and the hob and gear are of the same hand, the hob spindle is not inclined. This agrees with the rule just given, since the result of subtracting one angle from the other equals zero.

What fundamental principle is applied in cutting bevel gears by a generating process?

In bevel gearing, when the pitch-cone angle of one of the gears is 90 degrees, this gear is called a *crown gear*. In this case, there is, properly speaking, no pitch cone, but rather a pitch plane. The crown gear of bevel gearing is equivalent to the rack of spur gearing. Now in order to

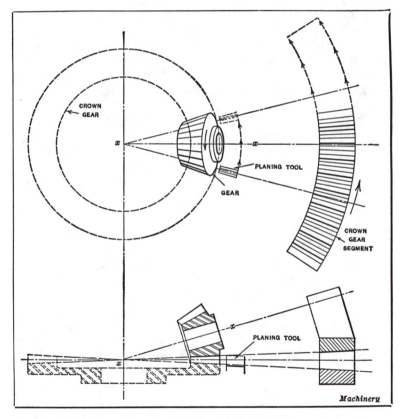

Fig. 9. Diagram Showing Action of Bevel-gear Generator when Generating Motion is Applied to Both Gear and Tool

illustrate the principle of generating bevel gear teeth, assume that a soft plastic bevel gear blank is rolled around a crown gear; this would form true bevel gear teeth and the machines for cutting bevel gears operate on this fundamental principle.

In connection with the cutting of spur gears by generating methods, it was explained that the straight-sided rack of involute gearing is represented either directly by the cutter used, or indirectly, as when a circular form of cutter is generated from the rack. Now the relation between a rack and spur gear is similar to that of a crown gear to a bevel gear; thus the pitch surface of a rack and also of a crown gear coincides with a plane. The teeth of a crown gear are also straight sided like those of a rack, although of converging form, and the inclination of each side corresponds to the pressure angle. The cutting tools of bevel gear generators, therefore, represent the crown gear, and when a bevel gear is being cut the tooth curves are derived by imparting to the work and to the cutting tool the same relative motion that would be obtained if the bevel gear being cut were rotating in mesh with the crown gear. In addition to this generating motion, provision must be made in a practical design of machine for giving the tool or tools a reciprocating motion for cutting, and an indexing movement to the work in order to cut equally spaced teeth around the entire gear.

In cutting bevel gears how are the tooth curves generated?

The generating motion for cutting bevel gears is obtained by rolling the gear being cut, relative to the cutting tool (representing a crown gear tooth) just as though this gear were finished and rolling around a stationary crown gear. A common type of bevel gear generator is so designed that the generating action is applied both to the work and to the cutting tools. In this case the action is similar to that of a crown gear rotating in mesh with the gear being cut, each gear revolving about a fixed axis. Fig. 9 represents this generating motion in diagrammatic form. While each gear tooth is being planed, the gear turns part of a revolution about a fixed axis x–x prior to each cut, just as though

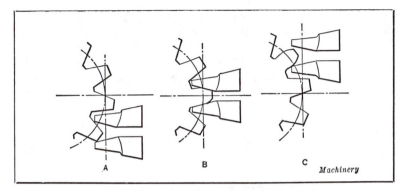

Fig. 10. Diagrams Showing Manner in which Gear and Planing Tools
Roll Together on Bevel-gear Generator

the imaginary crown gear were turning with it. The planing tool also swings around with the gear as though it were one of the crown gear teeth. Only one tool is shown on the diagram to simplify it, although two are used on the well-known Gleason machines. Because of the converging form of the space between two teeth, it is impossible to cut more than one side at a time with a single tool or cutting edge, although it is not only feasible but desirable to use two tools which operate on both sides of a tooth. The diagrams in Fig. 10 show how a roughed out tooth is formed as the gear and tools roll together. Diagram A shows the tools at the beginning of the generating motion, diagram B the central position, and diagram C the completion of the cut. The tools are first swung inward to the cutting position (diagram A) by a cam; then the rolling motion continues slowly until the particular tooth being planed rolls around out of contact with the tools, as shown by diagram C. During this generating movement, the two planing tools are taking a succession of cuts across each side of a tooth. Fig. 11 shows a "close up" of a bevel-gear generator and a finished gear in place.

On some machines the generating action indicated by Fig. 9 is controlled by a crown gear, which, as indicated by the diagram, swings with the tool-slide, and is in mesh with a master gear segment connected with the work-spindle. Such an arrangement causes the crown gear segment and the tools to move through an arc represented by the full and dotted lines, while at the same time the master gear rolls in mesh with the crown gear and causes the gear blank to turn in unison with the tools. Another method of regulating and controlling this generating motion is by using suitable combinations of change-gears instead of the crown and master gear segments. Thus, if gearing of the proper ratio is selected, the relative motion between the tools and work is the same as though a crown and a master gear segment were employed.

What are the general methods of cutting worm-gears?

The machines used for cutting worm-gears include ordinary milling machines, gear-hobbing machines of the type adapted to cutting either spur, helical, or worm-gearing,

Fig. 11. Bevel-gear Generator Set up for Cutting Teeth on a Gear

and special machines designed expressly for cutting worm-gears. The general methods employed are (1) cutting by using a straight hob and a radial feeding movement between hob and gear blank; (2) cutting by feeding a fly cutter tangentially with relation to the worm-gear blank; and (3) cutting by feeding a tapering hob tangentially. These three methods will now be considered.

When a straight or cylindrical hob is used, how are the worm-gear teeth formed?

When worm-gear teeth are generated by a straight or cylindrical hob, the latter is centered relative to the curved throat of the worm-gear blank, as shown by the end view of diagram A, Fig. 12; then as the gear blank and hob rotate together at the proper ratio, the hob is fed inward radially just far enough to form teeth of the right height. These worm-gear teeth on the mid-section x–x correspond to the teeth of an involute spur gear of the same pitch and diameter. The hob represents a standard rack for involute gearing, and generates the teeth the same as in hobbing a spur gear, except that a radial feeding movement is employed in order to form concave teeth.

When worm-gears are hobbed on an ordinary milling machine, the teeth are first formed roughly by a gashing operation; then a hob is placed in mesh with this gashed blank, and the rotation of the hob causes the blank to revolve as the hob is gradually fed in to the required depth. In using a gear-hobbing machine, this preliminary gashing is unnecessary, because the hob- and work-spindles are connected indirectly through gearing, so that the rotation of one relative to the other is positive and at the proper ratio. While worm-gear teeth can be formed by using a straight hob as just described, the taper hob method produces more accurate worm-gears. (This method will be described later.)

Is a worm-gear hob a duplicate of the worm that is to drive the worm-gear?

An ideal hob would have exactly the same pitch diameter and lead angle as the worm; repeated sharpening, however,

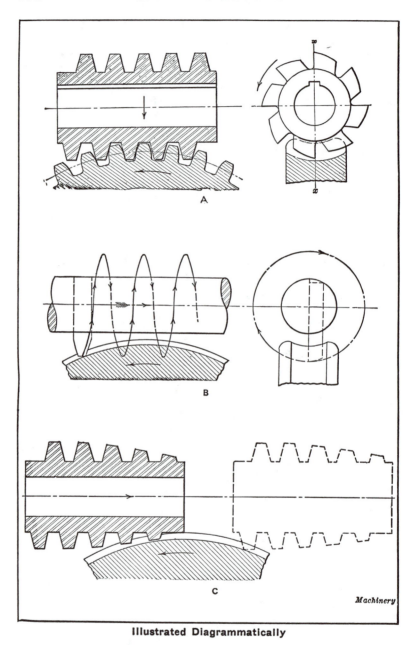

Illustrated Diagrammatically

Fig. 12. Different Methods of Cutting Worm-gears

would reduce the hob size because of the form-relieved teeth. Hence, the general practice is to make hobs (especially the radial or in-feed type) "over-size" to provide a grinding allowance and increase the hob life. An over-size hob has a larger pitch diameter and smaller lead angle than the worm, but repeated sharpenings gradually reduce these differences. To compensate for the smaller lead angle of an over-size hob, the hob axis may be set 90-degrees relative to the wheel axis plus the difference between the *lead* angle of the worm at the pitch line, and the *lead* angle of the over-size hob at its pitch line. This angular adjustment is in the direction required to increase the inclination of the worm wheel teeth so that the axis of the assembled worm will be 90 degrees from the wheel axis.

A second method is to make the worm diameter equal to the hob diameter as the latter is changed by sharpening. The worm wheel diameter may also be changed to maintain a given center distance. When this second method is employed, the hob's spindle remains in the 90-degree position.

What is a "fly-cutter" and how is it used in cutting worm-gears?

A fly-cutter is a simple type of formed milling cutter that is often used for operations that will not warrant the expense of a regular formed cutter. The milling is done by a single tool or cutting edge which has the required outline and is held on an arbor. The advantage of the fly-cutter is that a single tool can be formed to the desired shape, at a comparatively small expense.

Worm-gears are frequently cut by using a fly-cutter, which is shaped like a hob tooth of corresponding pitch. This fly-cutter is set to the full-depth position (unless allowance is made for a finishing cut) and it is given a tangential feeding movement relative to the worm-gear blank, as indicated by diagram *B*, Fig. 12, at the same time revolving at the proper ratio relative to the rotation of the worm-gear blank. A common method of obtaining this tangential movement is to mount the cutter-bar on a slide which is given a slow feeding movement, so that the cutter passes from one side of the gear blank to the other. If the worm-gear has, say, 40 teeth, then the cutter-bar will make 40 revolutions to one of the wheel, plus or minus a slight variation referred to later. As the fly-cutter begins to work on one side of the blank, forty shallow grooves are formed during the first revolution of the blank and these are gradually made wider and deeper as the fly-cutter feeds across the blank from one side to the other, thus completely forming the teeth, unless a light finishing cut is taken afterward.

What is the action of a fly-cutter in generating worm-gear teeth?

In order to cut worm-gear teeth of correct shape, the fly-cutter must be made to advance along a helix (see diagram *B*, Fig. 12), or as though it were moving along the thread of a worm similar to the one that the worm-gear is intended for. When a straight hob is used, as shown at *A*, each tooth of the hob rotates in a fixed plane, but the action of the hob is like that of a worm, and the successive cutting teeth keep in step with the worm-gear teeth as the hob and

gear are revolved together at the correct ratio. When a fly-cutter is used it is evident that it must perform the work done by a series of hob teeth. To accomplish this, the fly-cutter, as it feeds tangentially, is revolved at such a rate relative to the worm-gear, that the cutter gradually passes the positions which successive hob teeth would occupy; consequently, the same general effect is obtained with a fly-cutter as with a hob, although the worm-gear teeth are shaped somewhat differently, as compared with a straight hob.

To illustrate the action further, assume that successive hob teeth are designated by numbers 1, 2, 3, etc. Then when the fly-cutter is in the position occupied by hob tooth No. 1, it will perform the work of this hob tooth. Similarly, when the fly-cutter has advanced to coincide with the position of hob tooth No. 2, it performs the work of this tooth, and so on for the entire series. This feeding movement is necessarily very slow, because the fly-cutter has to take heavy cuts, especially in passing the center or full-depth position.

While the fly-cutter method is slow, as compared with hobbing, it has two decided advantages which account for its general use: First, a very simple and inexpensive cutter may be used instead of an expensive hob. This is of great importance when the number of worm-gears is not large enough to warrant making a hob. Second, with the fly-cutter method it is possible to produce worm-gears having more accurate teeth than are obtainable by the use of a straight hob, provided, of course, that the fly-cutter method is properly applied.

Is the taper-hob method of cutting worm-gears similar in principle to the fly-cutter method?

When a taper hob is used for cutting worm-gears, it is set to the full-depth position at one side of the blank, and fed tangentially across it as illustrated by diagram C, Fig. 12. The use of a taper hob makes it possible to cut worm-gears more rapidly than by means of a fly-cutter, and also very accurately, provided the hob itself is accurate. The taper hob method also increases the rate of production as compared with the use of straight hobs which are fed in radially.

In the taper-hob method, the rotation of the hob relative to the blank, as the hob moves tangentially, is such as slowly to advance or screw the hob along its own thread. The action of the hob is the same as that of a fly-cutter, and machines adapted for the fly-cutter method may also be equipped with taper hobs. The leading teeth on the hob are tapering, as indicated by the diagram, and they should be designed to increase progressively in width as well as in height from the small to the full size end to distribute the work of cutting. The tapering or leading end performs a roughing operation, whereas the full sized teeth take light finishing cuts, thus preserving their accuracy and insuring well-formed teeth. The tangential feeding movement continues until the large end of the hob passes out of contact on the side opposite the starting point, as indicated by the dotted lines of the illustration.

Taper hobs are especially adapted for cutting worm-gears that are to mesh with worms having large helix angles; they are also preferable for worm-gears having large face widths in proportion to the worm diameter. Worm-gear teeth are generated more accurately with a taper hob than with a straight hob that is given a radial feeding movement.

If a taper hob is used, a better bearing contact is obtained, especially for large helix angles and face widths, because such conditions are particularly unfavorable to the radial straight-hob method.

In hobbing worm-gears, what are the relative speeds of cutter and gear?

When a worm-gear is cut by using a straight hob and a radial feeding movement, the machine is geared so that the hob and work revolve according to the ratio of the number of threads in the worm and the number of teeth in the worm-gear. For instance, if a worm-gear has 50 teeth and is to mesh with a single-threaded worm, the machine will be so geared that the hob makes 50 revolutions to one of the worm-gear blank.

If a fly-cutter or taper hob is used, however, it is necessary to take into account the tangential feeding movement, and to so alter the ratio as to cause the cutter to follow a helical path. If the worm-gear is to have, say, 50 teeth, then the ratio must be a little greater than 50 to 1, assuming that the feeding movement of the cutter is against the direction in which the work is rotated, as shown by diagram B, Fig. 12. If the tangential feeding movement and the rotation of the work are in the same direction, then the ratio would be somewhat less than 50 to 1.

In order to illustrate the action more clearly, suppose the ratio is exactly 50 to 1 and that the tool is moved to the right until the point begins cutting a series of shallow grooves around the rim. It is apparent that if the tool is not moved farther to the right, it will simply rotate in unison with the blank and pass through the grooves without any further cutting action; but if the cutter-bar is again given a tangential feeding movement without changing the 50 to 1 ratio, the tool will no longer match with the grooves, but will begin to widen them by cutting away the right-hand sides in feeding to the right as indicated by the diagram. A continuation of this feeding movement would simply cut away the blank without forming teeth. This side-cutting action is due to the fact that a given tooth groove comes around to the same position each time the cutter makes fifty revolutions, but the cutter, owing to its advancing movement is continually arriving at different positions relative to the work. This may be prevented (assuming that the cutter is advancing against the direction of rotation) either by decreasing the speed of the work or by increasing the speed of the cutter, which produces the same result.

For instance, the cutter shown by diagram B should, for the movements of the tool and work indicated, revolve somewhat faster than the 50 to 1 ratio for cutting a 50-tooth worm-gear. On the contrary, if the feeding movement were in the same direction as the rotation of the work, the ratio should be decreased. This change in ratio varies according to the relation between the rotary and feeding movements of the cutter and the rotation of the work, and its practical effect is to cause the cutter to change its position successively relative to the work, the same as though it were moved to numerous positions along a helical curve. The proper relative movements of the work and cutter as a rule are controlled by means of change-gears.

How are gear tooth curves generated by grinding?

The generating process of finishing hardened gear teeth by grinding is similar in principle to generating methods as applied to gear cutting. A method of generating tooth curves by grinding is illustrated diagrammatically in Fig. 13. The grinding is done by the flat face A of the wheel, this face being perpendicular to the wheel axis and inclined from the vertical an amount equal to the pressure angle of the gear to be ground. Therefore, this flat side of the wheel face represents the side of a rack tooth the same as the teeth of certain gear cutters, like the hob, for example. In order to generate involute tooth curves, provision must be made for rolling the gear past the revolving grinding wheel, just as though an accurate gear were rolling along an accurate rack so located that the side of one tooth is in the same position as the grinding face of the wheel. A method of obtaining this rolling or generating motion is by the use of steel tapes in conjunction with a drum or disk having a radius approximately equal to the pitch radius of the gear. A master gear rolling along a rack is another method. The diamond for truing the flat grinding face

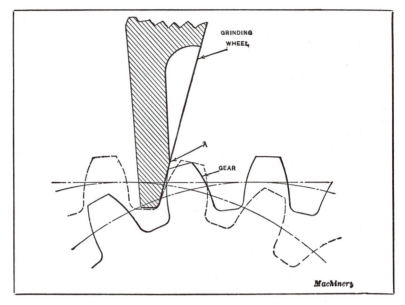

Fig. 13. Action of Gear Relative to Grinding Wheel on Machine of
Generating Type—The Grinding is done by the Flat Side of the Wheel

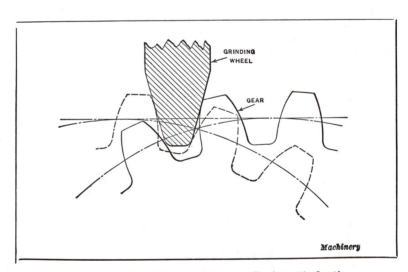

Fig. 14. Grinding Wheel which has Rack-tooth Section
and Grinds on Both Sides

is so mounted that it swings in a plane perpendicular to the axis of the wheel-spindle. The wheel face is maintained in one position because in dressing it the wheel is adjusted toward its truing diamond instead of adjusting the diamond toward the wheel. The grinding wheel is quite large in diameter, and does not have to be traversed parallel to the axis of the gear being ground. In other words, the wheel rotates about a fixed axis, but it covers the entire working surface of a tooth as the gear rolls past the grinding face. A slight arc or curved shoulder is formed at the bottom of the ground surface but this does not interfere with the tooth action provided it is confined to the clearance space and is not in contact with the ends of meshing teeth.

What type of machine is used to grind two tooth profiles simultaneously?

A method of grinding gears by utilizing both sides of a wheel is illustrated by the diagram, Fig. 14. The grinding faces of the wheel are shaped to represent a complete rack tooth instead of one side only. Since the grinding is not done by flat faces in this case, the wheel is given a traversing movement in a direction parallel to the axis of the gear. A reciprocating horizontal ram traverses the grinding wheel through the gear teeth, generating the adjacent sides of two teeth at the same time. The gear or stack of gears being ground is rolled past the reciprocating wheel by means of a master gear mounted upon the end of the work-spindle and engaging a rack. This master gear is a duplicate of the gear to be ground in regard to number of teeth, diametral pitch, and, consequently, pitch diameter. The grinding wheel makes as many passes per tooth as are necessary for the finish desired. The work is indexed automatically when it has rolled to one side out of engagement with the grinding wheel. This indexing takes place when the master gear is out of mesh with the master rack and in mesh with an indexing rack. In grinding spur gears, the center line of the gear is parallel to the ram movement, and the work-slide motion is at right angles. For helical gears, the work-slide is set to the helix angle of the gears to be ground. Means must be provided on a machine of this type for dressing both sides of the wheel to the pressure angle of

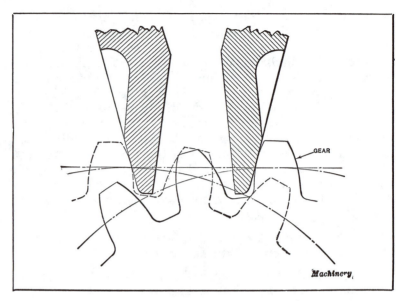

Fig. 15. A Method of Using Two Grinding Wheels
on Generating Type of Machine

the gears to be ground, and in addition the width must be
such that the right amount of stock is removed from the
sides of adjacent teeth when the wheel is in the full depth
position. The width or distance between the angular grinding faces depends upon the pitch of the gears to be ground.

Another method of grinding two tooth surfaces at the
same time consists in using two wheels which operate in
different tooth spaces as shown by the diagram, Fig. 15.
The grinding is done by using the flat side of each wheel,
as indicated by the diagram. This grinding side is inclined an amount equal to the pressure angle the same as
for the other generating methods referred to and the wheels
revolve in one position as the gear blank is given a rolling
motion for generating the tooth curves. The flat side of each
wheel corresponds in location to the side of an imaginary
rack tooth, and the generating action is the same as though
the pitch circle of the gear were rolling along the pitch line
of the rack, the motion being the same as with a single
wheel. After grinding each pair of tooth faces, the gear

is indexed a distance equal to the circular pitch. The amount removed in grinding is regulated by adjusting the distance between the wheels.

Fig. 16 shows a two-wheel type of machine grinding the teeth of a helical gear. As the slide which holds the gear is given a horizontal reciprocating motion, the gear rolls past the wheels. This rolling or oscillating motion is sufficient to bring the grinding faces of the wheels into contact with the entire tooth profile. The rolling motion is derived from the master gear seen at the end of the work-holding spindle. This gear meshes with and rolls along the stationary rack which may be seen above it. The work is indexed one tooth at each table reversal. The wheel-heads are adjustable for angle and spacing.

Fig. 16. Two-wheel Type of Gear Grinder in Action

Grinding Cylindrical and Tapering Parts

Grinding machines were used originally almost exclusively for truing tool steel parts which had been distorted by hardening, and are still indispensable for work of this class. The great improvements which have been made, both in grinding machines and abrasive wheels, however, have resulted in the application of the grinding process to the finishing of a great many unhardened parts. The work, as a rule, is first reduced to nearly the required size, by turning, milling, etc., and then it is ground to the finished dimension. After a part has been hardened, grinding is the only practicable method of truing it. On the other hand, unhardened pieces can be finished by other means, but grinding is preferable for most cylindrical work, because it enables parts to be finished accurately to a given diameter in less time than would be required by any other known method. Many different types of grinding machines have been developed for handling the various kinds of work to which the grinding process is applicable.

What are the main features of a cylindrical grinding machine?

Machines of the cylindrical type are intended primarily for grinding cylindrical parts, although they can also be used for taper work and other grinding operations, the extent of which may be increased considerably by the use of auxiliary equipment. All cylindrical grinding machines are equipped with a mechanism which enables the grinding wheel to be fed in automatically toward the work for taking successive cuts, and provision is made for varying the traversing movement of the work (or wheel) and the rotating speed of the work to suit different conditions.

314

The feed mechanism is so arranged that it can be set to stop the feed when the diameter has been ground to a predetermined size. This automatic cross feed is a great advantage, especially when grinding a large number of duplicate parts, as it prevents grinding them too small and makes it unnecessary for the operator to be continually measuring the diameter of the work. The automatic feed is also desirable because it moves the wheel inward an unvarying amount at each reversal. Fig. 1 shows a cylindrical grinding operation.

What is the difference between a plain and a universal grinding machine?

Cylindrical grinding machines, like milling machines, are divided into two general classes, known as *plain* and *universal* types. The first type is essentially a machine for manufacturing purposes. The general construction of a universal grinder is similar to that of a plain grinder of the same make, but it differs from the latter in having certain special features and auxiliary attachments which adapt it to a more general or universal class of work. The principal difference between the universal and plain types, as far as the construction of the machine itself is concerned, is as follows: The wheel slide of a universal machine can be swiveled with relation to the travel of the table as illustrated later; the headstock can also be set at an angle, and provision is made for revolving the headstock spindle for grinding parts that are held in a chuck or otherwise. With a plain machine, the wheel slide is permanently set at right angles to the table travel and the headstock cannot be swiveled.

In grinding a cylindrical part, what is the general procedure?

In grinding a cylindrical part, such as a rod or shaft, on a machine equipped with work-holding centers, it is mounted between the centers just as it would be placed between the centers of a lathe for turning; in fact, the same center-holes upon which the shaft was rough-turned are used when grinding. (For some classes of work, grinders of the centerless type are used, as explained later.) The work is rotated rather slowly upon the centers by a driving dog, and the surface is ground cylindrical by a disk-shaped wheel (see Fig. 1). This wheel rotates rapidly, and the grinding is done ordinarily either by traversing the rotating part past the face of the wheel or by traversing the wheel along the work. Some cylindrical grinders operate in one way, and some the other.

The revolving wheel is fed inward a slight amount at each end of the work and the latter is accurately ground to the required diameter. The wheel can be fed by hand or automatically, the latter method being generally em-

Fig. 1. An Example of Cylindrical Grinding

ployed, except when adjusting the wheel or starting a cut. The lateral movement of the work (or wheel) per work revolution, is always somewhat less than the full width of the grinding wheel-face, in order to secure a smooth surface free from ridges. This side traverse, as well as the rotative speed of the work, is varied to suit conditions.

Grinding wheels are composed of innumerable grains of some hard abrasive material which is held together by an adhesive bond. These grains or cutters, as they might properly be called, have sharp corners or edges which cut away the metal as the work traverses past the wheel-face, or vice versa. The relative rotation of the wheel and the part being ground is such that the grinding side of the wheel moves downward, and that side of the work being ground moves upward or in the opposite direction.

How much lateral feeding movement is required per work revolution?

In order to grind rapidly, it is the modern practice to use a coarse side feed of the work or wheel; that is, instead of feeding a distance equal to only $\frac{1}{8}$ or $\frac{1}{4}$ of the wheel width per revolution of the work, the side feed is only a little less than the full width of the wheel-face. This method of traversing the wheel is applied to the grinding of duplicate parts in connection with manufacturing operations, rather than to fine tool-room work. Comparatively wide wheels are also used in modern manufacturing, so that a surface can be ground rapidly.

Suppose the work is rotated fast enough to give a surface speed of 25 feet per minute and the fastest side feed is engaged in order to determine by trial what combination will give the best results. When the wheel is brought into contact with the work, if it leaves coarse, spiral feed lines (as shown at A, Fig. 2), having a greater pitch than the width of the wheel, the side feed should be reduced until the wheel does not leave any unground surface. On the contrary, if the side feed only moves the wheel laterally a fraction of its full width (as indicated by the narrow feed lines at B), the feed should be increased until it is nearly equal to the wheel width. Owing to the rapid side feed, the wheel will pass over the surface being ground in a com-

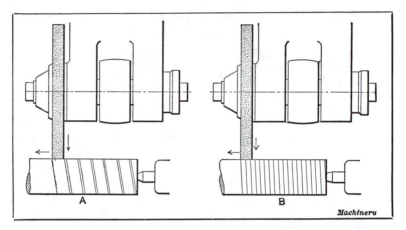

Fig. 2. (A) Traversing Movement Exceeds Wheel Width. (B) Traversing
Movement Much Less than Wheel Width

paratively short time, and by using a rather slow work speed, the wear of the wheel is minimized. This method of grinding is employed when using large machines which have sufficient driving power to enable such broad cuts to be taken and are rigid enough to prevent excessive vibration. When a small light grinder is employed, it is not feasible to use wide wheels and take such wide cuts, owing to the lack of rigidity and driving power. The depth of the cut, or the amount that the wheel feeds inward at each reversal, is also controlled by the power and rigidity of the machine used.

Why is the face of the grinding wheel flooded with cooling solution?

At the point where the wheel is in contact with the work, there is considerable heat generated; consequently, a cooling solution is very essential in order to maintain an even temperature. For example, in grinding a part held between centers, if a "coolant" is not used, the part being ground tends to bend toward the wheel, owing to the higher degree of heat and expansion on the grinding side; in other words, its axis will be continually changing, and, obviously, inaccuracy will be the result. The apparatus for supplying the coolant consists of a small pump and piping through which the cooling water is conveyed to the grinding wheel, and which plays upon that part of the work being ground at the contact point of the grinding wheel. The lower end of the nozzle spout and the deflector should be so shaped that the coolant is directed to cover the full face of the grinding wheel. If one or both edges of the wheel are allowed to cut dry, the surface finish may be marred by feed lines, so that, in order to secure good work, it is necessary to cover the entire cutting surface of the wheel.

The diagram, Fig. 3, shows some of the different types of nozzles used. At A is shown a plain pipe nozzle; the position of this nozzle has been reversed, so that it works unsatisfactorily and has a tendency to splash the operator. The reason for this is, that the water clings to the long side, so that it is thrown away from the wheel. The proper way of placing a nozzle of the type shown at A is shown at B. In this case it will be noticed that the angular face is not presented to the wheel but away from it; even with this arrangement, however, there is considerable splashing. One means of overcoming this is to use a deflector a, as shown at C and D, which can be adjusted so as to control the flow of the coolant. The diagram at E shows another type of adjustable nozzle. The deflector can be moved up and down by adjusting it along the pipe and it is also capable of an in-and-out movement that enables the operator to direct the stream of cooling lubricant where it is most needed. This nozzle is also so constructed and ad-

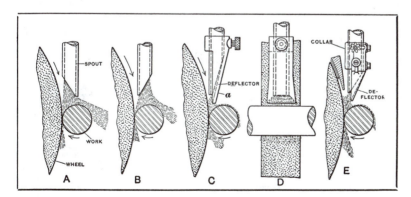

Fig. 3. Application of Water Nozzles to Cylindrical Grinding Machines

Fig. 4. The Work-table, with the Headstock and Footstock, is Swiveled to the Required Angle in Grinding Tapers

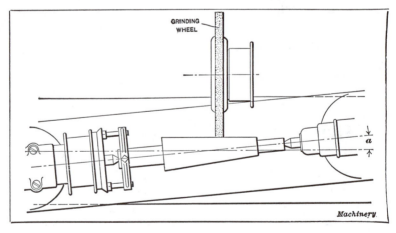

Fig. 5. Diagram Illustrating Principle of Taper Grinding Operation

justed that the stream of water or cutting lubricant is directed to cover the full face of the grinding wheel.

Grinding on a small variety of work in the tool-room is sometimes done dry, but for manufacturing grinding operations, dry grinding is not recommended.

What adjustments are required for taper grinding?

Taper parts are ground practically the same as those that are straight or cylindrical, provided the taper is not too steep or abrupt. The work is placed between the centers, as illustrated in Fig. 4, and the table with its headstock and footstock is set to the required angle a, Fig. 5, as shown by the graduations at one end. This adjustment locates the axis of the work at an angle with the line of motion of the table; hence, a taper is produced, the angle of which depends upon the amount that the table is turned from its central or parallel position. There are usually two sets of graduations for the swivel table, one reading to degrees and the other giving the taper in inches per foot. The taper should be tested before the part is ground to the finished size by using a gage, or in any other available way.

The diagram A, Fig. 6, shows how a taper surface is ground when the angle is beyond the range of the swivel table. The wheel slide e (which is normally at right angles to the table) is set to bring its line of motion parallel with the taper to be ground. The upper wheel-stand g is also set at right angles to slide e, to locate the wheel-face parallel with the taper surface. The table of the machine should be set in the zero position, so that the angular graduations on the wheel-slide base will give correct readings with relation to the axis of the work. After adjusting the table to the proper longitudinal position, the grinding is done by moving the wheel across the taper surface by using the hand cross feed, and the depth of each cut is regulated by slight longitudinal adjustments of the table. Evidently an operation of this kind must be done on a universal machine, because the wheel slide of a plain type does not have the angular adjustment.

Parts having a double or compound taper can be ground at one setting, provided one taper is within the range of

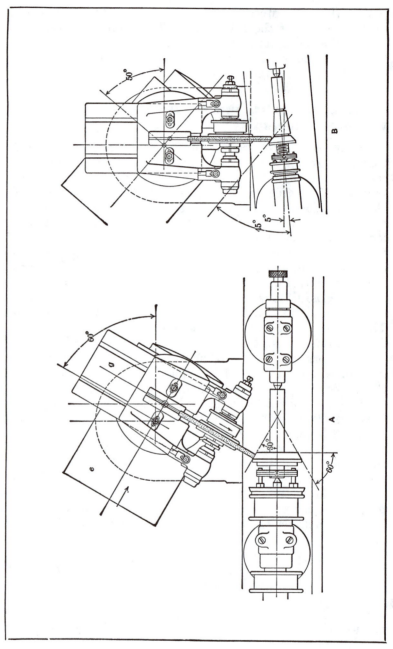

Fig. 6. (A) Grinding Abrupt Taper by Setting Wheel-slide of Universal Machine to Required Angle. (B) Grinding a Double Taper by Traversing both Plates and Wheel-slide Successively.

the swivel table. The latter is set for the smaller angle and the wheel slide for the greater angle, as indicated at *B*, Fig. 6. The wheel is set at right angles to the longest surface and one corner is beveled to suit the other surface. One part is then ground by traversing the table and the other by moving the wheel slide. The wheel-base, in this instance, should be set to an angle corresponding to the sum of the angles of both tapers, as measured from the axis. The sum of both angles, in the example illustrated, is 50 degrees.

How much material should be removed by grinding?

In modern practice, forgings and castings frequently are ground "from the rough." When this is done, the allowance for grinding is reduced as far as is practicable for forging and casting processes. If parts are turned or milled prior to grinding, the amount of metal to leave on the diameter of a piece for finish by grinding is governed by several conditions, such as: finish secured in lathe-turning operation; whether part to be ground is hard or soft; shape of work, as, for example, whether it is in the form of a solid shaft or in the form of a thin wall bushing; and length of portion ground in proportion to its diameter. If the work is turned too close to the finish size before grinding, it is evident that less material will have to be removed by the grinding wheel; but light cuts and fine feeds are not, as a rule, the most economical way of turning. On the other hand, if too large an allowance is left for grinding, this also is not economical because a large amount of stock could be removed more economically by rough-turning. The amount of stock that can be removed economically by grinding also depends, to a certain extent, upon the size and power of the grinding machine used.

There are two distinct methods recommended for roughing out the work previous to grinding: One is to turn the work carefully to within from 0.010 to 0.015 inch of the finished size, and then finish by grinding; the other is to rough out the work quickly, with comparatively coarse feeds and high work speeds, leaving about 1/32 inch to remove by grinding. As the tool "cuts a thread" or leaves a ridged surface on the work, only the high points have to be removed by the grinding wheel over and above the amount left for grinding when the work is turned smooth. There are advantages in both methods, depending on the class of work being machined. Where the work is large in diameter in proportion to its length and will stand heavy lathe cuts, the latter method can be used to advantage; but where the work is light or small in diameter in proportion to its length, the first mentioned practice is generally followed.

The grinding allowance for parts which have to be hardened naturally depends, to some extent, upon the shape of the part and the liability of distortion due to the hardening process. When parts are to be carburized or casehardened, it is important to finish them rather smoothly prior to hardening, so that the depth of the hardened case will be fairly

Fig. 7. Example of Cylindrical Grinding Showing Application of a Steadyrest

uniform over the entire surface. The allowance should also be small enough so that the hard case will not be removed entirely by grinding or be reduced excessively.

In cylindrical grinding, how are slender parts supported?

Most of the parts that are ground on centers should be supported by suitable steadyrests or back-rests, as their use will not only prevent chattering, when properly applied, but permit taking deeper cuts with coarser feeds and also increase the "sizing power" of the wheel. When grinding long and slender parts, such supports are indispensable, and, even for work which is short and rigid, steadyrests often are desirable to prevent vibration, which increases wheel wear and affects the quality of the ground surface. These supports are fastened to the table of the machine and are equipped with shoes of hard wood or metal which bear against the piece being ground. Fig. 7 illustrates the use of a single steadyrest for supporting the work half way between the headstock and tailstock. The number of steadyrests used depends on the form and diameter of the work. According to a commonly accepted rule, the distance between each steadyrest should be from six to ten times the diameter of the part being ground. Some recommend the use of as many rests as can conveniently be fixed in position.

What types of steadyrests are used?

Steadyrests are made in several different styles, and they may be divided into two general classes which differ in that one type is rigid and the other flexible. The rigid type gives a positive unyielding support, whereas the flexible steadyrest, as the name implies, can yield more or less, the supporting shoe being held against the work by springs. Most rigid steadyrests must be readjusted by hand as the diameter of the work is reduced by grinding, whereas the shoes of the flexible type adjust themselves automatically after being properly set. Then there is another form of steadyrest which has spring tension, but can be made rigid when desirable, and still another type is so designed that the supporting shoes are adjusted automatically, but the support is unyielding.

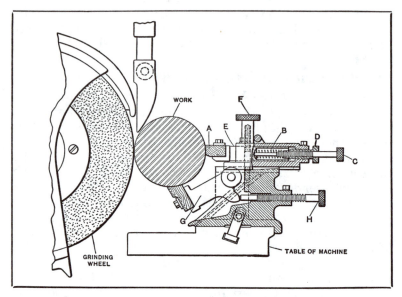

Fig. 8. Rigid Type of Steadyrest—One Design used on Cylindrical Grinding Machines

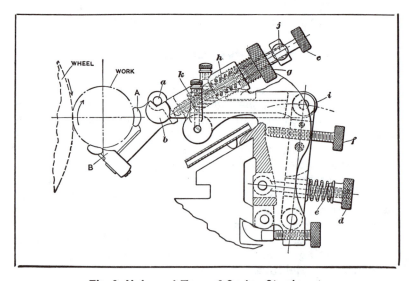

Fig. 9. Universal Type of Spring Steadyrest

Regarding the relative merits of the rigid steadyrest and the flexible or spring type, some advise the use of spring steadyrests for supporting light slender work, and the fixed or rigid form when grinding heavy stiff parts, whereas others advocate the use of rigid steadyrests for light as well as heavy work. Satisfactory results doubtless can be obtained with each type, under favorable conditions. When the work is light and flexible, the spring type is often used in preference to the fixed form. On the other hand, when a heavy rigid piece is being ground, solid unyielding steadyrests are commonly employed to provide as solid a support as possible, in order to absorb vibration and prevent chattering. Practically all spring steadyrests are capable of being locked, so that they are rigid enough, in most cases, to be applied to both heavy and light work.

How is the work supported by a fixed type of steadyrest?

The steadyrest is equipped either with one or two work-supporting shoes arranged to take backward and downward thrusts. There are various designs. The one shown in Fig. 8 has two work-supporting shoes which are operated independently. The horizontal shoe is carried in a plunger that is forced outward by means of a spring B and is backed up by an adjusting screw C. The tension of the spring is adjusted by means of screw D. Plunger E can be clamped rigidly in any position by means of the clamping screw F. The lower work-shoe support G is controlled by the adjusting screw H, which is operated to keep the work-shoe in contact with the work as the diameter of the latter is reduced by the grinding wheel. Where hardened work or large quantities of any particular part are being ground, the wooden blocks can be replaced by bronze or hardened steel blocks depending on the number of pieces to be turned out.

The spring with which the horizontal support is equipped is used only in the case of grinding light delicate work, and then the support is only controlled by the spring until the work has been trued up; after this the support is held rigid by the adjustable screw. When using the spring, it is adjusted with sufficient tension to give a light support to the work, and, at the same time, follow the work if it runs out of true without springing it out of line. Usually, when using the spring support, the work is spotted or ground true just at the point of support by feeding the wheel directly in without any traversing movement. After this the rest-jaw is brought up against the work rigidly by the independent adjustable screw for this purpose. The spring tension is controlled entirely independent from the rigid screw adjustment.

How is the universal spring type of steadyrest applied?

The back-rest or steadyrest shown in Fig. 9 is universal in all its movements and capable of delicate adjustment. This steadyrest may be equipped either with solid or adjustable bronze shoes. In using the steadyrest, first select a shoe of the size of the work being ground and hook the trunnion *a* in the V-support *b*; then turn screw *c* back far enough to allow the shoe to clear the work, loosen the nut *d* to relieve the pressure on spring *e*, and turn back screw *f*. Next turn screw *g* forward until a light pressure is given to spring *h*. Screw *c* should now be turned forward, and, if spring *e* is wholly relieved and screw *f* is back far enough, the shoe will come in contact with the work at both points *A* and *B*. A slight pressure should now be exerted on trunnion *i* to hold the shoe in contact with the work, and screw *f* should be carefully tightened, noting the slightest touch of the end of the screw against the stop, so that none of the parts will be moved. With this screw still in contact with the stop, the shoe should bear equally at both points *A* and *B*. Nut *d* should then be tightened to increase the pressure on spring *e*. The combined pressure on springs *e* and *h* should be just sufficient to resist the pressure of the wheel and prevent vibration of the work. After the proper adjustments have been made, clamping screw *j* should be tightened, to prevent screw *c* from loosening.

With the various screws properly adjusted, the springs of this steadyrest cannot push the work beyond the required size. When the work is finished to size, nut *k* and screw *f* should rest against the shoulder and stop, respectively, so that further pressure of the springs is impossible. The shoe and wheel will then be left in the proper position for sizing duplicate pieces. It is necessary, however, to get the relative pressures of springs *e* and *h* correct. When unground work is placed on the centers of the machine and in the shoe bearings, the nut *k* and screw *f* will be forced away from the shoulder and stop, thus compressing the springs *e* and *h*. Should the shoe bear unequally at points *A* and *B*, tighten screw *d* to increase the pressure at *A*, and screw *g* to increase the pressure at *B*. The combined pressure of springs *e* and *h*, however, should not be greater than is necessary to support the work, as long, slender work, although of uniform diameter, may not be straight when released from the shoes unless some allowance is made for the elasticity of the material. When adjustable bronze shoes are used to support pieces of various diameters, first loosen the screw which clamps the lower bearing shoe, and adjust it until the work bears centrally on both shoe surfaces, then retighten the set-screw.

What are the usual speeds for grinding wheels?

The peripheral or surface speed of a grinding wheel is usually somewhere between 5500 and 6000 feet per minute; speeds between 5000 and 6500 feet per minute are commonly employed, and, in some cases, wheels of hard bond are operated at speeds as high as 8000 feet per minute, although such high speeds are not recommended for general work. Grinding wheels made from corundum, alundum, aloxite, or other aluminous abrasives are generally operated at about 6000 feet per minute, whereas wheels made from carborundum, crystolon, carbolite, or wheels belonging to the carbide of silicon group are usually operated at about 5500 feet per minute.

As the wheel diminishes in size, it appears to get softer, even though the peripheral speed remains constant. This increase in wear is due to the fact that the abrasive grains are in contact with the work oftener as the size is diminished, owing to the increased number of revolutions necessary for maintaining the same surface speed. The speeds of grinding wheels of various diameters in revolutions per minute for obtaining a given surface speed may be obtained by the following rule.

Rule.—To obtain the wheel speed in revolutions per minute, divide its surface speed in feet per minute by 0.26 times the wheel diameter in inches. Note: For most purposes, the result will be accurate enough if the surface speed in feet per minute is divided by one-fourth the wheel diameter in inches.

A wheel which is perfectly adapted to grinding a certain kind of material will not work satisfactorily if the relative surface speeds of the wheel and work are not approximately correct. The work speed affects the wear of the wheel, which, when excessive, also affects the finish of the surface being ground. The amount of stock that the wheel removes for a given amount of wear can be increased or diminished by varying the work speed, the wheel wear being excessive when the speed is too high.

What factors govern work speed for cylindrical grinding?

As there are a number of factors, such as kind of material, finish desired, etc., which govern the proper work speed, it is impracticable to say just what this speed should be, and usually it is advisable to experiment somewhat when beginning to grind a new class of work, in order to determine the correct speed. A surface speed of 25 feet per minute might be correct for grinding a certain piece of steel, and not correct for another steel part having a different carbon content. The finish of a ground surface is also affected by the work speed. It is possible to grind a very rough or smooth surface by simply varying the speed, depth of cut, and side feed of wheel. For rough-grinding, the work speed would usually be about 25 feet per minute, and if this same wheel were used for finish-grinding, it would be necessary to reduce the work speed about 20 or 25 per cent to get the desired finish. If a harder and more compact wheel were used for finishing, the speed would be increased over that used for roughing.

In modern grinding practice, when rough-grinding, it is generally advisable to use a fairly coarse wheel of a soft enough grade to remain "sharp" and cut freely, and a comparatively slow work speed in conjunction with a coarse side feed of the wheel or work. This method can be used to the best advantage on machines that are heavily constructed and provided with sufficient driving power to enable these broad cuts to be taken without excessive vibration. With a light grinding machine, however, this coarse feed is not advisable, owing to the lack of rigidity and the low driving power, so that the question as to whether a high or low work speed for finishing is preferable depends, in most cases, upon the wheel used and to a certain extent upon the machine that drives it, as well as the nature of the grinding operation. As a general rule, the harder the bond and finer the grain of the wheel used, the higher should be the work speed.

In making grinding wheels, what general classes of abrasives are used?

Grinding wheels formerly were made from natural abrasives but nearly all grinding wheels in use today are made from artificial abrasives. Emery and corundum are natural abrasives; materials like alundum, carborundum, crystolon, aloxite, adamite and carbolite are produced artificially. *Emery* is a very tough and durable abrasive, but contains iron and other non-cutting elements, and is little used in grinding machines. *Corundum* is purer than emery and contains a much larger percentage of crystalline alumina, which is the element in both abrasives that does the cutting.

The abrasives used almost exclusively today are products of the electric furnace. They may be classified into two general groups—the *aluminum oxide abrasives* and the *silicon carbide abrasives*. The first group includes such trade names as "Aloxite," "Alundum," etc., and the second group, "Carborundum," Crystolon and various other names.

What is the meaning of "grade" as applied to a grinding wheel?

The term "grade" as applied to a grinding wheel, refers to the tenacity with which the bond holds the cutting points or abrasive grains in place and does not refer to the hardness of the abrasive. A wheel from which the abrasive grains can easily be dislodged is called "soft" or of soft grade, whereas, one which holds the grains more securely is called a "hard wheel." By varying the amount and composition of the bond, wheels of different grades are obtained. Also the hardness of the wheel is, to a certain extent, governed by the size of the grains; for instance, if two wheels have the same bond, the one composed of the finer grains of abrasive is the harder. In other words, a 120-grain wheel would be harder than a 24-grain wheel of the same bond. The combination of different size grains also has a certain effect on the hardness of the wheel.

What is the "grain" of a grinding wheel and how is it determined?

The grain or grit number indicates in a general way the size of the abrasive grains used in making a wheel, or the size of the cutting teeth, since grinding is a true cutting operation. It is generally assumed that grit number represents the number of meshes per linear inch through which the abrasive grains will pass; but, actually, this number is established with reference to several different sieve sizes.

To illustrate, take grit No. 24. All material must pass through the coarsest sieve—in this case the No. 16. Through the next to the coarsest sieve, termed the "control sieve"— in this case the No. 20—all material may pass, but not more than 20 per cent may be retained on it. At least 45 per cent must pass through No. 20, and be retained on No. 25 sieve, but it is permissible to have 100 per cent pass through No. 20, and remain on No. 25 sieve, the requirement being that the grain passing through No. 20, and retained on No. 25 and No. 30 must add to at least 75 per cent; consequently, if 45 per cent passed through No. 20 sieve and was retained on No. 24 sieve, then at least 30 per cent must be retained on the No. 30 sieve. Not more than 3 per cent is permitted to pass through the No. 35 sieve. The No. 25 sieve in this case has nearly 24 meshes per linear inch.

Standard Wheel Markings.—The grade, grain size, and other characteristics of grinding wheels are indicated by markings or a combination of symbols. The markings which have been adopted in the past by different wheel manufacturers vary more or less; however, a standard has been adopted by the Grinding Wheel Manufacturers' Association and is intended to replace, eventually, the individual systems. This standard consists of an abrasive letter, the grain size number, the hardness or grade letter, a structure or density number, and a bond or process letter, all written in the order named. The abrasive letter A represents aluminum oxide; C, silicon carbide. In either of these broad classes, some particular kind of abrasive may be designated by the manufacturer's own symbol used as a prefix (Example 38A). The grade letters from *A* to *Z* range from soft to hard. The bond letters are *V* for vitrified; *S* for silicate; *E* for shellac or elastic; *R* for rubber; *B* for resinoid; *O* for magnesite or oxychloride.

What principles govern the selection of grinding wheels?

The grade and grain of a grinding wheel depend largely upon the area of contact between the wheel and work, the kind of material to be ground, and its degree of hardness. A harder wheel should be used on soft machine steel than on hardened tool steel. The reason for this will be apparent if we think of a grinding wheel as a cutter having attached to its periphery an innumerable number of small teeth, for this is literally what the thousands of small grains of abrasive are. When the wheel is of the proper grade, these small teeth or cutting particles are held in place by the bond until they become too dull to cut effectively, when they are torn out of place by the increased friction. These grains or cutters will become dulled sooner in grinding hard than in grinding soft steel; hence, as a general rule, the harder the material, the softer the wheel, and vice versa, although soft materials, such as brass, are ground with a soft wheel, which crumbles easily, thus preventing the wheel from becoming loaded or clogged with metal, as would be the case if a hard-bonded wheel were used.

When a hard wheel is used for grinding hard material, the grit becomes dulled, but it is not dislodged as rapidly as it should be, with the result that the periphery of the wheel is worn smooth or glazed, so that grinding is impossible without excessive wheel pressure. Any undue pressure tends to distort the work, and this tendency is still further increased by the excessive heat generated. If the surface of the wheel becomes "loaded" with chips and burns the work, even when plenty of coolant is used, it is too hard. When a wheel is used which is too soft, the wear is, of course, greatly increased, as the particles of grit are dislodged too rapidly, and, consequently, the wheel is always "sharp." This means that the abrasive has not done sufficient work to become even slightly dulled and the result is a rough surface on the work. In regard to selecting the proper grinding wheel, as a general rule, materials of high tensile strength, such as soft and hardened steel, etc., require a wheel made from an aluminum oxide abrasive, whereas, for grinding materials of low tensile strength, such as cast iron, brass, bronze, etc., a wheel made from a carbide of silicon abrasive should be used.

Are wheels of fine grain always used for finishing cuts?

Generally speaking, coarse wheels are better adapted to most work, because the larger grains permit deeper cuts to be taken. When a very fine finish is required, particularly on a number of duplicate pieces, fine wheels are often used for finishing, after the work has been ground close to the required size with a coarse wheel. It is not necessary, however, to use a fine wheel in order to obtain a surface that is smooth enough, for ordinary machine parts, as a wheel of comparatively coarse grain will produce a finish fine enough for most purposes, if the work speed is reduced somewhat and the wheel is trued with a rather dull diamond just before taking the finishing cut; in fact, very fine surfaces can be obtained with a comparatively coarse wheel of the proper grade and grain, provided there is the correct relation between the surface speeds of the wheel and work and the wheel is carefully trued with a diamond tool. When roughing cuts are being taken, the cutting particles are constantly worn away or dislodged, so that the face of the wheel is kept rough or "sharp" and the ground surface is also comparatively rough. After the wheel-face has been trued with a diamond, however, light finishing cuts, in conjunction with a reduced work speed, will give a finish which is smooth enough for all practical purposes, even though a fairly coarse wheel is used.

Incidentally, it is not always the highly polished surface which represents the most accurate work, because this finish is sometimes obtained at the expense of accuracy, by using hard wheels that require so much pressure to make them grind that the work is distorted. In order to secure accuracy, the wheel must cut freely and without perceptible pressure. Sometimes, a coarse wheel will not cut after a surface has been finished to a certain point, because the cutting particles wear off somewhat and the ends become too large and blunt to enter the smooth surface. If this occurs, the wheel should be trued with a diamond or be replaced with one of finer grain. When grinding brass or soft bronze, the grain of the wheel must be as fine as the finish desired; in other words, it is not practicable to use a coarse wheel for finishing these metals.

Why are grinding wheels made in various shapes and sizes?

Grinding wheels are made in many different shapes and sizes to adapt them for use in different types of grinding machines and on different classes of work. Plain disk-shaped wheels similar to those illustrated at A, B, and D, Fig. 10, are the kind generally used in connection with cylindrical grinding operations. Wheels of this form vary greatly in size, the diameter and width of face naturally depending upon the class of work for which the wheel is used and the size and power of the grinding machine. In modern practice, quite large wheels are used on heavy work. In connection with form grinding, the wheel-face is frequently made wide enough to cover the entire surface of the work; that is, the wheel is dressed to conform to the shape required on the work and the latter is ground by feeding the wheel straight in, there being no lateral or transverse movement.

Very thin or narrow wheels C are sometimes used in connection with cutter or reamer grinding, or for cutting off stock. The form of wheel shown at E is intended for grinding up to a large shoulder. The wheel is mounted on the end of the spindle and is "dished" at the center so that the retaining nut on the spindle will not project beyond the side of the wheel and strike the shoulder. Wheel F is especially adapted for facing the ends of bushings or small shoulders. When the wheel is used for end facing, the grind-

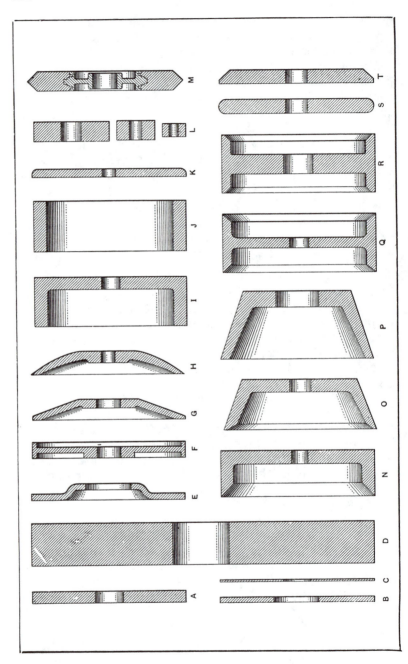

Fig. 10. Various Shapes of Grinding Wheels

ing is done on the side which is recessed to reduce the contact area to a narrow surface. The saucer or dish-shaped wheels *G* and *H* are extensively used for grinding formed milling cutters, etc., especially on regular tool- and cutter-grinding machines. The cup-wheel *I* is used for grinding flat surfaces by traversing the work past the end or face of the wheel. The cylindrical or ring-wheel *J* is also used for producing flat surfaces, the grinding being done by the end the same as with a cup-wheel. The latter is attached directly to the spindle, but the ring-type of wheel is held in a special chuck. The special form of wheel shown at *K* is used for thinning the points of twist drills when, as the result of repeated grinding, the web at the point becomes too thick, thus increasing the pressure required to force the drill through the metal. The small wheels shown at *L* are used in connection with internal grinding. These wheels, like all of those shown in the illustration, vary considerably in size, the diameter in any case depending upon the size of the work. The wheels illustrated at *M, N, O, P, Q,* and *R* are employed in connection with a variety of grinding operations on tool- and cutter-grinding machines, whereas wheels *S* and *T* are employed for saw-gumming. Grinding wheels are made in other shapes but most of them are modifications of the forms illustrated.

Have grinding wheel shapes and sizes been standardized?

Grinding wheels in various standard shapes, and in a range of sizes for each shape, have been standardized in the Simplified Recommended Practice R45 proposed by the Grinding Wheel Institute. These shapes (shown by the accompanying charts, Figs. 11 and 12) represent practically all of the forms used on standard makes of grinding machines. This standardization not only simplifies the stocking of wheels but enables the user to identify accurately any wheel by giving the type number and the important dimensions. The No. 1 or straight types are shown in diameters from 1/4 inch to 54 inches and in thickness from 1/64 inch to 12 inches. For internal grinding, this type of wheel is shown in diameters from 1/4 inch to 4 inches and in thicknesses from 1/4 inch to 2 inches. Complete tables giving

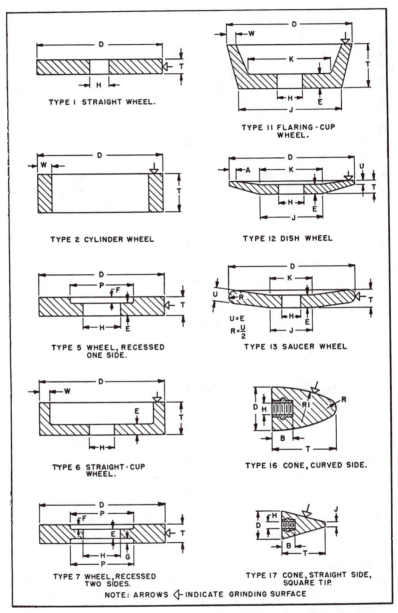

TYPE 1 STRAIGHT WHEEL.

TYPE 2 CYLINDER WHEEL

TYPE 5 WHEEL, RECESSED ONE SIDE.

TYPE 6 STRAIGHT-CUP WHEEL.

TYPE 7 WHEEL, RECESSED TWO SIDES.

TYPE 11 FLARING-CUP WHEEL.

TYPE 12 DISH WHEEL

TYPE 13 SAUCER WHEEL

TYPE 16 CONE, CURVED SIDE.

TYPE 17 CONE, STRAIGHT SIDE, SQUARE TIP.

NOTE: ARROWS ◁⊢ INDICATE GRINDING SURFACE

Fig. 11. Standard Types of Grinding Wheels Approved by the Grinding Wheel Institute

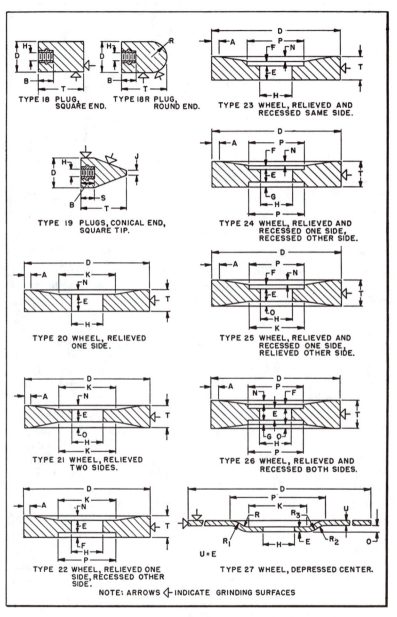

TYPE 18 PLUG, SQUARE END.

TYPE 18R PLUG, ROUND END.

TYPE 23 WHEEL, RELIEVED AND RECESSED SAME SIDE.

TYPE 19 PLUGS, CONICAL END, SQUARE TIP.

TYPE 24 WHEEL, RELIEVED AND RECESSED ONE SIDE, RECESSED OTHER SIDE.

TYPE 20 WHEEL, RELIEVED ONE SIDE.

TYPE 25 WHEEL, RELIEVED AND RECESSED ONE SIDE, RELIEVED OTHER SIDE.

TYPE 21 WHEEL, RELIEVED TWO SIDES.

TYPE 26 WHEEL, RELIEVED AND RECESSED BOTH SIDES.

TYPE 22 WHEEL, RELIEVED ONE SIDE, RECESSED OTHER SIDE.

TYPE 27 WHEEL, DEPRESSED CENTER.

NOTE: ARROWS ◁─ INDICATE GRINDING SURFACES

Fig. 12. Standard Types of Grinding Wheels Approved by the Grinding Wheel Institute (Continued)

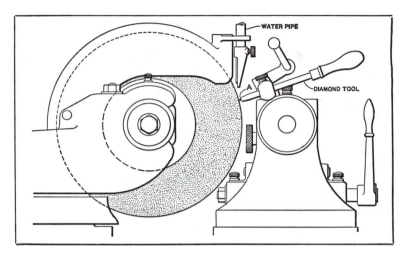

Fig. 13. Truing Face of Grinding Wheel by Use of Diamond Tool

shapes and dimensions of grinding wheels are shown in Simplified Practice Recommendation R45 available from the Grinding Wheel Institute or the U.S. Government Printing Office.

How are worn grinding wheel faces trued?

A grinding wheel should run true and have an even bearing on the surface of the work, the face of the wheel being parallel with the surface being ground; it is especially important to have a true, even wheel-face for taking a finishing cut. The only satisfactory method of truing a wheel, for machine grinding operations, is by the use of a diamond tool. This tool may be clamped to the footstock of the machine (as shown in Fig. 13) or in a special holder attached to the table, and the machine stroke is adjusted so that the diamond point will clear the wheel-face on each side. The wheel, which should revolve at the speed required for grinding, is then trued by bringing it into contact with the diamond as the table (or wheel) is traversed. Very light cuts should be taken and plenty of water used to keep the diamond cool. The diamond tool should be held with the point quite close to the clamp, or point of support, in order to reduce vibration and give a smooth accurate wheel surface. Diamond tools usually have round shanks, to permit clamping them in different positions, so that the wear on the diamond will not be confined to one or two points. The number of times that the wheel has to be trued depends upon the character of the work and the kind of wheel used. If it is necessary to remove considerable stock, the wheel may have to be trued before taking each finishing cut, provided the roughing and finishing operations are performed successively. When a number of duplicate parts are ground, this is avoided by first rough-grinding them all and then truing the wheel once for finishing the entire lot, or as many parts as the wheel will grind satisfactorily.

What kinds of diamonds are used for wheel truing?

For the truing and sharpening of wheels used on grinding machines, the most satisfactory tool to use is the diamond. The diamonds generally used in tools for truing grinding wheels are of two kinds, the carbon or black diamond and bort. The black diamond rarely has any visible crystallization; its color varies, but it is often of a dark purple brown. The bort is a semi-transparent stone or an imperfect "brilliant" and it is not as hard as the black diamond. The Brazilian bort or brownstone is sometimes called South African "premiers." Bort is considerably cheaper than the black diamond, but opinions differ as to the relative merits of the bort and the black diamond. In selecting a diamond, care should be taken to see that no seams appear, as such stones are liable to crack along the seams. The smooth "skin" stone is the one most likely to prove satisfactory. The more points a stone has, the better it is adapted to the truing of grinding wheels.

Is the diamond truing tool held in a radial position?

In truing the average grinding wheel, and especially when using a comparatively hard-bond wheel on a cylindrical grinding machine, it is necessary to maintain new cutting facets or points on the diamond to present to the grinding wheel. In order to do this, it is necessary to present the diamond point at a certain angle to the face of the wheel. The diamond holder is held at such an angle to the face of the grinding wheel that, by continuing the line representing the axis of the diamond holder, it coincides with an arc of from 1 to 1/2-inch radius scribed from the center of the wheel-spindle. By holding the diamond in this position, it is possible to secure the new cutting facets. For truing soft-bond wheels, it is sometimes advisable to use a diamond that has a rounded point and, in this case, the diamond need not be presented at an angle but can be held in a radial position.

What is the difference between a "glazed" and a "loaded" grinding wheel?

The wheels used for machine grinding not only require truing but "re-sharpening," when they will not cut effectively, owing to glazing and loading; this re-sharpening is done by removing the dull cutting points or grains of abrasive, a diamond tool generally being used. The difference between a glazed and loaded wheel is as follows: The cutting face of a loaded wheel has particles of the metal being ground adhering to it, the openings or pores of the wheel-face having been filled up with metal, thus preventing the wheel from cutting freely. A glazed wheel is one having cutting particles that have become dull, or worn down even with the bond, the bond being so hard that it does not allow the dulled cutting particles to be torn out of the wheel. Continued work with a wheel that glazes increases the smoothness of the wheel-face and decreases its cutting capacity.

On cylindrical grinding, loading may be caused by using a wheel of too hard a bond and running it too slowly. It may also be caused by taking cuts that are too deep and by not using the right cutting fluid. When a hard wheel is revolved at too slow a speed with plenty of power, the grains withstand the excessive cuts without breaking away, and thus pick up the metal. The remedy for loading is to increase the speed of the wheel or use a softer wheel. Glazing takes place when a wheel is too hard or revolves at too fast a speed. The remedy for glazing is to decrease the speed or use a softer wheel. Loading and glazing of the wheel makes excessive truing necessary and this, of course, means greater wheel wear. For truing a hard wheel, a sharper diamond is used than for truing a softer wheel. The reason for this is that a sharp diamond leaves the surface of the hard wheel with more clearance, and, consequently, this surface will cut for a greater length of time. Increasing the work speed also makes a wheel of hard bond or one having a tendency to glaze cut more freely.

What method of grinding is employed for parts of irregular shape?

Irregular or special shapes may be ground by *form grinding*. Parts are ground by using a broad wheel which is shaped to conform to the shape required, and without traversing either wheel or work laterally. The wheel is wide enough to cover the surface to be ground, and, for round work, is fed straight in, thus grinding the entire surface at the same time, without a traversing movement such as is common to ordinary cylindrical grinding. Form grinding bears practically the same relation to grinding by means of a traversing wheel as forming tools do to the ordinary single-point turning tool. In the case of forming tools, however, the sharpening is done on the flat face of the tool and the formed surface is not changed throughout the life of the tool, whereas, with a form grinding wheel, the sharpening is done on the formed surface which may be either straight, curved, or irregular. For ordinary shapes, this truing of the wheel is done without difficulty, and simply requires a special truing fixture which serves to guide the wheel truing or forming tool mechanically.

How are crankpins machined by grinding?

In grinding crankpins, it is necessary to hold the main crankshaft in an offset position so that the axis of the crankpin to be ground will coincide with the axis of rotation or centerline of the grinding machine. The amount of offset is, of course, equal to the throw of the crankshaft. If there are a number of crankpins on a crankshaft, as in the case of automobile engine crankshafts, it is also necessary to provide means of indexing the crankshaft so that each crankpin can be located successively in the central or

Fig. 14. Grinding the Crankpins of Tractor Crankshafts

Fig. 15. Another Crankpin Grinding Operation

grinding position. In automotive or other plants where large numbers of duplicate crankshafts are ground, the machines used are designed especially for crankshaft grinding. One of these machines (a Norton) is shown in Fig. 14 grinding the crankpins of large tractor crankshafts. Another crankpin grinding machine (a Landis) is shown in Fig. 15. These machines are equipped with special workholding fixtures for readily and accurately locating each pin in the grinding position. The grinding is done by feeding the wheel straight in. This wheel, which may have a diameter of 36 inches or more, is wide enough to grind the full width of the pin without a traversing movement.

Arnold Grinding Gage.—Both of the crankshaft grinders illustrated are equipped with an Arnold gage. This gage has contact points tipped with tungsten carbide. These points are in contact with the pin as it is being ground. The reduction in diameter is shown by a dial gage. In some cases, a single dial gage is used and in others two gages are employed. Sometimes two gages are used when considerable stock must be removed, and roughing and finishing cuts are indicated separately. For example, the first indicator graduated to 0.001 inch may be used for the roughing cuts; then a second indicator, graduated to 0.0001 inch, begins to register for the final finishing cut. With another arrangement, one indicator is for pin diameter and a second for pin width.

These special crankpin grinding machines are equipped with means for accurately locating the wheel opposite each pin to be ground. For example, the machine shown in Fig. 15 has a bar containing notches spaced to suit each pin and these notches are engaged by a plunger for accurately locating the work-table as it is traversed from one grinding position to the next.

Are cylindrical parts always supported on centers for external grinding?

Many cylindrical parts may be ground on machines of the centerless type. These machines may also be applied to taper parts. In centerless grinding the work is supported on a work-rest and between the grinding wheel and a regulating wheel which usually is made of material similar to that of the grinding wheel. The work-rest is equipped with suitable guides for receiving and supporting the work. The grinding wheel forces the work downward against the work-rest and also against the regulating wheel. The latter imparts a uniform rotation to the work which has the same peripheral speed as the regulating wheel, the speed of which is adjustable.

Through-feed Method.—There are three general methods of centerless grinding which may be described as through-feed, in-feed, and end-feed methods. The through-feed method is applied to straight cylindrical parts. The work is given an axial movement by the regulating wheel and passes between the grinding and regulating wheels from one side to the other. The rate of feed depends upon the diameter and speed of the regulating wheel and its inclination which is adjustable. It may be necessary to pass the work between the wheels more than once, the number of passes depending upon such factors as the amount of

Fig. 16. Centerless Grinding Pistons of Automobile Engines

stock to be removed, the roundness and straightness of the unground work, and the limits of accuracy required.

Fig. 16 shows a centerless grinding operation on automobile engine pistons. These pistons pass through two centerless grinding machines, arranged in tandem, so that they are automatically advanced to the second machine as they leave the first. These machines grind off about 0.015 inch of stock on the diameter, this dimension being held to size within plus or minus 0.0005 inch.

In-feed Method.—When parts have shoulders, heads or some part larger than the ground diameter, the in-feed method usually is employed. This method is similar to the "plunge cut" form grinding on a center type of grinder. The length of the section or sections to be ground in any one operation is limited by the width of the wheel. As there is no axial feeding movement, the regulating wheel is set with its axis approximately parallel to that of the grinding wheel, there being a slight inclination to keep the work tight against the end stop. An example of the in-feed or plunge cut method is shown in Fig. 17. The operation consists of finishing sandblasted cold-drawn steel tubes, 5 1/2 inches in diameter, for a length of 7 inches on each end. Stock to a depth of 1/16 inch on the diameter is ground off within an accuracy of 0.001 inch. The operation is performed in one plunge cut of the grinding wheel.

End-feed Method.—The end-feed method is applied only to taper work. The grinding wheel, regulating wheel, and the blade or work-rest are set in a fixed relation to each other and the work is fed in from the front mechanically or manually to a fixed end stop. Either the grinding or regulating wheel, or both, are dressed to the proper taper.

Automatic Centerless Grinding.—The grinding of relatively small parts may be done automatically by equipping the machine with a magazine, gravity chute, or hopper feed, provided the shape of the part will permit using these feeding mechanisms.

Rates of Production.—Rates of production vary widely according to the character of the work, the material, the accuracy and finish required, and other factors. As a general rule, parts ground by the through-feed method require two passes; however, when an extra-fine finish and extreme accuracy are essential, as for piston-pins, etc., the number of passes is increased. Most work is ground either by the through-feed or in-feed methods. The rate of production with the through-feed method depends chiefly upon the amount of stock to be removed, whereas, with the in-feed method, the production rate is limited to a considerable extent by the time required for loading and unloading.

When is a lathe used for grinding operations?

When parts require greater accuracy or a better finish than are obtained by ordinary turning, the turned surfaces usually are finished by grinding. This grinding, as a general rule, is done on a regular grinding machine and there are types designed for external cylindrical grinding, internal or hole grinding, and for the grinding of flat surfaces.

Fig. 17. Grinding Ends of 5 1/2-inch Tubes on a
Centerless Grinding Machine

Fig. 18. External Grinding in Lathe Equipped with Portable Grinder

Sometimes the lathe is used for grinding, as, for example, when a regular grinding machine is not available or possibly when the quantity of work is not large enough to warrant transferring the job to another machine. There is also an advantage in some cases in grinding a turned surface right after turning and without removing the work from the lathe, especially if it is held in a chuck or attached to a faceplate.

Portable electrically driven grinders are used in conjunction with a lathe. The grinder is mounted upon the lathe carriage and the lathe feeding movements can be utilized for traversing the grinding wheel the same as in turning. Fig. 18 shows how a Dumore portable precision grinder is used for grinding the cylindrical surface of a hydraulic hoist plunger. Internal grinding is also done on the lathe as explained in the section on Internal Grinding.

Surface Grinding and Types of Machines Used

The grinding of plane or flat surfaces is known as surface grinding. Several different types of machines are used for this work. The surface grinder is indispensable for truing parts that have been distorted by hardening and for producing fine accurate surfaces in connection with precision tool and die work, etc. Many of the surface grinding machines built at the present time are also efficient for producing flat surfaces in connection with manufacturing operations. Ordinarily, the surface grinder is used for finishing parts which have been milled or planed approximately to size, although many pieces are ground from the rough on the large machines used for manufacturing purposes. Surface grinders vary both in regard to the form of the grinding wheels used and the movement imparted to the work-table when grinding. For instance, the work-tables on some machines operate with a reciprocating motion, whereas others rotate; the grinding is done on some machines with a disk-shaped wheel, whereas other machines have a cylinder or ring-wheel. Some of these grinders are comparatively small in size and light in construction and are designed more particularly for tool-room use, whereas others are large and powerful, and are employed for grinding duplicate parts in connection with manufacturing operations.

How are flat surfaces ground by using the cylindrical face of a wheel?

A common method of grinding a flat surface and one that is generally used on tool-room work is shown by the diagram A in Fig. 1. The work a is traversed beneath the grinding wheel b, as indicated by the dotted line, and either the wheel or work is fed laterally (see end view) at each end of the stroke, so that the periphery or face of the wheel gradually grinds the entire surface.

Another method of producing flat surfaces is shown at *B* in Fig. 1. In this case, it will be noticed that the wheel *b* is slightly greater in width than the work and covers the entire surface to be ground. When using this type of wheel, the grinding must be done wet, as the surface contact of the wheel on the work is greatly increased. When the grinding is done wet, accurate work can be secured by this method, provided the wheel is properly trued.

How is the cylinder type of wheel applied to surface grinding?

The diagram at *C*, Fig. 1, shows one method of producing flat surfaces by using a wheel *b* of the cylinder or ring type. The vertical surface *c* is ground by being traversed past the face of the wheel; hence, this is often called *face grinding*. This method of grinding is used quite extensively in the grinding of comparatively large castings such as crankcase covers, gear housings, crankcases, and similar work.

The diagram shown at *D* illustrates the operation of a vertical surface grinder. The grinding is done either by a cup or cylinder wheel *b* which revolves about a vertical axis. The work *a* is held on a reciprocating table by means of a magnetic chuck or fixture and is traversed beneath the grinding wheel. The wheelhead remains stationary, as far as lateral motion is concerned, and is fed down gradually at the end of each stroke until the desired amount of material has been removed.

Diagram *E* illustrates the operation of another type of vertical surface grinding machine. In this case, the work-table has a rotary instead of a reciprocating movement, and the head carrying the cylinder wheel *b* is fed down a certain amount for each revolution of the work-table. This type of machine is suitable for grinding piston rings, facing the sides of ball bearing race rings, and for many other parts. It can also be used for the grinding of the sides of saws, the required clearance being obtained by setting the axis of the wheel-spindle to an angle less than 90 degrees with the top surface of the work-table. There is another

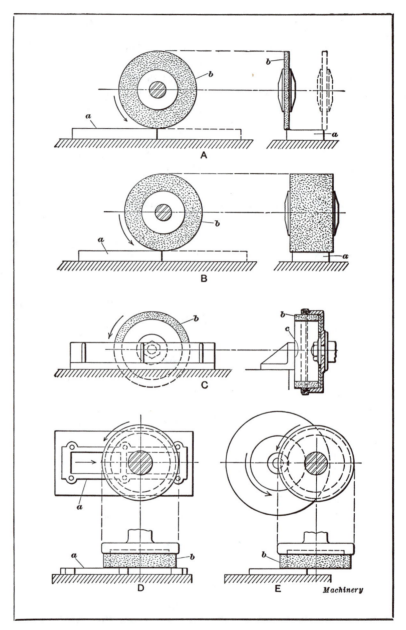

Fig. 1. Diagrams Showing Different Methods of Grinding Plane Surfaces

type of rotary surface grinder which differs from the type indicated at E in that a disk-shaped grinding wheel is used and the wheel is mounted on a horizontal spindle. Fig. 2 shows a machine of this design grinding one side of a small thrust collar.

Why are magnetic chucks used for holding parts on grinding machines?

Magnetic chucks are used because they provide a convenient accurate method of holding steel and iron parts. A large number of small parts may be held at one time and magnetic chucks are adapted for a wide range of work. They are made in a variety of sizes and shapes, the form depending upon the type of grinding machine and the shape of the work. The use of magnetic chucks for internal grinding is limited somewhat; their greatest field of usefulness is on surface grinding machines of the reciprocating and rotary types. The magnetic chuck is a special form of electromagnet which is connected by wires and a control switch with an electric power circuit. The top surface, against which the work is held, has a series of positive and negative poles which are separated by an insulating material. When the chuck is in use, the work is held by magnetic force when the current is turned on. (Figs. 2 and 3 show magnetic chucks.)

The chuck face is demagnetized so that work can easily be removed after the grinding operation by simply reversing the current through the chuck coils, momentarily, until the residual magnetism is removed. This does not demagnetize the work itself, but this is necessary for certain classes of work, because some materials become more or less permanently magnetized; hence, they attract small particles, which is sometimes quite objectionable. When the work must be demagnetized, a special apparatus called a demagnetizer is used.

When are work-supporting blocks used on magnetic chucks?

When the parts to be ground have relatively small surfaces in contact with the chuck face, and especially if the grinding operation is likely to displace the work, steel sup-

Fig. 2. Grinding One Side of a Thrust Collar to Obtain the Desired Fit in Assembling a Machine Unit. The Thrust Collar is Held in a Fixture which is Held by a Magnetic Chuck

Fig. 3. Grinding the Ends of Parts which are Held between Parallel Blocks on a Magnetic Chuck. Auxiliary Supporting Blocks are used on a Magnetic Chuck when the Parts being Ground are High in Proportion to the Area in Contact with the Magnetic Chuck Face

porting blocks may be required. The job shown in Fig. 3 is an example. The steel parts being ground are 3⅛ inches high and are finished on both ends. Because these pieces are high in proportion to the area in contact with the chuck, it is necessary to provide supplementary work-supporting blocks, as shown, to hold them in place on the magnetic chuck. After the parts are finished on one end, they are reversed for finishing the opposite end. They are ground at a table speed of 15 feet per minute.

Are magnetic chucks ever used for holding non-magnetic parts?

Some non-magnetic parts can be held by placing them adjacent to steel blocks or strips which are held by the magnetic chuck and prevent the non-magnetic parts from

Fig. 4. Grinding Non-ferrous Parts on a Magnetic Chuck by "Nesting" Them between Steel Pieces that can be Held in Position Magnetically

shifting during the grinding operation. As an example, the grinding of the sides of bronze shoes for clutch mechanisms (within plus or minus 0.002 inch) is illustrated in Fig. 4. These shoes are 2½ inches long by ¾ inch wide. Thirteen shoes are ground at one time. Since these parts are made of a non-ferrous material, they cannot be held on the chuck magnetically, and so they are nested between steel backing pieces. The magnetic chuck holds the steel pieces firmly in place, and they, in turn, prevent the work from slipping either endwise or sidewise. As the grinding pressure is always downward in this operation, a wide-face grinding wheel is used. About 0.005 inch of stock is ground off each side of these parts with the table operating at a speed of 40 feet per minute.

What is disk grinding?

The disk grinding process derives its name from the fact that the grinding is done by a disk of abrasive. Formerly it was general practice to use cloth or paper abrasive disks cemented or glued to a steel disk-shaped wheel. Now comparatively thick abrasive disks with some form of steel reinforcement are used extensively. These modern disks are held in place by bolts or screws and may readily be replaced. Regular ring wheels are also used for many disk grinding operations. This method of grinding is very efficient for many classes of work and is employed principally for truing plane surfaces. Most disk grinding is done without rigidly clamping the work; the casting or forging is pressed against the side of the disk wheel while held in the hands or in a suitable fixture. Rigid clamping should be avoided, if possible, not only because of the time required but also because it is desirable to present the work to the wheel in such a way that it will have a free "floating" movement. In some cases, clamping is necessary, but when the work is held in a fixed position, more material, as a rule, must be removed, in order to produce a true surface.

Internal Grinding Including Centerless Method

Internal grinding is done both on universal cylindrical grinding machines and on special types of machines designed exclusively for handling internal work. The use of the universal grinding machine for internal grinding is restricted ordinarily to tool-rooms for making jig bushings and similar parts, or to miscellaneous jobs in the shop. Internal grinders commonly are employed for producing duplicate interchangeable parts in manufacturing. Since the universal grinding machine is designed for external grinding, face grinding, and internal grinding, it is not as efficient for internal grinding as a machine which is designed and built for that special purpose. When internal grinding is done on a universal grinding machine, an internal grinding attachment is used. This consists of a bracket which holds the internal grinding spindle. When the attachment is in use, it is mounted on the base carrying the main grinding spindle.

What are the main features of internal grinding machines?

Machines designed primarily for internal grinding have a work-holding spindle which rotates the part to be ground, a rapidly revolving grinding wheel spindle and means for traversing the wheel or work. On one type of internal grinding machine, the work head which holds the rotating work spindle is stationary and the wheel-spindle is traversed. The wheel-spindle of another type is supported by a stationary portion of the frame, and the work head is mounted on the table of the machine and travels to and from the grinding wheel. The former type is used to a large extent in machines especially designed for internal grind-

Fig. 1. Internal Grinding Operation in Airplane Engine Cylinder Barrel

Fig. 2. Internal Grinding Ball-race Ring for Airplane Engine

ing. Figs. 1 and 2 show close-up views of typical internal grinding operations. In the design of internal grinders much attention has been given to the spindle and its mounting in order to obtain the high-speeds essential to efficient grinding.

For grinding cylinders or other parts which cannot be rotated readily, if at all, a special type of internal grinding machine has been developed in which the work does not have to be rotated, as is the case with the ordinary type of internal grinding machines. The grinding wheel not only rotates about its own axis but has a circular or planetary movement so that it follows around the walls of the hole being ground as the work is fed in a lengthwise direction.

One type of internal grinding machine designed particularly for high production on duplicate work is equipped with a dial type of size-indicating device, and automatic control for various important movements. The position of this sizing unit can be adjusted for various sizes of holes.

The application of the centerless grinding principle to internal work is another important development.

Why are relatively soft wheels used for internal grinding?

The wheels used for internal grinding should generally be softer than those employed for other grinding operations, because the contact area between the wheel and work is comparatively large. A soft wheel that will cut with little pressure should be used to prevent springing the spindle. The grade of the wheel depends upon the character of the work and the stiffness of the machine, and, where a large variety of work is being ground, it may not be practicable to have an assortment of wheels adapted to all conditions. By adjusting the speed, however, a wheel not exactly suited to the work in hand can often be used. If the wheel wears too rapidly, it should be run faster, and, if it tends to glaze, the speed should be diminished.

The rigidity of a machine has about as much to do with the selection of the grade and grain of the wheel as any other one factor. The fact is often overlooked that a machine which is not rigid should use a harder wheel than one which is stiff and rigid. The greater the rigidity of

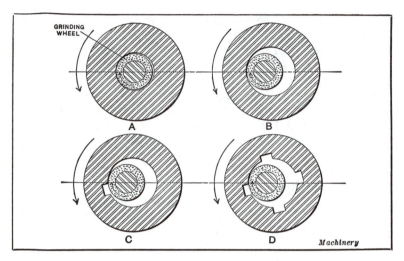

Fig. 3. Examples Illustrating Points to Consider when
Selecting Wheels for Internal Grinding

the machine, the softer the grade and coarser the grain of
the wheel should be. Of course, the rigid machine also has
the advantage of grinding more rapidly and accurately.

Should an internal grinding wheel be as large as possible?

The larger the diameter of the wheel in relation to the
hole in the work, the larger is the arc of contact and the
greater the tendency to glaze and heat the work. In addition to a greater arc of contact, a wheel used for internal
grinding has less body or volume to carry out the heat
generated and will heat up and glaze much quicker than a
wheel used externally. Wheels for internal grinding are
generally used until completely worn out. Wheels usually
are ordered as large as possible, in order that the greatest
possible number of pieces may be turned out with one wheel,
thus reducing wheel expense. From a theoretical stand-
point, however, the smaller the diameter of wheel in relation to the hole in the work, the less will be the arc of
contact and the higher will be the cutting efficiency of the
wheel.

If the hole is less than 1 inch in diameter, for economical

reasons, the wheel may be nearly as large as the hole, as shown at *A* in Fig. 3. The arc of contact of wheel and work is large, and, consequently, a much softer wheel must be used than if the wheel were small in relation to the diameter of the hole. Where the wheel is small, a harder bond and finer grain wheel can be used. The lower the wheel speed, the harder the bond should be, other conditions remaining the same. When a hole is plain, as shown at *B*, a softer wheel should be used than if the hole were keyseated, as shown at *C* and *D*. Slots or keyseats have a shaving action on the wheel-face and quickly tear out the grains; hence, for keyseated work, a harder wheel should be used than on plain hole work, and it should also have a wider face.

What factors determine the widths of internal grinding wheels?

The widths or thicknesses of internal grinding wheels depend upon the diameter and the kind of grinding operation. After several years of experimenting with different widths of wheels, a prominent manufacturing concern found that, for ordinary plain internal grinding on holes larger than 1 inch, a grinding wheel with a ¾-inch face gives the best results. This width of face of wheel is used exclusively on all internal work over 1 inch and under 3 inches in diameter that is not provided with grooves or keyways. For a considerable length of time, wheels of ½-inch face were used, but it was found that these did not have sufficient cutting surface or width of face to remove the metal efficiently. On the other hand, it was found that, with a wheel wider than ¾ inch, there was too much wheel surface in contact with the work and trouble was experienced with heating of the work.

For grinding holes in which keyways are cut, a narrow face wheel cannot be used with success, because the small wheel surface reduces the life of the wheel. The greater the number of splines or keyways, as a rule, the harder the wheel must be because a keyway has a tendency to tear out the grains from the bond of the wheel. For this class of work, the concern previously referred to uses a wheel with a 1⅛-inch face on holes up to 2 inches in diameter, and, for larger work, a wheel of 1½-inch face. Such wide wheels are used only on internal grinding machines especially designed for manufacturing.

Should the hole or exterior of a bushing be ground first?

There are two ways of grinding bushings; one is to grind the interior first and then slip the bushing on an arbor and finish the outside on a plain grinding machine; by this method, accurate work is secured and the exterior and interior holes can be made truly concentric with each other. The other method is to grind the exterior of the bushing first, and then slip this into a holder or bushing chuck fitted to an internal grinding machine, and grind the hole last. While this may seem to be a very satisfactory method, in practice, it is not, for the following reasons:

1. It seems to be true that a bushing will change shape more when the inside is ground last than when ground first.

2. It is more difficult to grind the outside of a bushing to accurate limits, when the hole is rough and cannot be satisfactorily supported on a grinding arbor, than it is to grind the hole in the bushing first (the bushing being held by clamping endwise), and then slip the bushing on an arbor and finish the outside; the bushing, in this case, fits the arbor perfectly and has a good solid support for the grinding wheel at all points.

3. In fitting up a bushing chuck, it must be put onto the internal grinding machine spindle and be ground to size after being mounted, in order that the hole will run true. After one lot of bushings is finished, the chuck will be removed from the grinding machine and later replaced when another lot of bushings is to be handled. In making this replacement of the bushing chuck, the operator will find it very difficult, in placing the chuck back on the grinding spindle, to have the hole run "dead true"; therefore, he must grind out the hole again to make it run true, but then the bushings are not a good fit unless ground to a larger diameter to fit the new diameter of hole in the chuck. All of this difficulty is avoided when the hole is ground first and then the bushing is finished externally.

How should thin bushings be held to prevent distortion?

In grinding thin bushings, a special form of chucking device should be used in which the bushing is *clamped end-wise*, because any radial pressure, such as would result from using a three-jaw chuck or similar arrangement, would distort the piece; consequently, after it is removed from the grinding machine, the hole will not be truly cylindrical. Very little clamping pressure is required to hold a bushing endwise, and these chucks can be made so that they are simple and quickly operated, thus producing work accurately and efficiently.

What are the causes of "bell-mouthed" holes?

A hole may be ground "bell-mouthed" or slightly over-size at the ends by traversing the wheel clear through the hole. In adjusting the machine for grinding a hole, the length of the stroke should be regulated so that the wheel will only travel beyond the ends of the hole one-fourth or one-half its width. The wheel should never be withdrawn from the work except when it is necessary to remove it to gage the hole. While the power traverse is being used for advancing and withdrawing the wheel, the wheel should never pass clear through the work, but about one-half of the wheel surface should still remain in the hole. The reason for this is that, when the wheel is withdrawn, any slackness in the spindle tends to throw it out to one side, because the strain of cutting has been removed; consequently, the hole is made bell-mouthed. Bell-mouthed holes may also be caused by looseness in the spindle.

What is the proper allowance for internal grinding?

The allowance depends upon the size of the hole, the rigidity and power of the grinder and the accuracy readily obtained in chucking the work. With hardened parts, there is also the question of distortion, because distortion makes it necessary to leave more stock than would otherwise be necessary. One reason why allowances on similar work in different shops vary so greatly is because of the difference in the way the work comes to the grinding machine, so far as the previous machining operations are concerned, and, the difference in methods of holding the work.

The following is an example which illustrates the relation of the amount of material removed to wheel wear. An extended experiment was made in the grinding of a tapered hole in a pinion blank. These pinion blanks were made from vanadium-steel drop-forgings, heat-treated, and were drilled and reamed so that 0.015 inch was left on the diameter for grinding. Various grains and grades, as well as makes of wheels, were tried, and the greatest number of pieces that could be obtained from one wheel was sixty. The limit of allowance was then reduced to 0.010 or 0.008 inch on the diameter and the production from one wheel increased to eighty. The limit was again reduced to 0.005 inch on the diameter, whereupon the production increased to 200 pieces from one wheel. Such reductions in the grinding allowance would be impracticable for some classes of work, but the job referred to shows that tests and experiments often are required in order to ascertain what is the best practice.

Are coolants used for internal grinding?

Internal grinding is generally done dry in the tool-room, but when grinding duplicate parts in connection with manufacturing, a coolant is often used. The general practice, in regard to the use of a coolant for internal grinding, is to grind soft and hardened steel work wet, and to grind cast iron, bronze, and brass dry. There are, however, exceptions to this rule, and sometimes work that is hardened is of such a shape that the use of coolant is either inconvenient or not necessary. For instance, when the surface to be ground is narrow and a small amount of stock is to be removed in proportion to the size of the work, the grinding can be done without any appreciable change in temperature. As a general rule, however, soft and hardened steel parts should be ground wet. Wet grinding produces a smoother and truer hole than dry grinding, and also facilitates gaging. Where a part has been heated to a considerable extent by grinding and a cold gage is placed in it, the gage is likely to "freeze," and difficulty will be experienced in removing it. A further reason for wet grinding is that it is impossible for the operator to know just what size the hole will be at normal temperature if the temperature has increased 20 or 30 degrees during the grinding. On interchangeable work and where accurate limits must be maintained, wet grinding is imperative.

There are cases where cast-iron parts should be ground wet. When the work must be ground to very accurate limits, a coolant is used to keep down the temperature, in order to prevent any change of diameter or shape. This applies more particularly to the grinding of thin, cast-iron bushings. It has been found that a cast-iron bushing, ground dry, will spring out of shape, to some extent, after it is taken out of the machine. By keeping the part at a uniform temperature, by means of a coolant, any change in diameter and shape is avoided. The pipes for conveying the coolant or compound to the grinding wheel may be seen in Figs. 1 and 2.

How is the planetary principle applied to internal grinding?

The planetary type of internal grinding machine, which is used for grinding cylinders, etc., is so designed that the work does not have to be rotated as is the case with the ordinary type of internal grinding machines. The grinding wheel not only rotates about its own axis but has a circular or planetary movement (as indicated by the diagram, Fig. 4) so that it follows around the walls of the hole being ground as the work is fed in a lengthwise direction. A table mounted on a traveling carriage is provided for supporting the work. The grinding spindle is carried in double eccentrics, or with one eccentric inside another. The two eccentrics may be rotated relatively to each other, to increase or decrease the eccentricity of the wheel-spindle, and thus give the wheel a circular path of the required diameter. The cylinder to be ground is traversed with relation to the grinding wheel. A machine of the planetary type is of especial value when the parts to be ground are of such a shape that it is impracticable to revolve them.

It is common practice, when taking the roughing cut, to rotate the head slowly and use a feed for the work-table equal to about one-half or three-quarters of the wheel width.

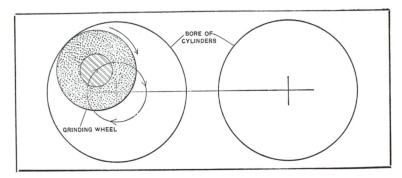

Fig. 4. Diagram Illustrating Method of Grinding Cylinders
on Machines of Eccentric-head or Planetary Type

After the stock has been removed to within about 0.0025 inch of the required size, the wheel should be trued carefully for the finishing cuts, and then the speed of rotation for the eccentric head should be increased and very light cuts taken until the hole is ground to size. When finishing, the wheel should be allowed to run through the work at least two or three times without any increase of cut, so that any possible spring in the spindle will be eliminated. Ordinarily, for finishing, the feed or table traverse is reduced, in order to remove the feed marks, although, by using the most rapid table feed for the final cuts, the time for the finishing operation may be reduced and the surface produced will be smooth enough for practical purposes.

A wheel with a ¾-inch face is used in many shops. When roughing, inasmuch as the advancing edge of the wheel does most of the work, the alternate use of first one-half the wheel-face and then the other half, by reversing the position of the wheel on the spindle, tends to cause the wheel to wear evenly. If there is surface dirt or oil in the cylinder, this should be removed before attempting to grind, as, otherwise, a soft free-cutting wheel is liable to cut and glaze. The allowance for cylinder grinding should ordinarily be about 0.010 inch.

How is internal grinding done on a lathe?

The lathe may be adapted to internal grinding by mounting a portable grinder upon the carriage. Fig. 5 shows the application of a Dumore precision grinder to a typical example of internal grinding on a lathe. A grinding attachment of this kind makes it possible to finish the hole accurately after the boring operation, without transferring the work to a grinding machine. While regular internal grinders are preferable for quantity production, frequently there is a decided advantage in adapting the lathe to internal and also external grinding, by means of these portable electrically driven grinders. Grinders of the small portable type have very high spindle speeds. For example, these speeds may range from 500 to 700 revolutions per second (30,000 to 42,000 R.P.M.), thus adapting them to light precision grinding operations in small holes.

What is internal centerless grinding?

The type of internal grinding machine known as the *centerless* type, utilizes a previously ground outer surface for rotating the work and holding it in proper relation with

Fig. 5. An Example of Internal Grinding on Lathe

Fig. 6. Grinding Airplane Engine Sleeves on Internal
Grinder of Centerless Type

Fig. 7. Rear View of the Centerless Grinder Illustrated in Fig. 6.

the grinding wheel. Fig. 6 shows a close-up view of an internal centerless grinder arranged for grinding the cylinder sleeves of airplane engines. The method of holding and rotating the work is more clearly shown by the rear view, Fig. 7. The sleeve is supported between three rolls as shown. The larger or "regulating" roll, seen at the left in Fig. 7, rotates the work and also prevents it from attaining too high a speed due to its contact with the rapidly revolving grinding wheel. The upper roll at the right exerts sufficient pressure on the work to hold it in proper relation with the regulating roll and against the lower supporting roll. As the grinding of each part is completed, the pressure roll is lifted, usually by automatic means, to permit ejecting the work over the top of the regulating roll. As a general rule, bores can be ground to a given diameter within 0.0003 inch, and concentric, straight and round within 0.0001 inch.

Centerless grinding was applied first to external grinding, and, as the name indicates, the centerless method differs from the usual method of external grinding in that the part being ground is not supported at the ends upon work-holding centers. Internal grinding has, of course, always been centerless, but this term has been applied to the internal method just described evidently because it is based upon fundamental principles which are similar to those employed in external grinding.

The indicator seen at the front of the machine in Fig. 5 registers readings taken on the diameter of the work by moving a diamond-pointed finger along the surface being ground. This finger can be withdrawn entirely from the machine during loading and can be withdrawn partly from the surface of the work during rough grinding. It is brought into contact with the ground surface by operating the lever that may be seen mounted on the indicator unit.

Is internal centerless grinding applicable to taper bores and "blind" holes?

Taper holes or bores, and also holes that are "blind" or closed at one end, can be ground by the centerless method. The outer surfaces may also be tapering or have shoulders. In grinding taper holes, the entire work-supporting unit is merely adjusted to whatever angular position is required. If the part has a double taper, as in certain bearing races, both taper surfaces are ground by merely rechucking and reversing the work after one taper surface has been finished. If the outer work-supporting surfaces are tapering or have shoulders, work-supporting rolls of special form are used. For example, a roller bearing race having a tapered outer surface and also two shoulders on the outside, would be supported on the taper surface by rolls that are also tapering. The rolls in this case would be relieved to clear the shoulders on the work. Internal centerless grinding can be applied to quite a range of diameters and lengths. If the part is quite small, a narrow blade is used for supporting it on the under side instead of the roll that is used for medium size and large work.

Grinding Milling Cutters and Reamers

Milling cutters and reamers should be ground whenever there is any indication of dullness. Dull cutters are inefficient, and reconditioning extremely dull cutters requires the removal of so much material in obtaining sharp edges, that the life of the cutter is greatly reduced. The old adage about a stitch in time saving nine, applies to milling cutter and reamer grinding.

Cutter grinding is often done on cylindrical grinding machines of the universal type, and even in the lathe by the use of a grinding attachment, especially in small shops, but it is preferable to use a machine designed especially for this class of work. These special machines are so arranged that they may be used for grinding plain cylindrical cutters, angular cutters, end-mills, side-mills, formed cutters, reamers, circular forming tools, saws for cutting-off machines, and a variety of other tools. The universal tool- and cutter-grinders made by different manufacturers vary more or less as to details, but they are similar in their general arrangement and operate on the same general principle.

What are the general methods of sharpening cutters?

The general methods of sharpening milling cutters differ in regard to (1) the type or form of wheel used; (2) the part or surface of the tooth that is ground; (3) the direction of grinding wheel rotation relative to the tooth being ground. Three different types of grinding wheels are extensively used in cutter grinding. These are (1) the "straight" or disk-shaped wheel; (2) the cup type in either the straight or flaring form; and (3) the dish type. (These standard types or forms are illustrated in the section on Grinding Cylindrical and Tapering Parts—see Figs. 11 and 12.)

371

Milling cutters are sharpened and given the required amount of clearance, by grinding the "lands" of the teeth, excepting formed cutters, which will be dealt with later. The diagrams A and B, Fig. 1, illustrate the meaning of "land." The land is that narrow clearance surface which is back of and adjacent to the cutting edge. While the diagrams show the land at the top of a tooth, it may also extend along the sides of radial cutting edges, as, for example, in the case of side milling cutters. The clearance angle is measured from the surface of the land. The clearance at the top of a tooth is the angle between the land and a line tangent to the cutter circumference at the cutting edge.

The width of the land varies for cutters of different types and sizes. Repeated sharpening of the cutter increases the land width, and, consequently, the amount of metal which must be removed in sharpening. To reduce excessive land width, a secondary clearance angle may be ground on the backs of the teeth, as illustrated by the diagram B.

When a straight or disk type of wheel is used, the land

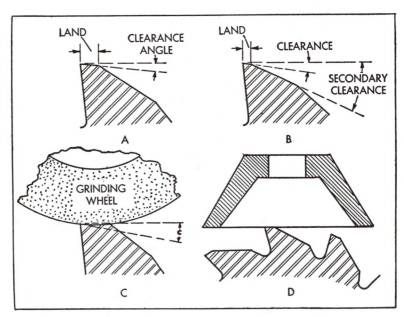

Fig. 1. Diagrams Illustrating Tooth Forms and Terms
Used in Cutter Grinding

is slightly concave in form as shown by diagram *C*. The actual clearance angle *c* applies to that part of the land adjacent to the cutting edge; hence, the actual clearance may be greater than the apparent clearance. A cutting edge having a concave land is not as strongly supported as one that is backed up by a flat land such as would be produced by the flat face of a cup-wheel, as illustrated by diagram *D*; however, when the grinding is done by using 6-inch or 8-inch straight or disk wheels, which are common sizes, the resulting concavity is extremely slight, especially on narrow lands, and of little or no practical importance in most cases.

In what direction should the grinding wheel rotate relative to the cutter?

The wheel rotation relative to the tooth being ground may be either from the back to the front of the tooth (as shown by diagrams *A* and *C*, Fig. 2), or vice versa, as shown by diagrams *B* and *D*. When the rotation is as shown at *A* or *C*, the thrust of the grinding operation is directly against the tooth-rest which bears against the front face of the tooth being ground, thus locating each successive tooth as it is brought around to the grinding position. When wheel rotation is in the opposite direction or from the front of the tooth to the back (diagrams *B* and *D*), the thrust of the grinding operation is away from the tooth-rest and serious damage might result if a cutter should revolve while grinding a tooth; however, this method is often preferred, because a keener edge will be obtained without forming a burr, and there is also less danger of drawing the temper. Care must be taken to hold the work securely against the tooth-rest when grinding as shown at *B* and *D*, as otherwise the wheel may draw the cutter away from the rest and score the tooth. There is also danger of the wheel being broken. Rotating the wheel as shown at *A* and *C* is the safer method, as the wheel then holds the tooth against the rest.

How is a cylindrical cutter traversed past the grinding wheel?

There are two general methods: The cutter may be traversed along a close-fitting stationary arbor, or the cutter (on a tight fitting arbor) may be traversed by moving the table of the cutter grinder. When the cutter is traversed along an arbor held between the centers of the machine, it should have a free sliding fit so that it can be traversed without play. When the arbor fits the cutter tightly and the table is traversed while grinding, the straightness of the cutter depends upon the accuracy of alignment between the axis of the arbor and the line of table movement. To avoid unnecessary refinement in aligning the table, it is common practice to traverse the cutter along the arbor; this insures a straight or cylindrical cutter, because the distance be-

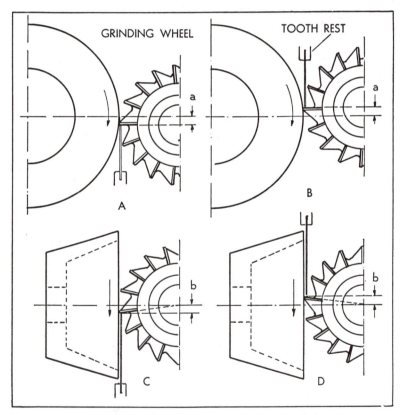

Fig. 2. Illustrations Showing Grinding Wheel Rotation Relative to Cutter, and Methods of Obtaining Clearance when Using Disk Wheels as at A and B or Cup-wheels as at C and D

tween the grinding face of the wheel and the stationary arbor remains unchanged; hence, the entire surface of each tooth is ground to a uniform radial distance from the cutter axis. On the other hand, if the table were traversed, any angularity between the axis of the cutter and the ways of the table would cause the cutter to be ground tapering. Instead of mounting cylindrical cutters directly on an arbor, they are sometimes held on sleeves which are adapted for cutter holes of different diameters, so that it is not necessary to have an accurate fitting arbor for each cutter.

What form of wheel is used for grinding cylindrical cutters?

Ordinary milling cutters may be sharpened either by using the periphery of a disk wheel as shown at *A* and *B*, Fig. 2, or the face of a cup-wheel as at *C* and *D*. The cup-wheel has the advantage of grinding flat lands. The periphery of a disk wheel leaves the teeth slightly concave, thereby causing them to become dull sooner than they would if the lands were flat; however, if the disk wheel is not under 3 inches in diameter, the teeth will not be concaved enough to cause trouble; moreover the disk wheel will grind faster.

Grinding wheels for cutter sharpening should be of a medium-soft grade and not too fine—never finer than 60 grit. Fine wheels cut slowly and tend to burn the teeth. Wheels for sharpening cutters made of high-speed steels can be a little coarser than those used on carbon steel. If the wheel is too soft, it will wear rapidly, which makes it difficult to keep the cutter round while sharpening it. This difficulty can be overcome by using a wheel at least $3/4$-inch wide, instead of the $3/8$- or $1/2$-inch wheels commonly used. A wide soft wheel will last as long as a narrow one of harder grade, and being softer, it tends to eliminate the danger of burning. For sharpening ordinary milling cutters, a wide wheel will not be especially inconvenient, as generally there is plenty of room.

Why are light cuts taken in cutter grinding?

Care should be taken not to remove too much stock at one cut; if crowded, the tendency will be to "burn" the work or draw the temper, and also cause the wheel to wear away too rapidly. The rapid wearing of the wheel will leave the cutter considerably out of round. Under favorable conditions a cut of 0.003 inch can be taken without burning the teeth or wearing the wheel away enough to cause the cutter to be out of round more than 0.001 inch. A slight error of this kind is corrected in the finishing cut. The operator should finish up with a light cut, say 0.001 inch deep, and before removing the cutter, he should make sure that it is round. This is easily determined by the manner in which the teeth "spark" on the wheel. The first tooth sharpened should be marked with chalk before starting.

Is the tooth-rest of a cutter grinder fixed relative to the grinding wheel or cutter?

In cutter grinding, a tooth-rest is used to support the cutter while grinding the teeth. When grinding a cylindrical cutter having helical teeth, the tooth-rest must remain in a fixed position relative to the grinding wheel. The tooth being ground will then slide over the tooth-rest (as illustrated later), thus causing the cutter to turn as it moves longitudinally, so that the edge of the helical tooth is ground to a uniform distance from the center, throughout its length. If the cutter is traversed along its arbor, the tooth-rest may be fixed to the machine table, and stops should be used to prevent the cutter from sliding off of the rest. If the table is traversed, the tooth rest must be attached to a stationary part of the grinder. When grinding a straight-fluted cutter, it is also preferable to have the tooth-rest in a fixed position relative to the wheel, unless the cutter is quite narrow, because any warping of the cutter in hardening will result in inaccurate grinding if the tooth-rest moves with the work. The tooth-rest should be placed as close to the cutting edge of the cutter as is practicable, and bear against the face of the tooth being ground. When the tooth-rest is fixed relative to the wheel, it should be somewhat wider than the wheel face so that the cutter will have a support before it reaches the wheel and also after it has been traversed past the wheel face. The end of the tooth-rest should have an even bearing upon the tooth being ground. Narrow tooth-rests may be used when they are attached to the table, and remain fixed relative to the work.

Fig. 3. Grinding a Milling Cutter Having Helical Teeth. The Machine is a Universal Type of Tool Grinder

What adjustments are made to obtain the required clearance angle?

The clearance angle is regulated, when grinding, by setting the grinding wheel center above or below the center of the cutter, or by adjusting the tooth-rest below or above the center, depending upon the type of wheel used. When grinding with a plain disk wheel, the clearance may be obtained by setting the center of the grinding wheel above the center of the work as shown at *A*, Fig. 2, or below the work center when the position of the cutter is reversed as at *B*. When using a cup-wheel, the clearance may be obtained as indicated at *C* or *D*, but this method should not be employed for grinding angular cutters. The grinding wheel and work centers are at the same height at *C* and *D* and the tooth-rest is set below or above the centers as shown. The amount of offset *a* and *b* for disk and cup-wheels, respectively, may be determined by the following rules:

Rule for Disk Wheel.—Multiply the grinding wheel diameter, in inches, by the clearance angle, in degrees, and multiply this product by 0.0087 to obtain the amount of offset *a* (Fig. 2) in thousandths of an inch.

Rule for Cup-Wheel.—Multiply the cutter diameter, in

inches, by the clearance angle, in degrees, and multiply this product by 0.0087 to obtain the amount of offset b in thousandths of an inch.

Milling cutters usually have from 4 to 7 degrees of clearance. The practice in some shops, where special roughing and finishing cutters are used, is to give the roughing cutters a clearance of 7 degrees and finishing cutters, 5 degrees. Excessive clearance causes chattering when milling, and the teeth become dull quickly. The Brown & Sharpe Mfg. Co. recommends a clearance of 4 degrees for plain milling cutters over 3 inches in diameter, and 6 degrees for those under 3 inches. The clearance of the end teeth of end mills should be about 2 degrees, and it is advisable to grind the teeth about 0.001 or 0.002 inch low in the center so that the inner ends of the teeth will not drag on the work.

How are plain or cylindrical milling cutters ground when the teeth are helical?

In grinding cylindrical cutters having helical teeth, either a cup-wheel (as shown in Fig. 3) or a plain disk wheel may be employed. A cup-wheel generally is preferred, and its position relative to the cutter and tooth-rest is shown by the diagram, Fig. 4. The cutter may be mounted upon a close-fitting arbor, and it should be free to slide in a lengthwise direction along the arbor to permit traversing the cutter past the grinding face of the wheel. The tooth-rest remains in a fixed position relative to the wheel. This tooth-rest should be somewhat wider than the wheel face, so that the cutter will be supported at the beginning and end of the tooth grinding operation. The top of the tooth-rest should be inclined as shown, so that it has an even bearing on the tooth face and it should be centered relative to that side or face of the wheel which is to be used. Arbor stops set to regulate the cutter stroke are preferable.

The particular tooth-rest illustrated has a flexible part A to facilitate indexing the cutter from one tooth to the next. In sharpening a tooth, as at B-C, the grinding begins at B and continues throughout the length of the tooth until point C has passed the face of the grinding wheel. The traversing movement of the cutter is then reversed, and when end B is over the flexible part A, the cutter is turned or indexed, thus springing part A which clears the wheel

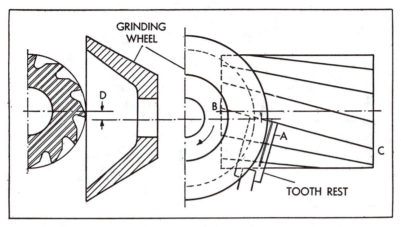

Fig. 4. Arrangement of Wheel and Tooth-rest when Grinding
Cylindrical Cutter Having Helical Teeth

and snaps back after passing the face of the next tooth.
As the cutter is traversed longitudinally by hand along the
arbor, it also rotates a part of a revolution as the helical
tooth face slides over the tooth-rest; hence, the land or
top of the tooth progressively moves past the central or
grinding position. The center of the cutter is located above
the center of the grinding wheel an amount D, depending
upon the degree of clearance desired. If the cutter is long
enough to reach the non-grinding side of the wheel face
during the traversing movement, the wheel-head should be
inclined about ½ degree from the right-angle position relative to the cutter axis, in order to provide a little clearance
for the non-grinding side. If a plain or disk type of grinding wheel is used, the general procedure is the same as
just described.

What is the general method of grinding the radial teeth of side milling and face cutters?

A cup-wheel is used ordinarily and the cutter is mounted
on an arbor held in a head or attachment having angular
adjustments in both the vertical and horizontal planes. The
adjustment in the vertical plane, as indicated by the left-
hand diagram, Fig. 5, is to provide clearance which usually
ranges from 2 to 4 degrees. The adjustment in the hori-

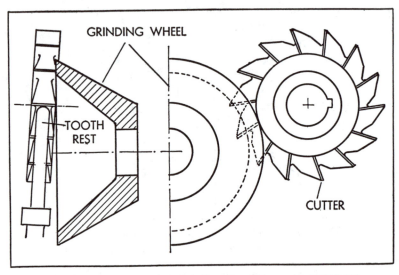

GRINDING WHEEL

TOOTH REST

CUTTER

Fig. 5. Arrangement for Grinding the Side or Radial Teeth
of a Side Milling Cutter

zontal plane of about ½ degree only is for grinding the
teeth 0.001 to 0.002 inch lower at the inner ends, so that
they will have this slight amount of clearance in a radial
direction. The tooth-rest is set to the same height as the
center of the cutter. As each successive tooth land is
ground, the cutter is indexed for positioning the next tooth
by merely turning it backward, thus causing the flexible
tooth-rest to spring outward until the point or top passes
the face of the next tooth to be ground. Fig. 6 shows the
grinding of the radial faces of a 16-inch face mill.

In grinding angular cutters, what is the proper position for the tooth-rest?

In grinding or sharpening angular cutters, the tooth-rest
should be set to the same height as the center or axis of
the cutter while grinding the teeth on the conical face (see
diagram, Fig. 7). In other words, the edge of the particular tooth being ground should lie in the same horizontal
plane as the axis of the cutter, so that the cutter will be
ground to the required angle as indicated by the graduations on the machine. For example, in sharpening a

Fig. 6. Grinding the Faces of the Teeth on a 16-inch Face Mill,
Using a Cutter and Tool Grinder

60-degree single-angle cutter, the work-holding head is adjusted to hold the axis of the cutter at an angle of 30 degrees from the wheel face, as shown by the plan view. This angle, however, would not be obtained on the cutter if the centers of the wheel and cutter were in the same plane and clearance were obtained by adjusting the tooth-rest as illustrated by diagrams *C* or *D*, Fig. 2. When the tooth being ground is inclined relative to the horizontal plane or the plane of angular adjustment, as shown by these diagrams *C* and *D*, the resulting change in angle is due to the same fundamental cause as the error in taper turning when the cutting point of the tool is above or below the horizontal plane intersecting the axis of the work. In using a disk wheel, the clearance is obtained either by an offset adjustment *a*, Fig. 7, of the wheel center or by adjusting the work-holding head (depending upon the design of the machine) in order to obtain whatever clearance is required. If a cup-wheel were used, the clearance would have to be obtained by inclining the grinding face of the wheel an amount equal to the clearance angle desired.

Grinding the End Teeth.—When grinding the end teeth or flat face of a single-angle cutter, a flaring type of cup-wheel is used and the procedure is similar to that illustrated by Fig. 5 which shows the general arrangement for grinding a side milling cutter. Clearance for the lands of the teeth is obtained by inclining the work-holding spindle in a vertical plane to whatever angle is required. The cutter usually is also inclined one or two degrees in a horizontal plane, so that the inner ends of the teeth will be 0.001 or

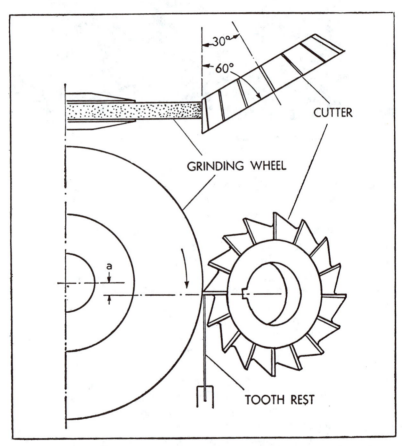

Fig. 7. Grinding Teeth on Conical Face of 60-degree Single-angle Cutter. In Grinding the Teeth on the Side or End, the Procedure Is Similar to that Shown in Fig. 5

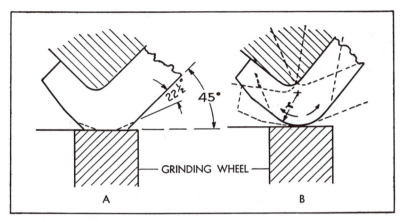

Fig. 8. (A) Beveling Corner of Face Mill Tooth. (B) Rounding Corner of Face Mill Tooth

0.002 inch low to provide clearance in a radial direction. The tooth-rest is attached to the table and the tooth being ground is traversed past the wheel face for the cut. A disk wheel is sometimes used instead of a cup-wheel.

Why and how are the corners of face milling cutters rounded or beveled?

Some milling cutters such as end-mills, angular cutters, etc., must have sharp corners in order to form square or angular corners on the work. Sharp or delicate corners, however, should be avoided on face mills or any cutters of this general type designed for roughing operations. Corners are often beveled to an angle of 45 degrees, as illustrated in the diagram A, Fig. 8. The corners may be further eliminated by beveling each side to an angle of 22½ degrees, as illustrated by the dotted lines, thus approaching more nearly a rounded form. It is preferable, however, to eliminate all corners, as illustrated by diagram B. This corner-rounding operation requires the use of a radius attachment. This attachment is mounted upon the table of the grinder and provides means of swinging the cutter about a vertical axis so that the corner will be ground to any required radius r. Fig. 9 shows a Cincinnati radius grinding attachment set up for rounding the corners of the teeth of a face mill. This attachment has a gage for setting the cutter in a zero radius position. Then, by turning two adjusting screws located at right angles to each other, the attachment can be set for grinding an arc that is tangent both to the face and periphery of the cutter.

Attachments of this general type are also used for grinding convex and concave cutters or any other circular forms. Fig. 10 shows the grinding of a convex cutter on a Landis machine equipped with a radial grinding attachment. The lower slide of this attachment swivels on a central stud and the adjustable slides are graduated to permit setting the attachment for a given radius.

In grinding formed cutters, how is the original shape retained?

All formed cutters are given clearance at the time the cutter is made, by a relieving operation, so that the original shape of the cutter can be retained as the front faces of the teeth are ground back in resharpening; hence, in sharpening formed cutters, the relieved or formed surfaces are not touched and only the front faces are ground. Some

Fig. 9. Grinding a Rounded Corner on a Milling Cutter by Means of a Radius Grinding Attachment

**Fig. 10. Grinding Convex Cutter by Means of
Radial Grinding Attachment**

formed cutters have radial faces and others have rake to
improve the cutting action. The radial form will be considered first. Formed cutters of the radial-face type should
be re-ground so that all of the tooth faces lie in planes
passing through the axis of the cutter, as shown by the
diagram *A*, Fig. 11. The teeth should also be uniform in
height so that each one will do an equal amount of work
when the cutter is used. When setting up a grinder for
sharpening formed cutters, the grinding side of the wheel
should run true and be in line with the center-line or axis
of the work-holding arbor. If the wheel is set off-center
and the teeth are not ground radially, the contour of the
cutting edge will be changed and the cutter will not mill to
the shape for which it was intended. Grinding the faces so
that they lie either ahead or back of the radial line changes
the projected shape, and as it is that which defines the outline of the cut, the result is a departure from the established
tooth shape.

The cutter must also be ground so that the height of
all of the teeth will be equal. This could be accomplished

readily by using an indexing fixture, if the cutter were round with evenly spaced teeth; these conditions are seldom realized, however. Changes of shape in hardening throw the teeth out of even spacing, and, if ground on an accurate indexing fixture, some of the teeth will be so short that only a few do the cutting. The result is rough work and heating of the cutter which reduces the quality and quantity of the work. A cutter-grinder that employs the following principle of indexing will enable accurate results to be obtained. The teeth should be indexed by a finger resting on the *formed* part of the tooth, an indicator being provided which enables the operator to gage the position of each tooth with reference to its own shape as well as to the axis of the cutter. With this arrangement, each tooth-face may be ground truly radial and all the teeth to the same height.

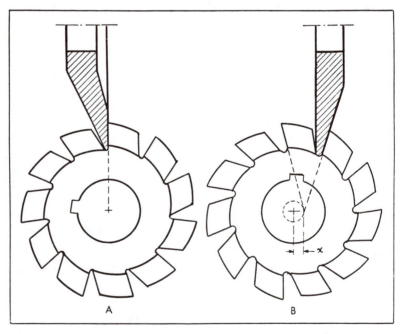

Fig. 11. (A) Grinding Radial Tooth Faces on Formed Cutter, Using Flat Side of Wheel. (B) Grinding Formed Cutter Having Rake. Beveled Side of Wheel is Used when Formed Cutter or Hob Has Helical Flutes

If a formed cutter has rake, how is the original shape retained when sharpening it?

If the teeth of a formed cutter have rake or are of the "hook type," this means that the front faces lie in planes that are offset relative to the cutter axis, as shown by the diagram B, Fig. 11. The amount of offset x depends upon the rake angle. In resharpening formed cutters having rake, the grinding wheel is set relative to the cutter axis, so as to retain this original offset x, as illustrated by the diagram. This offset should be stamped on the cutter so that the original shape may be retained in resharpening.

The rake of formed cutters or the inclination of the front faces of the teeth from a radial line, usually varies from 5 to 10 degrees, depending upon the character of the milling operation. This rake angle not only improves the cut-

Fig. 12. Cutter and Tool Grinder Arranged for Grinding Hob Having Helical Flutes

ting action but makes it possible to use the cutter for a longer period without sharpening. The relieved surfaces in back of the cutting edges of cutters having rake are formed by tilting the relieving tool upward or so that the upper face of the tool is in a plane passing above the center of the cutter. The angle of the relieving tool face equals approximately the rake angle of the cutter. If a formed cutter relieved in this manner were ground so that all of the front faces were radial, this would, of course, change the contour of the cutting edge and the shape of the surface produced by the cutter.

What is the general method of sharpening or grinding hobs?

Hobs for cutting spur gears, helical gears, and worms are sharpened by grinding the front faces of the teeth. Since these teeth are formed or relieved, it is necessary to retain the original shape by setting the grinding wheel so that the ground faces will be either in a radial position or inclined to provide hook or rake, depending upon the type of hob. The tooth faces must also be an equal distance apart so that all teeth will have the same height, thus insuring an even distribution of work or uniform cutting. It is also essential to retain the lead of the grooves or flutes. When grinding helical flutes, the bevel side of the wheel should be used, as illustrated by the diagram B, Fig. 11, representing a formed cutter. If the flat side were used, there would be interference or lack of clearance between the wheel and the teeth at points away from the mid-grinding position. The larger the helix angle of the flute, the larger the angle of the bevel required on the grinding wheel in order to avoid this interference. When the helix angle is exceptionally high, as in the case of multiple-threaded worm-wheel hobs, it may be necessary to use wheels having curved profiles instead of the straight bevel form.

As the hob is traversed past the wheel, there are two general methods of causing it to rotate so that the grinding wheel will make proper contact with the helical flute. One method is illustrated in Fig. 12 which shows a Cincinnati cutter and tool grinder. A master form or guide is mounted on the work-holding arbor, and it has grooves

Fig. 13. Hob-grinding Attachment with Covers Removed to
Show Gearing which Rotates Helically Fluted Hobs

or flutes of the same lead as those on the hob. This master form is in contact with the stationary finger against which the form is held as the hob is traversed past the wheel. The work is fed to the grinding wheel by a micrometer adjustment on the tooth-rest. Another method of rotating the hob to suit a given lead is illustrated in Fig. 13, which shows a Landis universal and tool grinder equipped with a hob-grinding attachment. In this case the work-spindle is rotated through a combination of change-gears by means of a pinion which meshes with a stationary rack attached to the front of the machine. As the machine table is traversed, the work-spindle is given a rotary movement in whatever ratio to the table movement is required to obtain the correct lead. By using suitable change-gears, hobs having either very short or very long leads may be sharpened.

In cutter grinding, when is the cutter rotated?

Some tool and cutter grinding machines are so designed that the cutter can be rotated, as, for example, in connection with face grinding. Fig. 14 illustrates how the side of a metal-slitting saw is ground slightly concave to provide clearance. The saw is mounted upon an expanding arbor

Fig. 14. Face Grinding a Metal-slitting Saw

or collet and is further supported by the rotating faceplate, while the side is being ground concave. Another example of face grinding is shown in Fig. 15, which illustrates the sharpening of a Fellows gear shaper cutter. For some face grinding operations the work is held in a chuck attached to the headstock spindle. Hand reamers for steel are sometimes ground cylindrically. The reamer is mounted between the centers of the grinding machine and is rotated so that the heel of the blade comes into contact with the wheel first. The lands of these reamers are very narrow, and grinding from the back to the front of the tooth results in a very slight springing action which provides the clearance. (Figs. 16 and 17 show additional examples of cutter grinding.)

How are straight reamers ground?

The grinding of reamers is a more delicate operation than the grinding of milling cutters. The lands frequently are very narrow and the cutting action of the reamer is more closely related to the exact amount of clearance than in the case of a milling cutter. Either disk or cup-wheels may be used. As the lands of hand reamers may be only 1/64 inch or less for steel and 1/32 inch or less for cast iron, a straight or disk type of wheel forms a land that is practically flat.

Fig. 15. Face Grinding a Fellows Gear Shaper Cutter

Diagram *A*, Fig. 18, illustrates the application of a straight or disk wheel. The tooth-rest is set to the same height as the reamer axis, and the clearance is obtained by locating the center of the reamer slightly below the wheel center.

The diagram *B* illustrates the application of a cup-wheel. In this case, the clearance is obtained by lowering the tooth-rest. The axis of a cup-wheel should be set about ½ degree from the 90-degree position so that there will be a slight amount of clearance for the non-grinding side. When grinding reamers, as in all grinding, very light cuts should be taken to avoid excessive heating and burning of the cutting edges. The tooth-rest should be in contact with the face of the particular tooth being ground in all cases. If a reamer has helical teeth, the grinding procedure is practically the same as described in connection with helical milling cutters. The tooth-rest is fixed relative to the grinding wheel. It should have a straight edge and be somewhat wider than the wheel face, and this edge against which the reamer tooth bears should be inclined to suit the helix angle, thus obtaining an even bearing throughout its width.

Clearance for Reamer Teeth.—Too much or too little clearance on the teeth of a reamer will tend to produce unsatisfactory results. Too much clearance causes a reamer

Fig. 16. Grinding Teeth of Metal-slitting Saw

Fig. 17. Grinding End Teeth of a Shell End-mill

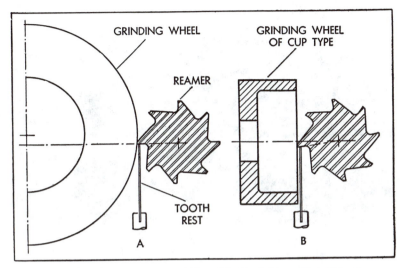

Fig. 18. Grinding Reamer (A) with Straight or Disk Wheel, and (B) with Cup-wheel

to chatter. Too little clearance will cause excessive reamer wear. There is also a tendency for the reamer to bind in the hole, thus injuring the hole as well as the reamer. The amount of clearance that a reamer tooth should have depends somewhat upon the class of reamer, and its size. A suitable clearance for reamer teeth, expressed in degrees (the face of the tooth being radial), is not a fixed amount, but varies for different diameters—the smaller the reamer, the greater should be the angle of clearance. Hence, if the cutter and reamer grinder are set to grind the right amount of clearance on a large reamer, a small reamer would not have suitable clearance at that setting.

How is a taper reamer set up for grinding?

In grinding a taper reamer, the tooth-rest must be set to the same height as the center of the reamer. A straight or disk wheel is used and the clearance is obtained by locating the center of the reamer slightly below the wheel center when the wheel rotation is from the back to the front of the tooth (diagram A, Fig. 18). The cutting edge of the tooth being ground must lie in the same horizontal plane

Fig. 19. Grinding a Taper Hand Reamer

as the center of the reamer in order to grind the reamer to the taper or angle to which the machine is set, as explained in connection with the grinding of angular milling cutters. Fig. 19 shows a Cincinnati cutter and tool grinder set up for grinding a taper hand reamer.

The taper of a reamer may be tested by inserting it into a standard collet or taper gage. Another final test is to actually ream a hole and then test this hole with a standard taper plug gage. In reamer grinding, as in cutter grinding, the rotation of the wheel should, as a general rule, be from the back to the front of the tooth, so that the thrust of the grinding operation is against the tooth-rest. If the wheel rotation is reversed, a somewhat smoother cutting edge will be obtained, but there is always danger of rotating the reamer away from the tooth-rest unless a special holding device is provided. The slight burr or roughness along the edge which may result from grinding from the back to the front of the tooth can readily be removed by rubbing an oilstone along the face of the tooth.

Why are some reamers made with helical flutes and teeth?

Reamers in both straight and tapering forms may have helical or spiral flutes and teeth. There are two possible advantages: One object is to provide a shearing cut and thus improve the cutting action and reamed surface. Another reason for using reamers of the helical type on some classes of work, is to bridge over keyways, grooves, or other openings in the reamed hole, thus preventing binding and chatter which might result if straight teeth are used. Helical-fluted reamers may have either right-hand or left-hand flutes. The right-hand flutes and teeth tend to draw the reamer inward, whereas left-hand teeth react in the opposite direction. When reamers are held securely and have a positively controlled feeding movement, there is no danger of right-hand flutes drawing the reamer forward. When hand-feeding movements are employed, a reamer with right-hand flutes might tend to feed in too rapidly. This would not apply, however, to end-cutting reamers and such reamers should preferably have right-hand flutes to improve the cutting action, especially at the ends of the teeth. Left-hand flutes are commonly applied to taper reamers. Teeth which wind around to the left tend to force the chips ahead, which is a good feature when reaming holes which extend through the work, particularly when such reaming is done on a vertical type of machine. Some steep taper reamers have right-hand helical cutting edges in order to secure a slight drawing-in tendency and thus improve the cutting action. A taper reamer known as the "duplex spiral" type has varying angles of flutes in order to eliminate any tendency to chatter and leave a smooth surface.

What kind of wheel should be used in cutter grinding?

In grinding high-speed steel cutters, use an *aluminum oxide* abrasive (see page 331) and a *vitrified bond*. The preferable *grain size* is medium (about 60), the *grade* is medium (about J), and the *structure* or grain spacing is also medium (No. 5). If applied properly, a cutter may be ground dry with the type of wheel mentioned, without excessive heating or wheel wear. It is important to use a wheel that will retain its size while grinding a cutter, in order to obtain a uniform cutter diameter or (in case of a face mill) edges which lie in the same plane. The surface speed of the wheel should be 4500 to 6000 feet per minute.

Lapping and Other Precision Finishing Processes

Certain classes of work, in tool and gage making and also in manufacturing, require highly refined finishing processes for obtaining exceptionally true surfaces and, usually, very accurate dimensions as well. The quality of the finish obtained by these refined processes is not mere polish or apparent smoothness. Highly polished surfaces often are deceptive and if seen through a microscope may consist of minute ridges and valleys which would be objectionable when truly cylindrical and flat forms are essential.

While it is possible to produce very true cylindrical and flat surfaces by ordinary machine grinding, some classes of work require even greater refinement as to quality of finish and degree of accuracy. For example, in making precision tools and gages, grinding is often followed by "lapping" (as described later) to obtain an extremely fine quality of finish and a true surface conforming to a given dimension within very close limits. In some cases, the quality of the finish may be more important than the actual dimension of the part, or this order may be reversed. Whenever extremely accurate dimensions are essential, true surfaces are, of course, also necessary since the dimension relates directly to them.

In the practical application of these refined processes, the equipment used depends more or less upon the shape and size of the work and the exact methods employed often require considerable skill resulting from repeated experiences with similar classes of work. It is possible, however, to explain the fundamental principles involved.

What is lapping as applied in precision tool and gage making?

Lapping is a refined abrading process which may be applied in finishing either cylindrical or flat surfaces requiring great precision. As applied in precision tool and gage making, it is based upon the practice of the *lapidary*, who cuts and polishes precious stones by means of diamond dust charged or embedded in a copper disk. The ease with which the lapidary found it possible to cut a substance as hard as a diamond paved the way for the handling of hardened steel in the same manner: that is, by having a soft surface in contact with a hard surface, with abrasive material between the two. As in the case of cutting and polishing precious stones, the abrasive material embeds itself into the softer substance, and scores or cuts the harder material as the lap is manipulated manually or mechanically.

Lapping not only produces a very smooth surface, but it is adapted for minute reductions in size, as, for example, in finishing a plug gage to the right diameter. It is not only applied to cylindrical external surfaces, but to the finishing of holes requiring great precision. Plain or flat surfaces may also be finished by lapping, in which case the cbject is to correct local errors wherever necessary to produce a flat surface. The lapping tool or lap is made of some soft metal, such as cast iron or brass, and it is "charged" with an abrasive which is embedded into its surface. The grade or coarseness of the abrasive depends upon the finish required and the amount that must be removed by lapping, but all lapping abrasives are fine. The form of the lap naturally depends upon the shape, size, and location of the surfaces upon which it is used. While the main essential points of the art of lapping can be described, it is necessary that the workman shall do considerable lapping before he can become proficient. There are certain motions, touches, sounds, refinements, etc., which the skilled workman acquires by practice, that are impossible to enumerate and describe in a way that would be intelligible to an inexperienced man.

What materials are used for laps?

The material of which the lap is made is one of the most important factors, especially when the work is soft. It is very important that this material should be softer than the work. Laps are usually made of soft cast iron, copper, brass, or lead, although soft steel is used for a certain class of laps that will be referred to later. In general, the best material for laps to be used on very accurate work is soft, close-grained cast iron. If the grinding, prior to lapping, is of inferior quality, or an excessive allowance has been left for lapping, copper laps may be preferable. They can be charged more easily and cut more rapidly than cast iron, but do not produce as good a finish. Whatever material is used, the lap should be softer than the work, as, otherwise, the latter will become charged with the abrasive and cut the lap, the order of the operation being reversed.

A common and inexpensive form of lap for holes is made of lead which is cast around a tapering steel arbor. The arbor usually has a groove or keyway extending lengthwise, into which the lead flows, thus forming a key that prevents the lap from turning. When the lap has worn slightly smaller than the hole and ceases to cut, the lead is expanded or stretched a little by the driving in of the arbor. When this expanding operation has been repeated two or three times, the lap usually must be trued or replaced with a new one, owing to distortion. The tendency of lead laps to lose their form is an objectionable feature. They are, however, easily molded, inexpensive, and quickly charged with the cutting abrasive.

A more elaborate form of lap for holes is composed of a steel arbor and a split cast-iron or copper shell, which is sometimes prevented from turning by a small dowel-pin. The lap is split so that it can be expanded to accurately fit the hole being operated upon. For hardened work, some toolmakers prefer copper to either cast iron or lead. For holes varying from ¼ to ½ inch in diameter, copper or brass is sometimes used; cast iron is used for holes larger than ½ inch in diameter. The arbors for these laps should have a taper of about ¼ or ⅜ inch per foot. The length of the lap should be somewhat greater than the length of the hole. The length of an external lap should at least equal the diameter of the work, and might well be longer.

What kinds of abrasives are used for lapping?

Both natural and artificial abrasives are used for lapping. Flour of emery is a natural abrasive that has been used extensively. Artificial abrasives such as alundum and carborundum are also adapted to lapping. For very fine work, diamond dust may be used.

As a general rule, there is no advantage in using a lapping abrasive coarser than No. 150. Lapping abrasives may be graded as to fineness by the precipitation method. To illustrate, a quantity of, say, flour emery is placed in a heavy cloth bag, and the bag gently tapped. The finest emery will work through first, and should be caught on a piece of paper. When sufficient emery is thus obtained, it is placed in a dish of lard or sperm oil. The largest particles of emery will rapidly sink to the bottom, and in about one hour the oil should be poured into another dish, care being exercised that the sediment at the bottom of the dish is not disturbed. The oil is now allowed to stand for several hours, say, over night, and then is decanted again, and so on, until the desired grade of abrasive is obtained.

What kind of diamond dust is used for lapping and how is it graded?

The diamond dust used for very precise lapping operations is made from Brazilian bort or imperfectly crystallized diamonds. Brazilian bort, in a pebbly form, is crushed in a suitable mortar, and graded to suit requirements, and it is particularly applicable to hand or form laps, or laps for delicate or sharp corners. Diamond dust is commonly used for lapping or grinding small precision work in tool-rooms, watch factories, etc., where great accuracy is required.

The grades of diamond dust used for charging laps are designated by numbers, the fineness of the dust increasing as the numbers increase. The diamond, after being crushed to powder in a mortar, is thoroughly mixed with a light oil. According to one system of grading, olive oil is used and the mixture is allowed to stand 5 minutes and then the oil is poured into another receptacle. The coarse sediment which is left is removed and labeled No. 0. The oil poured from No. 0 is again stirred and allowed to stand 10 minutes, after which it is poured into another receptacle and the sediment remaining is labeled No. 1. This operation is repeated until practically all of the dust has been recovered from the oil, the time that the oil is allowed to stand being increased finally to several hours (as shown by the following table) in order to obtain the smaller particles that require a longer time for precipitation:

To obtain No. 1—10 minutes. To obtain No. 4—2 hours.
To obtain No. 2—30 minutes. To obtain No. 5—10 hours.
To obtain No. 3—1 hour   No. 6—until oil is clear.

The No. 0 or coarse diamond which is obtained from the first settling is usually washed in benzine, and recrushed unless very coarse dust is required. This No. 0 grade is sometimes known as "ungraded" dust. In some tool-rooms the time for settling, in order to obtain the various grade numbers, is greater than that given in the table.

Very light oils are sometimes used for grading to reduce the time required. If the oil is comparatively heavy, like machine oil, sometimes weeks are required to eliminate all but the finer particles. The process can be hastened by the use of a liquid of less specific gravity than machine oil, such as kerosene, benzine, ether, or alcohol. In these lighter liquids, the particles settle very rapidly, and it is possible to accomplish in a few minutes what would take weeks with a heavier liquid.

Diamond dust usually is used either with a cast-iron, machine-steel or copper lap. In the charging operation, it is only necessary to take a few drops of the liquid containing the abrasive, smear it over the lap, and begin working with the hardened steel, flint, glass, or whatever substance is to be cut. The diamond dust immediately embeds itself and the work may proceed.

How are laps "charged" with abrasive?

The method of "charging" a lap or embedding in its surface minute particles of abrasive depends upon the shape of the lap. To charge cylindrical laps for internal work, spread a thin coating of prepared abrasive over the surface of a hardened steel block, preferably by rubbing lightly with a cast-iron or copper block; then insert an arbor through the lap and roll the latter over the steel block, pressing it down firmly to embed the abrasive into the surface of the lap. For external cylindrical laps, the inner surface can be charged by rolling-in the abrasive with a

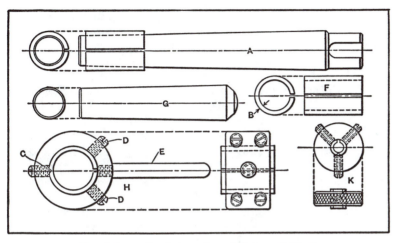

Fig. 1. Laps for Internal and External Work

hardened steel roller that is somewhat smaller in diameter than the lap.

To charge a flat cast-iron lap, spread a very thin coating of the prepared abrasive over the surface and press the small cutting particles into the lap with a hardened, flat steel block. There should be as little rubbing as possible. When the entire surface is apparently charged, clean and examine for bright spots; if any are visible, continue charging until the entire surface has a uniform gray appearance. When the lap is once charged, it should be used without applying more abrasive until it ceases to cut. If a lap is over-charged and an excessive amount of abrasive is used, or more than can be embedded into the surface of the lap, there is a rolling action between the work and lap which results in inaccuracy. The surface of a flat lap is finished true, prior to charging, possibly by scraping and testing with a standard surface plate.

In lapping holes, what form of lap is used?

A common form of lap for holes consists of a split sleeve and a tapering holder or arbor (*A*, Fig. 1) for expanding the sleeve to fit the hole. The lap is traversed through the hole, thus gradually finishing it to the required size. The included angle of the taper of this arbor should be about 2 degrees. The lap proper, or the part that is in contact with the work, may be made of copper or of cast iron and is shown in detail at *F*. The lap is split as shown, to allow it to expand as it becomes worn. The length of the lap should be somewhat greater than the length of the hole to be operated on, and thickness *B* should not be more than one-sixth nor less than one-eighth of the diameter of the work. When making these laps, especially small ones, a hardened swedging plug (shown at *G*), ground to the same taper as the arbor, can be used to advantage for tapering the hole through the lap before it is turned and slotted. If in the operation of lapping, the hole becomes "bell mouthed," that is, enlarged at the ends, this is caused by the introduction of sharp abrasive from time to time as the hole is being lapped. To obviate this, the lap should be cleaned of all loose abrasive and expanded by driving the arbor farther into it. The hole is then dry lapped by using only the abrasive that sticks or is charged in the lap. This process must be repeated occasionally until the proper size is obtained. If the operator is careful to see that the abrasive used is not too coarse, and the lap is kept expanded to fit the work at all times, the result will be a straight hole.

In lapping cylindrical plugs, what form of lap is used?

One form of lap for external cylindrical surfaces is shown at *H*, Fig. 1. The proportions of the lap proper should be the same as were given for inside laps. The same method of procedure described for inside work should also be followed, *viz.*, the lap must be freed from oil and loose abrasive, from time to time, as the work progresses. The pointed screw *C* keeps the lap from slipping out of place, and the adjusting screws *D* compress it to fit the work. A handle *E* should be used on all laps of large size, as it will be found much more convenient than a lathe dog, which some workmen use for moving the lap across the work. At *K* is illustrated an outside lap and holder for small work, say, less than ½ inch in diameter. Laps of this size are not provided with a handle, but are knurled on the outside as shown.

How is a hole finished by lapping?

The exact method of lapping a hole may depend upon its size and the accuracy required. The general type of lap shown at A, Fig. 1, is often used. The bushing forming the lap frequently is made of cast-iron. This bushing is split on one side to allow for expansion as it is forced onto the taper arbor, to compensate for the gradual enlarging of the hole being lapped. The actual lapping operation consists in traversing the lap in and out so as to cover the entire surface of the hole. The lap should be about three times the length of the hole it is intended to be used in. The same rules regarding abrasive, speed, etc., apply as in the lapping of plug gages. Care should be exercised to avoid a too generous application of the abrasive as the process nears completion, for, if applied too lavishly, the particles have a tendency to crowd under the edges and cause a "bell-mouth" effect or enlargement of the hole at the ends. This latter trouble is sometimes eliminated by making the rings with a small extension collar on each end which is ground off after the rough lapping has been completed; but this is somewhat expensive, and, except for master gages, hardly necessary. In the making of small ring gages which do not allow the insertion of a substantial cast-iron lap, a tool-steel lap charged with diamond dust can be used.

When are holes finished by using a rotating lap?

The rotary type of lap is used for accurately finishing very small holes, as in hardened steel. While the operation is referred to as lapping, it is, in reality, a grinding process, the lap being used the same as a grinding wheel. Laps employed for this work are made of mild steel, soft material being desirable because it can be charged readily. A form commonly used is shown in Fig. 2. The grinding is done by the enlarged end, and the lap is held in the spindle of a grinding attachment by the taper shank. A bench lathe is used and the rotating spindle of the attachment is traversed by hand.

Charging is usually done by rolling the lap between two hardened steel plates. Diamond dust and a little oil are

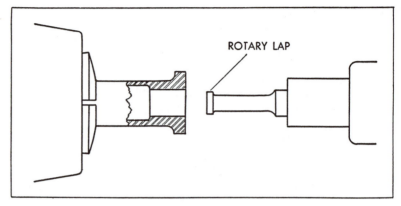

Fig. 2. Rotary Lap which is Used Like a Grinding Wheel

placed on the lower plate, and, as the lap revolves, the diamond is forced into its surface. After charging, the lap
should be washed in benzine. The rolling plates should also
be cleaned before charging with dust of a finer grade. It
is very important not to force the lap when in use, especially if it is of a small size. The lap should just make
contact with the high spots and gradually grind them off.
If a diamond lap is lubricated with kerosene, it will cut
freer and faster. These small laps are run at very high
speeds, the rate depending upon the lap diameter. Soft
work should never be ground with diamond dust because
the dust will leave the lap and charge the work.

Is lapping applicable to tapering or conical holes?

Tapering holes may be lapped but a rotary lap is preferable. The following method is recommended: First grind
the hole to size, plus the allowance for lapping; then, without disturbing the position of the slide-rest or grinder head,
change the grinding wheel for a lap made of copper (of the
same shape as the wheel but wider) and lap in the same
manner as the hole was ground, taking care not to "crowd"
the lap by attempting to lap or grind too fast.

Taper or conical holes are sometimes lapped by using
cast-iron plugs or laps having the same taper as the hole

in the work. This method of finishing taper holes is liable to be unsatisfactory. In the first place, it is difficult to secure a smooth surface, because the conical lap cannot be moved back and forth across the surface being lapped; moreover, the lap tends to cut annular grooves into the work, as it remains in one position, and imperfections in the hole will, to some extent, be transferred to the lap. Taper laps should be charged by rolling the abrasive into the surface, in the same way that cylindrical laps are charged. When loose abrasive is used, a conical lap tends to produce a more abrupt taper in the hole being lapped, because the abrasive is gradually carried outward toward the mouth of the hole by the action of centrifugal force which increases as the diameter of the hole increases. At least one roughing and one finishing lap should be used and, if a smooth hole is necessary, several laps may be required. Slight errors in the taper are sometimes corrected by charging the roughing lap in accordance with the error; for instance, if the taper is slightly greater than it should be, the small half of the lap only is charged.

What is the general procedure in lapping a cylindrical plug gage?

As an example of cylindrical lapping, assume that a 1-inch plug gage is to be lapped to size. Such a gage needs only about 0.001 or 0.0015 inch for lapping. For plug gages, cast iron is generally considered the best lap material. Although cast iron cannot be charged with abrasive as readily as copper or lead, it gives better results and wears less. Laps for lapping plugs are made from disks ranging from 3/16 to about ½ inch in thickness; they are drilled and reamed to a sliding fit on the ground plug, and split on one side to allow for adjustment. The type of lap and lap-holder shown at H in Fig. 1 is adapted to work of this kind. The piece to be lapped should be running at the speed required in grinding, which varies according to diameter, and the lap adjusted at all times to grip firmly on the surface, but sufficiently free to allow its being held by the fingers. In the case of large work, a wood clamp may be used. As the piece revolves, the lap is slowly drawn along it from end to end, and this traversing movement should continue while the plug is in motion. A very fine grade of abrasive is used. It is mixed with sperm or lard oil, to the consistency of molasses, and applied sparingly to the surface being treated, from which it is taken up by the lap, which becomes charged as it passes over after each application. When the operation is almost completed, use only a drop or two of oil charged with the finest particles of abrasive. This is obtained by sifting about a tablespoonful of abrasive into a tumbler of lard oil. After standing an hour, the oil should be poured off, and will be found charged with the very finest abrasive, the coarser particles having settled at the bottom. This abrasive is applied, a drop at a time, from the end of a small pointed stick or wire. The lap requires constant adjusting, to compensate for the wear of the lap and reduction in size of the gage. This adjustment is effected by the screws in the holder. When the gage has been lapped to within 0.0002 inch of the finished size, allow it to thoroughly cool and then, by hand, lap lengthwise of the gage to the finished size. By so doing, all minute ridges that are caused by circular lapping are removed, thereby leaving a true surface and also imparting a silvery finish. A gage should never be lapped to size while warm (heated by the friction of the lap), because the gage expands when heated, and if then lapped to size it will contract enough to spoil it.

What kind of lap is used for flat surfaces?

Laps for producing plane surfaces are made of cast iron. In order to secure accurate results, the lapping surface must be a true plane. Many toolmakers claim that a flat lap that is used for roughing or "blocking down" will cut better if the surface is scored by narrow grooves. These are usually located about ½ inch apart and extend both lengthwise and crosswise, thus forming a series of squares similar to those on a checker-board. A flat cast-iron lap is shown in Fig. 3. As will be seen, this particular lap is provided with ribs across the back the same as a small surface plate. The first requisite of perfect lapping is a perfect lap. To make a flat lap, it should be carefully planed, strains due to clamps being avoided, and then it should be carefully scraped to a standard surface plate. This is done by rub-

Fig. 3. Cast-iron Lap for Lapping Plane Surfaces

bing the face of the lap on the standard surface plate to obtain the bearing marks, and scraping down the high spots until a plane surface is obtained.

How is a flat lap charged with abrasive?

To charge a flat cast-iron lap, spread a very thin coating of the prepared abrasive over the surface and press the small cutting particles into the lap with a hard steel block. There should be as little rubbing as possible. When the entire surface is apparently charged, clean and examine for bright spots; if any are visible, continue charging until the entire surface has a uniform gray appearance. When the lap is once charged, it should be used without applying more abrasive until it ceases to cut. If a lap is over-charged and an excessive amount of abrasive is used, or more than can be embedded into the surface of the lap, there is a rolling action between the work and lap which results in inaccuracy.

How are flat surfaces lapped?

After the lap is charged, all loose abrasive should be washed off with gasoline, for fine work, and, when lapping, the surface should be kept moist, preferably with kerosene. Gasoline will cause the lap to cut a little faster, but it evaporates so rapidly that the lap soon becomes dry and the surface caked and glossy in spots. When in this condition, a lap will not produce true work. The lap should be employed so as to utilize every available part of its surface. Gently push the work all around on its surface, and try not to make two consecutive trips over the same place on the lap. Do not add a fresh supply of loose abrasive to a lap, as is frequently done, because the work will roll around on these small particles, which will keep it from good contact with the lap, causing inaccurate results. If a lap is thoroughly charged at the beginning, and is not crowded too hard and is kept well moistened, it will carry all the abrasive that is required for a long time. This is evident, for if a lap is completely charged to begin with, no more abrasive can be forced into it. The pressure on the work should only be sufficient to insure constant contact. The lap can be made to cut only so fast, and, if excessive pressure is applied, it will become "stripped" in places, which means that the abrasive which was embedded in the lap has become dislodged, thus making an uneven surface on the lap.

Are flat surfaces ever lapped by the rotary method?

For some classes of flat lapping, a rotating disk-shaped lap is used. Sketch A, Fig. 4, illustrates how a lap of this type may be used for lapping the jaws of a snap gage. The lap is made of cast iron and the sides are relieved as shown, leaving only a narrow edge or flange on each side to bear against the jaws. The gage is clamped in a vise attached to the machine table, and the circular lap is mounted on an arbor inserted in the spindle. As the table is traversed back and forth, the lap passes over the entire surface of the jaw, grinding it down in the same manner as would be done with a cup grinding wheel. Care must be taken to clamp the gage in the vise so as not to spring it.

A lap should be turned on the arbor on which it is to be used, as it is almost impossible to put a lap back on an arbor after it has been removed, and have it run true. Therefore, the lap should be recessed quite deeply, to allow for truing up each time the lap is placed on the arbor.

Diagram B, Fig. 4, shows the operation of charging a

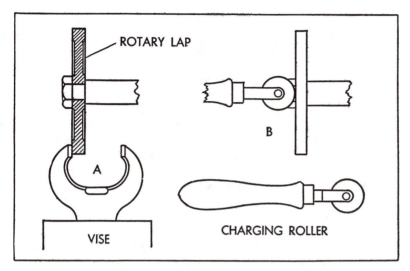

**Fig. 4. (A) Lapping Snap Gage with Rotary Disk Lap.
(B) Charging the Lap**

circular lap, using a roller mounted in a suitable handle for the purpose. The abrasive is rolled in under moderate pressure. It is good practice to make the roller of hardened steel, and, after charging the lap, all loose abrasive should be thoroughly washed off.

Rotary disk laps, similar to the form shown at *A*, are also made of soft steel, the sides being relieved so that there is a narrow lapping surface. These steel laps are made quite narrow and are often mounted on arbors and used in a bench lathe or surface grinder for lapping out sharp corners, etc., on fine hardened work. A charged steel disk can be used for fine work that could not be done with an abrasive or grinding wheel.

Why should the temperature of the work and lap be about equal in lapping plane surfaces?

To prevent the distortion and inaccuracy resulting from decided temperature differences. With an accurate plane surface-lap, it is possible to produce surfaces that are either plane, convex, or concave, the form depending on the temperature of the lap with relation to the work. If the work is warmer than the lap, the result will be a convex surface on the work. If the work is colder than the lap, a concave surface will be produced on the work. If the temperatures are the same, the surface produced will be plane.

In the usual attempt to produce plane surfaces by hand-lapping, the operator holds the work in his hand. The lap is colder than blood temperature, and the work becomes warmed through contact with the hand; hence, this method results in warm work and a cold lap. The lower surface of the work is therefore chilled by contact with the lap, and contraction of the surface is inevitable. The remainder of the work is warm because of the heat absorbed from the hand. When moved across the surface of the lap, the work is abraded only on the edges or margins of the concave surface. Under these conditions lapping might be continued until the lap and work fit each other perfectly, and one might suppose that a plane surface would be produced; but when the work is removed from the lap and the temperature equalizes itself throughout the mass, the lower surface expands, resulting in a convex surface. The remedy is to allow the work to remain in contact with the lap until the two acquire a uniform temperature, and then lap the work without contact of the hands. In the application of this remedy, make a work-holder that will fit over the work without bending it. In the lapping of thickness gages, for instance, the work-holder should hold the work without exerting distorting pressure, and it must be so made that it may be moved about by means of a wooden handle, such as an old file handle.

How are precision gage-blocks lapped so as to eliminate practically all errors?

The manufacture of gage-blocks, which are flat, parallel and to a given size within a millionth of an inch—or at most a few millionths of an inch—is a great achievement in the art of precision lapping and measuring. In making Hoke precision gage-blocks, lapping is done mechanically and the lapping is accompanied by a systematic distribution and elimination of the errors as described later. The blocks, after proper "seasoning" and aging, are ground to within approximately 0.001 inch of the basic size. They are then lapped.

The special lapping machine has two flat horizontal laps of circular form. The lower lap is attached to the base of the machine, and the upper lap is secured to an arm by a connection which permits the lap to move freely in any direction but not to revolve. This arm is pivoted at one end so that the upper lap can be swung to one side to expose the lower lap and the work. When a machine is in use, one lap is above the other, and the gage-blocks are between them, so that both the upper and lower surfaces of the blocks are lapped simultaneously. When the blocks are to be removed or inserted in the machine, the upper lap is swung out of the way.

Between these two cast-iron laps there is a steel plate or "spider" which contains twenty-four holes into which the blocks to be lapped are inserted. These holes are slightly larger in diameter than the distance across the corners of the blocks, so that the latter are free to move and bear evenly on the faces of the laps. When the machine is closed and ready for use, this spider is about midway between the laps but does not come into contact with them. The function of the spider is to hold the blocks in position and impart to them the motion required for lapping.

Why does the gage-block lapping apparatus have combined rotary and planetary motions?

The multiple lapping motion (see diagram, Fig. 5) is to secure uniform lapping. The upper and lower laps remain stationary, while the spider receives a planetary motion which brings each block into contact with the entire surface of each lap. This motion of the spider is derived from three cranks A, B and C which are caused to rotate in unison, and as they revolve, the spider, which has gear teeth around its circumference, is carried around between pinions at A, B and C. The motion of the spider is partly indicated by the diagram, which shows it in the four positions corresponding to 90-degree movements to the cranks. As these cranks revolve, the spider itself is given a rotary motion about its own axis, while at the same time this axis follows a circular path having a radius equal to the crank radius. This combined motion carries the blocks inward and outward as the spider revolves. As it is very essential to distribute the wear on the laps as uniformly as possible, the machine is equipped with gears of such ratio that this gearing seldom occupies the same relative position. The result is that the spider and gage blocks do not follow the same path around the surfaces of the laps, during successive turns. In fact, the cranks make about 300,000 revolutions before a given block in the spider again follows the same path relative to the faces of the laps. The wear is also further distributed by reason of the fact that the blocks themselves revolve with more or less irregularity at first and then quite uniformly and in the same direction, as the surfaces become true.

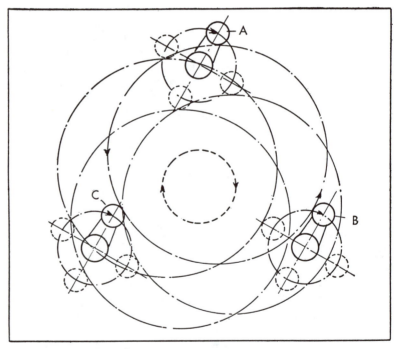

Fig. 5. Diagram Illustrating the Planetary Motion of the Plate or "Spider" which Moves the Gage-blocks around between the Stationary Lapping Surfaces

Why are the positions of the gage-blocks changed while lapping them?

The blocks are shifted around systematically to reduce them equally and obtain both flat and parallel surfaces. While it would be relatively simple to lap the upper and lower ends until they are flat, it is essential to so handle the work that both *flat* and *parallel* surfaces are secured either before or when the gage is finished to the required dimension. When the twenty-four blocks are first placed in the machine, there are sure to be very slight irregularities in regard to the heights of different blocks, as well as slight errors in the flatness and parallelism of their surfaces. As the upper lap is free to adjust itself, it naturally rests upon the three highest points or surfaces on the circular row of blocks, and as the blocks are carried around between the laps, these high areas are soon enlarged. Now if the lapping were continued without changing the blocks, the surfaces might be finished flat, but the plane of the upper surfaces would not be parallel to the plane of the lower surfaces, assuming that there was some slight variation in height to begin with, as would ordinarily be the case. The problem then is to lap all the blocks flat and parallel while they are being reduced to the same length or height. This is accomplished by changing the relative positions of the blocks at intervals during the lapping period.

How are the gage-blocks transposed in lapping to make them flat, parallel and the right size?

Just how the lapping is done will be illustrated by considering in detail the various steps connected with the actual lapping of blocks having a nominal size of 0.550 inch. In this particular case, approximately 0.0008 inch had to be removed by lapping. In order to have the lapping periods under accurate control, each machine is equipped with a combination revolution indicator and automatic stop which can be set to stop the machine after a given number of revolutions. In the lapping operation, the history of which is recorded on a card, the machine was first set to make 70 revolutions; then, after transposing the blocks, as explained later, it made 70 more revolutions. As the result of the lapping due to 140 revolutions, the blocks were reduced from plus 0.0008 or 800 millionths of an inch to plus 400 millionths. Sixty additional turns reduced the blocks to within 225 millionths of the required size, and then a number of other lapping periods followed, until the gage blocks were only 15 millionths over size. The machine in this case was then set to make 50 revolutions, which gave the additional reduction required.

The transposition of the blocks is a very essential part of the work, and if this were not done, it would be impossible to lap the opposite surfaces of the blocks flat and parallel at the same time. Wherever a set of rough blocks is placed between the laps, the upper lap is sure to rest upon the three highest points or areas, and if the upper lap were in an inclined position to begin with, it would ordinarily tend to remain out of parallel with the lower lap. To avoid this lack of parallelism, the minute errors in heights of different blocks are gradually eliminated by the following method: After the machine has made a certain number of turns, the upper lap is swung out of the way and every other block is moved 180 degrees, or to a position diametrically opposite. As the result of this shifting of the blocks, either the highest one in the set or the one next to it, is moved across the plate, and owing to these changes of position, height errors are reduced about one-half. Sometimes the blocks are shifted only 90 degrees, thus further dividing and reducing the error, and by repeatedly changing positions, the upper surfaces are finally brought into parallelism with the lower surfaces.

There is no fixed relation between the number of lapping machine revolutions and the amount of reduction by lapping. This is due to the fact that as the blocks are transposed and occupy different relative positions between the laps, the high areas, in some cases, are quickly reduced, whereas for other positions, the surfaces are more nearly in the same plane, and, consequently the laps are in contact with larger areas, and any reduction in the height of the blocks requires much more time. For instance, in one case, 20 revolutions served to reduce the blocks from 225 to 200 millionths, and then 30 revolutions more did not result in any appreciable change. The next lapping period, however, of 40 revolutions again reduced the size 25 millionths.

The laps used for these precision gages are charged with a fine grade of Turkish emery, which has been found superior to artificial abrasives for this particular work. A slight amount of additional abrasive is applied to the laps during the early stages of the lapping process. This abrasive is mixed with kerosene which is pumped through a pipe to the upper lap. This upper lap is provided with a channel extending around the outer edge, and holes through which the oil flows to the space between the laps. Much depends upon the accuracy of these laps; they are made of gray cast iron, which is homogeneous and entirely free from any porous or spongy places. In order to eliminate all internal stresses and prevent subsequent changes, the castings are thoroughly annealed and seasoned. They are next rough-machined and then are further seasoned. The laps are scraped to a master plate, and then ground together. This grinding-in process is similar to the well-known triple method of making surface plates and straight-edges, which is based on the principle that three surfaces cannot fit together interchangeably unless they are true planes. For example, laps Nos. 2 and 3 are first ground in with lap No. 1. Then the order is varied by grinding in lap No. 2 with lap No. 3, and then lap No. 1 with lap No. 2, and so on. In this way, errors which would be transferred from one plate to another if only two plates were made, are eliminated and plane surfaces are generated. The fine grade of Turkish emery used for grinding in the laps becomes embedded in the surface, so that when the lap is finished it is also charged and ready for use. The newly ground laps are used on the finishing machines, and those which have become worn slightly through use are applied to the rough lapping.

Why is the honing process applied to cylindrical surfaces?

The hone-abrading process is applied to the cylindrical surfaces of cylinder bores or other parts requiring extreme precision and also to some external cylindrical surfaces. In honing, several desirable results are combined in the process.　These include: (1) Generating a true cylindrical form by the rapid economical removal of metal necessary to eliminate whatever inaccuracies may have remained from a previous or preliminary machining operation; (2) obtaining a true cylindrical form of given diameter within extremely small dimensional limits; (3) securing a final surface finish of practically any desired quality or degree of smoothness needed for precision work.

The honing tool contains abrasive stones or sticks which

Fig. 6. Honing Machines

vary as to width, length, and thickness. These abrasive stones, as applied to a cylindrical bore, are expanded to bear evenly against it with a pressure that is varied to suit requirements. The honing machine imparts combined rotary and reciprocating motions to the hone which is self-centering in the bore. A true cylindrical form is generated by these combined motions in conjunction with positively controlled expansion and equalized pressure of the honing tool. The cutting action of the hone is also under control to secure required surface quality. Honing tools may be operated either mechanically or hydraulically.

The amount of metal removed by honing varies for different classes of work and also depends upon the accuracy of the preceding operation. For example, from 0.001 to 0.003 inch may be removed from a ground hole; 0.003 to 0.005 inch from a reamed hole, and 0.005 to 0.010 inch from a bored hole. These figures are merely by way of illustration, and much larger amounts can be removed readily when necessary. In some classes of gun barrel honing, as much as 65 cubic inches of material have been removed per hour in correcting, sizing, and finishing the bore. The surface speeds in honing may range from 10 to 250 feet per minute, depending upon the amount of stock to be removed and the final finish desired. An ample and continuous supply of clarified coolant should be used in honing. The kind of coolant depends upon the material to be honed.

Fig. 6 shows two machines arranged for honing the crank-bearing ends of connecting-rod and cap assemblies. Four assemblies are honed at one time by placing them above each other in the multiple type fixture shown, with the piston-pin bores of the individual rods located over separate plugs.

While most honing operations are internal, the process is applicable to external surfaces. An external honing machine is shown in Fig. 7. This machine is used for finishing the piston-rods of gun recoil mechanisms. Diameter and roundness are held within a tolerance of 0.0003 inch on rods, say, 60 inches long, the machine having a capacity for work up to 10 feet in length. About 0.005 inch of stock on the diameter is removed in the rough- and finish-honing operations. The piston-rods are of steel or Monel metal

The honing head floats in a fixed position in the center of the bed, and the work both revolves and reciprocates through the hone. As the hone is mounted in a floating holder, it is free to follow slightly curved rods. The hone is adjustable for diameter, but is of such a design that the abrasive stones are held solidly after being adjusted. Separate heads are used for rough- and finish-honing. There are six abrasive stones in each head.

What is the "superfinishing" process?

The term "superfinish" has been applied to a finish of fine quality obtained by a process developed in the mechanical laboratories of the Chrysler Corporation for application to certain automotive parts. This process may be used in finishing flat, round, concave, convex, and other surfaces. The finish is obtained mechanically and the machines used vary in design to suit the shape and size of the work. The

Fig. 7. External Honing Operation on a Steel Piston-rod 5 Feet Long. The Work Revolves and also Reciprocates through the Hone

finish is produced by stones, which are comparatively hard and of medium grit and operate with a "scrubbing action." Low abrasive pressures, low cutting speeds and constantly changing speeds are factors connected with this process. The object in superfinishing, as in honing, is to eliminate minute scratches and surface defects created by previous mechanical operations and produce a bearing surface in which any remaining scratches will be below the mechanically made bearing surface. This process may be applied to such parts as cylinder bores, pistons, crankshaft bearings, cams, etc.

In the process, ordinary abrasive stones of proper grain and hardness, acting in a suitable lubricant and under sufficient pressures progressively applied, are brought into contact with the metal surface to be superfinished. At least three motions are required to produce superfinish and five or more are desirable. Equipment is in use that has as many as ten motions operating simultaneously. As the result of this multi-motion scrubbing with abrasives, the superfinished surface need have no indentation deeper than a few millionths of an inch.

Superfinished surfaces, in a bearing, can move on each other with a very thin lubricating film and a minimum of clearance. Two metal surfaces are said to be lubricated when there is a sufficient number of oil molecules between them to maintain fluid friction and completely separate the bearing surfaces under a normal bearing load. Projections above the bearing surface of the metal will rupture the oil film, with a resulting bearing failure. Superfinished surfaces have no metal projections above the bearing surface to break the oil film so that complete lubrication can be assured. Superfinish is also applied to advantage on many non-lubricated surfaces.

Is there a precise method of indicating quality of finish?

What is known as the *root mean square* (*rms*) value is a definite indication of surface roughness. All metal surfaces which have been finished by turning, grinding, honing, or other means, consist of more or less minute irregularities. If. from a central reference line, the heights

and depths of these irregularities are measured at equally
spaced intervals, and if the sum of such measurements is
divided by the number of points at which measurements
are made, the resulting average might be used in comparing the roughness of this surface with that of another having an average value obtained in a similar manner. If these
measurements are in micro-inches (1 micro-inch equals
0.000001 inch), and if each measurement is squared, then
averaged as explained, the square root of this average yields
a figure indicative of surface roughness which is much more
convenient to use in practical work. The irregularities on
any surface vary in size and the root mean square average
value gives the larger profile values greater weight than
the smaller ones, which is desirable in a surface profile
roughness number. In recent years the *arithmetical average
(AA)* has largely replaced the use of *rms* values which had
been in use in industry a relatively short time. The difference between *rms* and *AA* values is slight. *Rms* values are
approximately 11 per cent greater than *AA* values for the
same surface roughness.

How is the finish of a surface checked or measured?

Various methods have been developed for checking or
measuring surface irregularities. Some of these are intended primarily for use in the research laboratory or inspection department, whereas others are particularly suitable for use in the shop. The working and even the inspection
tests may only approximate theoretical roughness values
such as the *rms*, in order to simplify the tests and permit
the use of methods suitable for shop use.

Comparison with Standard Specimens: In plants where
surface roughness values are specified, it is essential to
provide some method of checking which may readily be
applied in the shop. Comparison of a machined surface
with a standard finished sample or specimen may be made
by the sense of touch. A simple method is by dragging the
finger-nail first over the standard block and then the "lay"
or ridges of the machined surface. This simple check is
more accurate than might be supposed.

A number of sets of standard surface specimens or finish

samples have been placed on the market. These sets consist of blocks which have surfaces varying from the smoothest to the roughest likely to be required. Some of the more complete sets have from 18 to 25 standard surfaces with a series of nominal *rms* values. Some of these specimen blocks are made of stainless steel.

Comparoscope: With this instrument the finish of an accepted standard for a given class of work is compared with the work itself by optical means. The more numerous and wider the surface irregularities, the brighter the surface will appear to the eye when the work is compared with the master. The images of both appear through the eyepiece of the instrument which contains a dual microscope and identical illumination for both master and work.

Profilometer: In measuring surface roughness with this instrument, the surface irregularities are measured and averaged as a diamond tracer point moves slowly across the surface, and a meter shows by direct reading the average deviation in microinches. The profilometer may be applied to flat, cylindrical or curved surfaces. Different types of tracers are obtainable to permit checking various kinds of surfaces, such as small holes, narrow slots, etc. In measuring unusually smooth surfaces, odd shapes, or any place where only a short stroke can be taken, the tracer point is moved mechanically.

Brush Surface Analyzer: This instrument produces on a graduated chart a graphic ink-line record of surface irregularities showing not only the amplitude, but the form of these irregularities greatly magnified. The surface is traced or explored by a stylus having a point radius of 0.0005 inch. The displacements of this stylus by the surface irregularities are finally recorded in ink on a moving paper chart. The recorded deflections are directly proportional to those of the stylus, but are greatly magnified.

Projection Method: In applying this method, a small piece of clear plastic film is softened by the use of a solvent and then is pressed into the surface, thus transferring the unevenness to the film. The surface pattern on the film is enlarged to 100 diameters or more if required, by means of a projector, thus permitting comparison with another film.

Broaching Internal and External Surfaces

The broaching process may be applied in machining holes or other internal surfaces and also to flat or other external surfaces. Some machine parts are finished by broaching because it is the only practical method. In other cases, broaching is selected in preference to other methods because for certain classes of work, especially in interchangeable manufacture, it is more rapid, and, consequently, less expensive. Broaching, when properly applied, is also very accurate and leaves a finish of good quality. Generally speaking, broaches are expensive tools, but they often make it possible to machine either internal or external surfaces in a few seconds. This explains the extensive use of the broaching process in automotive and other plants where duplicate parts must be produced in large quantities and frequently to given dimensions within small tolerances.

What is the principle of the broaching process?

Broaching consists in cutting away metal to obtain a given form, size and finish by using a broach (or several successive broaches in some cases) having a series of teeth which progressively increase in size or height from the starting end, so that each tooth takes a light cut and thus, by a succession of cuts, forms a surface quickly and accurately. In other words, the shape of the machined surface is a reproduction of the shape of the final cutting edges on the broach. Broaching is applied to many different classes of work. A simple example of internal broaching consists in forming a hole of square, hexagon, or other form from a drilled hole. Originally, broaching was restricted to internal work of this kind and to the cutting of keyways; but now many flat or other external surfaces are machined by this process.

To what general classes of work is internal broaching applied?

Internal broaching is applied in forming either symmetrical or irregular holes, grooves, or slots in machine parts, especially when the size or shape of the opening, or its length in proportion to the diameter or width, makes other machining processes impracticable. The diagrams, Fig. 1, show just a few typical examples of internal broaching.

Broach *A* produces a square hole. Prior to broaching square holes, it is usually the practice to drill a round hole having a diameter *d* somewhat larger than the width of the square. The teeth of the broach gradually cut away the metal, each tooth doing a small part of the work. Broach *B* is for finishing round holes. Broaching is superior to reaming for some classes of work, because the broach will hold

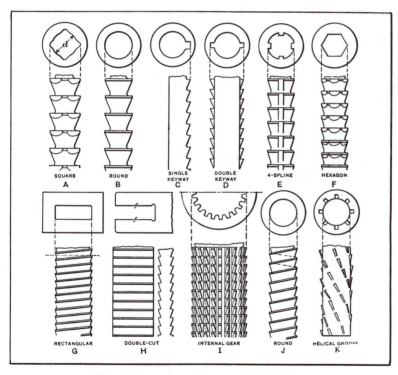

Fig. 1. Examples of Internal Broaching

its size for a much longer period, thus insuring greater accuracy. Broaches *C* and *D* are for cutting single and double keyways, respectively. The former is of rectangular section and, when in use, slides through a guiding bushing which is inserted in the hole. Broach *E* is for forming four integral splines in a hub. The broach at *F* is for producing hexagonal holes. Rectangular holes are finished by broach *G*. The teeth on the sides of this broach are inclined in opposite directions, which has the following advantages: The broach is stronger than it would be if the teeth were opposite and parallel to each other and it takes shearing cuts; thin work cannot drop between the inclined teeth, as it tends to do when the teeth are at right angles, because at least two teeth are always cutting; the inclination in opposite directions neutralizes the lateral thrust. The teeth on the edges are staggered, the teeth on one side being midway between the teeth on the other edge, as shown by the dotted line. The broach shown at *H* is for finishing both sides *f* of a slot. (Usually broaches for flat surfaces have inclined teeth to obtain shearing, overlapping cuts.) Broach *I* is used for forming the teeth in internal gears. It is practically a series of gear-shaped cutters, the outside diameters of which gradually increase toward the finishing end of the broach. Broach *J* is for round holes but differs from style *B* in that it has a continuous helical cutting edge. Some prefer this form because it gives a shearing cut. Broach *K* is for cutting a series of helical grooves in a hub or bushing. Either the broach or the work turns slowly to form the helical grooves, as the broach is pulled through. In addition to the typical broaches shown, many special designs are now in use for performing more complex operations.

What general types of broaching machines are in use?

Broaching machines may be divided into horizontal and vertical designs, and they may be classified further according to the method of operation, as, for example, whether a broach in a vertical machine is pulled up or pulled down in forcing it through the work. Horizontal machines usually pull the broach through the work in internal broaching. External surface broaching is also done on some machines of horizontal design, but usually vertical machines are employed for flat or other external broaching. As a general rule, in broaching holes, the broach is pulled through the work which is securely held in a fixed position. A second method is to push the broach instead of pulling it. This is applicable to certain internal broaching operations which can be performed with comparatively short rigid broaches. A third method that is employed in some cases is to traverse the work past a stationary broach or series of broaches.

What are the principal operating features of broaching machines?

The general function of a broaching machine is to supply the power required for broaching and provide whatever stroke and speed adjustments may be needed. The machine must also be equipped with a suitable work-holding fixture and with means of supplying a cutting fluid to the broach. Modern broaching machines, as a general rule, are operated hydraulically rather than by mechanical means. Hydraulic operation is efficient, flexible in the matter of speed adjustments, low in maintenance cost and the "smooth" action required for fine, precision finishing may be obtained. The hydraulic pressures required, which frequently are 800 to 1000 pounds per square inch, are obtained from a motor-driven pump forming part of the machine and connected with the cylinder containing the broach-operating piston or plunger. Broaching machines for general use are so designed that the length of the stroke can be adjusted to suit the length of the broach. The broach length depends upon the number of teeth needed to remove a given amount of metal. The cutting speeds of broaching machines may be varied for different materials and operations. These speeds frequently are between 20 and 30 feet per minute, and the return speeds often are double the cutting speed or higher, to reduce the idle period.

What is the meaning of the term "surface broaching"?

Broaching originally was utilized for such work as cutting keyways, machining round holes into square, hexagonal, or other shapes, forming splined holes, and for a large variety of other internal operations. The development of broaching machines and broaches finally resulted in extensive application of the process to external, flat, and other surfaces. While all broaching is, of course, applied to surfaces of some kind, external broaching commonly is referred to as *surface broaching* to distinguish it from internal broaching. This application of the term surface broaching may originally have been due to the fact that the finishing of *external* flat or irregular surfaces, and in competition with other machining methods, was a novel feature of the process when external broaching was introduced. External surface broaching is closely allied to the development of hydraulically operated machines. Most surface broaching is done on machines of vertical design but horizontal machines are also used for some classes of work. The part to be surface broached should be heavy enough to withstand the cutting pressures and of such a shape that it can be held rigidly in a fixture.

How is a vertical single-slide surface broaching machine operated?

The exact method of operation varies somewhat with different machines and may also depend upon the type of work-holding fixture. One machine of typical design has the work-holding fixture mounted upon a horizontal table which is moved into the broaching position and then out to a convenient loading position. In operation, the part to be broached is first placed in the fixture. Then the operator, by means of a lever, moves the table in to the broaching position. At this point a valve opens automatically, thus starting the broach slide on its downward stroke. As soon as the broach has completed its work, the table automatically moves out to the loading position. The broach slide then returns to its starting position where it remains until the cycle of movements is repeated. The broaches used on flat surface broaching machines have inclined teeth to provide continuous shearing cuts. The lower teeth take the roughing cuts and remove most of the metal and the upper teeth do the final finishing. The broaches usually are designed to suit a given job.

When two or more parts are broached simultaneously, how is this done?

The method employed depends upon the kind of broaching operation and possibly upon the shape of the part. Multiple broaching is often applied to the smaller classes of work. In some cases, two or more parts are so shaped that they can readily be placed one upon the other in stack formation, so that all are broached at the same time. Another method consists in using two or more parallel broaches, so that an equal number of parts are broached simultaneously. Then the broaches are moved in unison by the slide or ram, and the fixture is designed to locate each part in alignment with its broach.

Figs. 2 and 3 show how two parts are broached during each cutting stroke. The fixture is tilted upward for loading (see Fig. 2). After the operator loads the fixture, he locks the parts in place by means of a cam-lever and presses a push-button. The fixture then rocks downward to center the work pieces over pilots in front of the broaches. This tilting and positioning of the fixture is accomplished hydraulically. Fig. 3 shows the position of the fixture during the cutting stroke, the operation having been almost completed. At the end of the stroke, the fixture swings upward, ready for reloading, and while the reloading is being done, the broaches return to the starting position.

Why are some broaching machines equipped with two vertical slides which operate alternately?

The vertical duplex type of surface broaching machine has two slides or rams which move in opposite directions and operate alternately. While the broach connected to one slide is moving downward on the cutting stroke, the other broach and slide is returning to the starting position, and this returning time is utilized for reloading the fixture on that side; consequently, the broaching operation is practically continuous. Each ram or slide may be equipped to perform a separate operation on the same part when two operations are required. These duplex machines may have receding work-tables or swiveling fixtures which automatically move into the cutting position and out to the reloading position in unison with each ram movement.

Fig. 2. Broaching Fixture which Tilts up for Convenient
Loading and Unloading of the Work

Fig. 3. Hydraulically Operated Fixture Shown in Fig. 2
in the Broaching Position

Fig. 4 shows a close-up view of a duplex machine tooled up for connecting-rods. The fixture at the left is seen in the broaching position, and the fixture at the right in the loading position. When the left-hand ram descends, the right-hand ram rises, and as it reaches its top position, the right-hand fixture swings into place for broaching the newly loaded connecting-rod with the next downward movement of the ram. In the meantime, as the left-hand ram reaches the end of its downward stroke, the left-hand fixture swings approximately 15 degrees to the right to facilitate reloading and to permit the left-hand ram to ascend without interference during the next cycle of the machine. Each cycle consists of a downward movement of one ram and an upward movement of the other. A foot-pedal must be depressed for each cycle. The broach rams are hydraulically operated, and this is also true of the two fixtures.

Each connecting-rod is automatically clamped by block *A* as it is indexed into the broaching position. This block is tapered, so as to enable the rod to slide beneath it, and it swivels to permit a firm seat on the I-beam section of the rod. Sufficient pressure is exerted by the spring-actuated lever *B* to hold the connecting-rod for the operation.

What is the difference between "slab" and "progressive" broaching?

When a surface is broached by the slab method, each broach tooth cuts the full width of the surface being machined, and each tooth is slightly higher than its predecessor, with the result that every tooth takes a light cut over the entire width of the surface. In the progressive method the broach teeth are much narrower than the surface to be finished. The line of teeth is inclined at an angle with respect to the length of the broach, and the direction of its movement, so that the entire width of the work surface is machined as the full length of the broach is traversed past the work; also, in this broaching method, each tooth takes a cut of full depth, so that the work surface is finished at one pass of the broach, as in most broaching operations. Whether the slab or the progressive broaching principle should be adopted for finishing a surface depends upon conditions. Slab broaching is employed when the surface to be

machined is so wide as to prohibit progressive broaching because of the excessive broach length that would be
required.

What are the chief features of the "pull-up" type of broaching machine?

Vertical hydraulically operated machines which pull the
broach or broaches up through the work are used for internal broaching of holes of various shapes, for broaching
bushings, splined holes, small internal gears, etc. A typical
machine of this kind is so designed that all broach handling
is done automatically. When the broaches are in the downward or starting position, they are completely disengaged
from the pulling head. There may be a single broach or
several parallel broaches for multiple broaching. Assuming that two or more broaches are used, each part is placed
over the shank of its broach; then the operator starts the

Fig. 4. Broaching Connecting-rods on Vertical Machine of Duplex Type

hydraulic broach elevator which raises the broaches up to the point where they automatically are connected with the pulling head. After the broaches have been pulled up through the parts being broached, the work drops down and may be deflected into a discharge chute. The broaches next move downward to a point where they are automatically disconnected from the pulling head. Then they are lowered farther by the broach handling mechanism, until the shanks are in position for receiving more unbroached parts. Reversal of the main ram at the top of the stroke may be automatic or under manual control, depending upon the broaching operation.

How are parts broached on a vertical "pull-down" type of machine?

The various movements in the operating cycle of a hydraulic pull-down type of machine equipped with an automatic broach-handling slide, are the reverse of the pull-up type. After the work is in place, the broach-handling slide moves downward to the point where the broaches are connected automatically with the main pulling ram. The upper ends of the broaches may be guided during the cutting stroke until the finishing sections are approaching the work. Then the upper broach-holders are released automatically. When the broaching stroke is completed, the main ram or slide returns the broaches to the point where they are connected with the broach-handling slide, which continues to raise them until there is room beneath for reloading the fixture. The broaches for a pull-down type of machine have shanks on each end, there being an upper one for the broach-handling slide and a lower one for pulling the broaches through the work.

Another design of pull-down machine is not equipped with the automatic broach-handling mechanism. The operator inserts the broach through the work and into the pulling head. After the broached part is removed, the ram is returned to its starting position.

How are splines broached in tapering holes?

Work of this kind may be done by holding the tapering part at an angle and indexing according to the number of

Fig. 5. Helical Splines are Accurately Broached on Machine Equipped with a Long Helically Splined Shaft that Imparts a Rotary Movement to the Broach

splines required. A vertical hydraulic machine designed for this class of work has an automatic indexing mechanism consisting of a rotary work-table, a table slide, and a change-gear box driven by an electric motor. The work-table can be tipped upward to any desired angle to suit the tapered holes in which splines are to be broached.

The machine is equipped with a hardened and ground broach guide above and below the work. The broach travels up and down in the guide during an operation. At the end of each cut, the work is moved away from the broach by a hydraulic cylinder connected to the table slide. At the same time, the indexing mechanism is actuated, indexing the rotary table during the return stroke of the broach. At the end of each upward stroke, the work is again moved toward the broach against a fixed stop, and the broach is pulled downward to cut the next spline. This cycle is repeated automatically until the operation is finished. The fixture is adjustable to suit different diameters, tapers, and lengths of hole, and it can be indexed any required number of times.

In broaching helical grooves or splines, how is the lead of the helix controlled?

The broach has teeth which extend along a helical path and either the broach or the work itself is rotated to suit the lead of the helix. Fig. 5 shows how a horizontal machine is used for broaching helical splines in sliding clutch sleeves. The machine is fitted with a shaft A that is splined to correspond with the lead required. A gear in head B, at the end of the broach ram, slides along the stationary shaft A which imparts a rotary movement to the gear and also to a second gear on the broach ram; consequently, the broach is rotated an amount corresponding to the desired lead as it is pulled through the work. Vertical machines are also equipped with master splined shafts for helical broaching.

Is broaching applicable to external cylindrical surfaces?

One type of broaching machine is so designed that the work revolves about its axis during the broaching operation. This type of machine has been applied, for example, in machining the main bearings of automobile engine crankshafts.

As the crankshaft revolves, the hydraulically operated slide traverses the broaches tangentially past the bearings. There is a broach for each bearing and both the roughing and finishing operations are completed at one stroke. This type of machine is called a "broaching lathe."

How are parts broached by drawing them over stationary broaches?

Although parts usually are broached by traversing the broach itself, some machines are designed to hold the broach or broaches stationary during the actual broaching operation. This principle has been applied both to internal and surface broaching. The description which follows applies to a vertical hydraulically operated machine for internal work. This machine may be equipped with one or more broaching tools to permit either rough- or finish-broaching at the same time, or multiple broaching. Assume that two parts are to be broached simultaneously. The two pieces are placed on the work-table or in a fixture which centers them approximately under the broaches. The starting pedal is then depressed, causing the broaches which are above the work to descend. The shanks at the lower end pass through the parts and engage in mechanical locking sockets. The work-table then automatically rises, thus drawing the work over the stationary broaches. At the top of the stroke, the upper locking sockets disengage the broaches and the table continues upward until the work clears the broach shanks. An ejector bar then moves forward and removes work from the fixture, and the arrangement may be such that the parts fall into a chute or conveyor. The table now descends quickly to the loading position. The upper broach sockets also descend, engage the broaches, and lift them to the starting position. This completes the cycle. During the actual operation, the broaches are securely held at both ends.

How is the rotary principle applied in broaching with stationary broaches?

A surface broaching machine, known as the rotary type, has stationary broaches and a rotating work-table for traversing the parts past these broaches. The broaching tools are mounted in three holders located on the outer rim

of the circular base of the machine. The first holder contains roughing broaches; the next, broaches for an intermediate cut; and the last holder, broaches for finishing.

What is the operating principle of a continuous or chain type of surface broaching machine?

This type of machine is equipped with stationary broaches
and several work-holding fixtures which are held on an end-
less chain. As the part to be broached passes through a
horizontal "tunnel" containing the broach or broaches, the
machining operation is completed. The operator merely
places the work in the fixtures at a loading station. The
work clamping and unloading is automatic. As each fixture
passes through the broaching tunnel, it is rigidly supported
by hardened steel guides. This special type of machine is
for very high production and is applicable either to flat or
to irregular surfaces which can be broached by straight-line
movement.

In surface broaching cylinder blocks, what methods are employed?

One hydraulically operated horizontal type of machine is
designed to broach the bottom face and also the bearing cap
seats. (This is the operation illustrated in Fig. 6.) Another
machine of this same general type is used to broach the top
or cylinder head surface. The large broaching tool used is
of the inserted-blade type and it has tungsten-carbide finishing blades. About 3/16 inch of metal is removed and the
broach leaves a flat smooth and accurate surface.

Another special machine is designed to broach the upper
and lower surfaces of cylinder blocks. The hydraulically
operated broaching ram is equipped with two sets of
broaches which operate in opposite directions to broach first
the bottom or pan-rail surface of a cylinder block held in
the right-hand fixture, and then the top or cylinder-head
joint surface, the valve-chamber cover pad (at right angles
to the top), and an angular pad on one side of a second casting held in the left-hand fixture. Both the fixtures swing
through 180 degrees to carry the work from the loading
position to the broaching position and vice versa. Between
the two fixtures is a power-operated mechanism that re-

ceives the casting from the first-operation fixture and turns it through 180 degrees to bring the top uppermost, into the proper position to enter the second-operation fixture. Each fixture is alternately swung into and out of the broaching position in synchronism with the ram movements, so that broaching is practically continuous. While the casting in one fixture is being broached, the other fixture is being reloaded. The time per cylinder is only about one minute.

Why are hydraulic presses used for many broaching operations?

Broaching is often done on vertical hydraulic presses which are also designed for other operations such as assembling bushings or other tight-fitting parts which must

Fig. 6. Special Machine for Broaching Surfaces of Cylinder Blocks

be pressed or forced together. These presses are simple in design, less expensive than regular broaching machines, and may be preferable for the simpler classes of push broaching or where a press is needed both for broaching and assembling. Hydraulic presses ordinarily are used for internal broaching, but they may also be equipped for surface broaching. A typical design of broaching and assembling press has a hydraulically operated ram that is controlled either by hand-lever or foot-pedal. When the foot-pedal control is used, the operator's hands are free for inserting and removing work.

In making broaches what are the important features?

Broaches are designed especially for a given class of work, as a general rule. The depth of cut per tooth, the pitch or spacing of the teeth, shape of the teeth, the kind of steel used for the broach and its heat-treatment, are all very important. Broaches are made from carbon steel and from certain types of high-speed or alloy steels. An important feature of broaching is that the depth of cut per tooth is very light and may only be 0.001 or 0.002 inch, depending upon the material. The cutting or roughing teeth usually are followed by several finishing teeth which take extremely light cuts so that broaches ordinarily will finish a great many parts accurately before sharpening is required.

The pitch of the teeth must be large enough to provide room for the chips formed during one stroke; but unnecessary pitch or spacing will unduly increase the length of the broach, and, consequently, the length of the working stroke and the broaching time. In designing a broach for a given job, the number of teeth is governed by the total amount of metal to be removed and the depth of cut allowed per tooth. The over-all length must, of course, include the shank which engages the pulling head and a pilot or plain section at the finishing end.

The relation between the length of the broached part and the pitch should be such that from three to six teeth will be cutting at the same time. In broaching very thin parts, several may be held in a stack, thus increasing the total length so that it spans two or more teeth. (This stacking of parts may also be done primarily to reduce the broaching time per piece.) Most broaches are made from a solid piece of steel, but the sectional type is preferable for some applications. A number of sections may be attached to a holder or bar. A worn broach frequently can be restored by adding a new section to the finishing end, advancing all the other sections and then regrinding the entire broach. Broach-making is a highly specialized business and it should be done only by those who have had wide experience in the application of broaches to various classes of work.

Why are burnishing broaches used?

Broaches for burnishing holes have polished rings of rounded cross-section, and these rings compress the metal instead of cutting it. The burnishing operation, which is for final finishing only, is applied to certain classes of bronze sleeves or bearings and also to "white metal" bearings, in order to produce a dense, smooth and durable surface. Burnishing broaches having rings of solid tungsten carbide have proved to be much more durable than those made entirely of tool steel and chromium plated. One broach of this type has eight tungsten-carbide rings. The first six tungsten-carbide rings do the actual swaging, each succeeding ring being ground 0.0015 inch larger in diameter than the preceding ring. The last two tungsten-carbide rings, as well as tool-steel rings which follow, "size" the hole. This burnishing broach expands the bearings 0.009 inch, but the hole contracts 0.001 inch after the burnishing operation.

How are broaches sharpened?

As previously pointed out, broaches are so designed that each tooth takes light cuts; consequently, broaches frequently will produce a large number of duplicate parts before sharpening is required. Dull broaches should not be used because they require an excessive amount of power for the broaching operation and may also leave a poor finish on the work. The front faces of broach teeth, which are the surfaces against which the chips bear as they are being severed, usually have some rake, which may be designated either as the "face angle" or "hook angle." The top surface adjacent to the front face is known as the "land." This land usually has a flat or straight section that is parallel to the axis, especially at the finishing end of the broach. This straight section is very narrow and may not be over 1/64 to 1/32 inch wide originally. It is followed by a relief or clearance angle of 1½ or 2 degrees.

Broaches for internal work should be sharpened usually by grinding on the front face only so that the diameter, width, or other size is retained, at least until all of the flat land at the finishing end has been ground away. When internal broaches are applied to metals which are abrasive in character, the width of the straight land may increase so much due to wear, that it cannot be reduced sufficiently by grinding the front faces. In that case, the land width is further reduced by grinding on the back of the tooth. Broaches for external surfaces are ground on the faces and also on the lands, when necessary. When the lands are ground as well as the front faces, the back surfaces or relief areas should also be reground to reduce the land widths. Special machines have been designed for broach sharpening. These machines are arranged to rotate internal broaches of circular cross-section. Flat broaches for external surfaces are sharpened by traversing the wheel across the teeth.

Chipping, Filing, Scraping and Hand Grinding

In producing machine parts, chipping and filing are not done as much as formerly because the improved tools now in use have largely eliminated such work, although at times even in a modern shop it is necessary to resort to the hammer, chisel and file, particularly in erecting and general repair work. Hand finishing by filing or grinding is often required in making certain classes of dies, molds, etc., especially when there are cavities or forms which cannot be finished entirely by mechanical means. Portable power-driven chipping, filing and grinding tools have been developed which greatly facilitate these hand-finishing processes.

What are the common types of metal-cutting chisels?

The various types of metal-cutting or "cold" chisels commonly used for chipping are shown in Fig. 1. The flat chisel is used for a general class of work. The cutting edge is either ground straight for light work or made slightly convex for heavy chipping to prevent the corners from breaking. The included angle at the end should be about 60 to 70 degrees, although a greater or less angle is advisable when the metal is either exceptionally hard or soft. The side chisel differs from the flat type in that it is ground and beveled on one side only, which permits it to be used on surfaces which could not be reached with a double-angle end; it is also used for chipping the sides of keyways, slots, etc. A cape chisel has a narrower cutting point than a flat chisel, and is used principally for cutting grooves, etc. The diamond point is adapted to chipping V-grooves, squaring corners, etc., while the grooving chisel is for cutting oil grooves or for similar work. The round point is another form of grooving chisel. The half-round gouge, as its shape indicates, is used on curved surfaces.

440

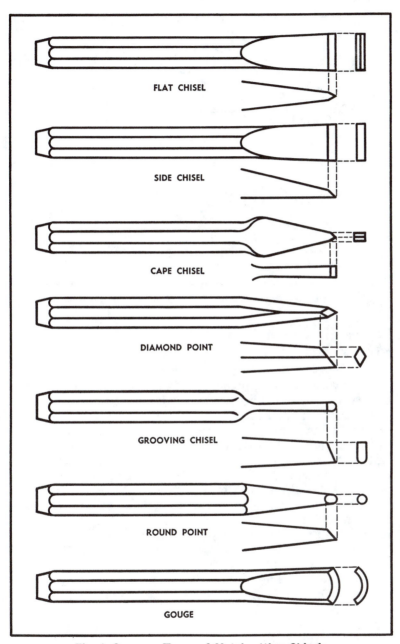

Fig. 1. Common Types of Metal-cutting Chisels

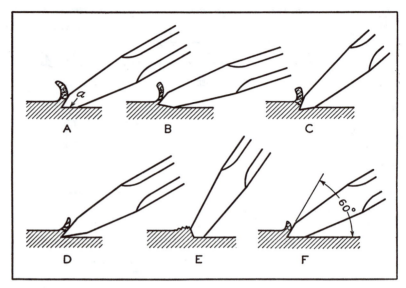

Fig. 2. Effect of Position and Form of Chisel on Cutting Action

In chipping with hammer and chisel what is the general procedure?

Most workmen who are not accustomed to chipping proceed very awkwardly, principally because of fear of striking the hand, and in order to avoid this, the eyes are continuously fixed on the head or striking end of the chisel. It is not necessary, however, to look directly at the chisel head in order to hit it; in fact, it is important to continually watch the cutting end, as the chip being removed determines whether the chisel should be raised or lowered to increase or decrease the depth of the cut. The hand holding the chisel should be quite close to the striking end to obtain better control, and it is well not to grip the chisel shank too firmly with the thumb and forefinger, as an accidental blow on the relaxed muscles will be less painful than when the chisel is gripped tightly by the entire hand.

The way in which the position of the chisel alters the depth of the cut is illustrated by the diagrams in Fig. 2. As each successive blow is struck, the chisel tends to move along the plane of the supporting surface a (diagram A), so that lowering the striking end causes the point to move upward as at *B*, and raising it has the opposite effect, as illustrated at *C*.

The angle of the cutting point illustrated by diagram *D* is too small for general use because this acute form is comparatively weak. On the other hand, the point shown at *E* is too blunt. While this blunt form is strong, it tends to push the metal rather than cut it and is similar in principle to a turning tool which has negative rake. For general all-around use, an included angle of 60 degrees, as shown by diagram *F*, is recommended.

It is advisable before chipping a surface that needs to be carefully finished to chamfer the corner where the cuts end to prevent the stock from breaking out below the depth line.

The ball peen hammers commonly used for chipping weigh from 1 to 1½ pounds. This weight does not include the handle, which is usually from 12 to 14 inches long and made of hickory.

Why are the battered "mushroom" heads of chisels dangerous?

The heads of cold chisels, sets, punches, and other percussive tools of the class driven by hammers and sledges often batter and break away under repeated blows. The bits of steel displaced and sent flying by the hammer blows are often causes of painful accidents. The eyes are especially vulnerable, and in foundries, steel works, and machine shops especially, many workmen have lost the sight of an eye by being struck with a bit of steel broken from a chisel head. If goggles with safety glasses were always worn, the danger to the workmen actually using the chisel would be practically eliminated, but mushroom heads are a source of danger to others near by. If the battered ragged heads were ground off regularly, the danger of steel flying off and causing accidents would be small. The danger of ragged heads can be eliminated by hardening and tempering the heads of chisels etc., as carefully as the cutting parts.

What are the advantages of chipping with a pneumatic hammer?

When a pneumatic hammer is used for chipping, the continuous cutting action makes it possible to take a smoother cut and to work more rapidly than is possible by hand chipping. Whenever there is considerable chipping to be done, a pneumatic hammer should be used if one is available. The chisel, with the shank inserted in the hammer, is held in the left hand and the hammer with the right. By regulating the admission of air to the cylinder, the speed of the hammer blows is controlled. When beginning a cut, the hammer should be started gradually so that the movement of the chisel may be more easily controlled. Both hammer and chisel should be held firmly against the work and at an angle which will depend upon the depth of cut desired. When the hammer is working at full speed, the cutting movement of the chisel is almost continuous, owing to the rapidity of the blows.

Air Chipper.—The small pistol-shaped "Kipp air chipper" shown in Fig. 3 may be used effectively, not only for tool and die work, but for metal and wood pattern making, and

Fig. 3. Kipp Air Chipper which Delivers 6000 Blows per Minute

for many special riveting, chipping, and grooving operations in the mechanical industry. This air-driven chipper delivers 6000 blows per minute to the chisel. The resultant rapid pulsating effect provides an even flow of power when the chisel is pressed against the work by the operator. The intensity of the blows is regulated by the pressure the operator exerts against the chisel, and also by the position of the trigger valve; that is, lighter cuts may be taken with less pressure on the handle and with less pressure on the trigger valve. The operator, therefore, has extremely close control of his work so that he may work to a line and stop at a line.

In making cold chisels, what kind of steel is used?

Chipping chisels have proved satisfactory when made from steel containing from 0.75 to 0.85 per cent of carbon; 0.30 per cent of manganese; 0.10 per cent of silicon; and with a percentage of sulphur not exceeding 0.025 per cent; and of phosphorous, not exceeding 0.025 per cent. A chisel which has given excellent service in practice, has been analyzed and was found to contain carbon, 0.75 per cent; manganese, 0.38 per cent; silicon, 0.16 per cent; sulphur, 0.028 per cent; and phosphorus, 0.026 per cent.

How are cold chisels hardened and tempered?

Chisels made from steel as specified in the preceding paragraph, should be heat-treated as follows: The chisels are carefully heated, preferably in a gas-fired or electrically heated furnace, to a temperature of about 1350 degrees F. When uniformly heated, the chisel is quenched to a depth of from 3/8 to 1/2 inch from the point, in water, and then the whole chisel is immersed and cooled off in a tank containing linseed oil. After this treatment, the chisel will be found to have a dead hard point and a tough structure for the remaining portion. The point is then tempered by immersing the chisel into an oil tempering bath, heated to about 420 or 430 degrees F. The recommended method is to first immerse a batch of chisels in a bath heated to 420 degrees F., and then to gradually raise the heat until a temperature close to 430 degrees F. is reached. This insures that the chisel will not be immersed or drawn at too high a temperature. When removed from the bath the chisels are permitted to cool in the air.

How are files classified?

Files are classified according to their shape and also according to the pitch or spacing of the teeth and the nature of the cut or method of forming the teeth. Various standard shapes of files are illustrated in Figs. 4 to 7. These illustrations give the names which are commonly used to designate different classes of files, and they also show the cross-sectional shapes as well as the form in a lengthwise direction.

What is the difference between a single-cut and a double-cut file?

There are three general classifications of files according to the cut—single-cut, double-cut, and rasp. The *single-cut* file (or "float," as the coarser cuts are sometimes called) has single rows of parallel teeth extending across the face at an angle of from 65 to 85 degrees with the axis of the file. This angle depends upon the form of the file and the nature of the work it is intended for. The *double-cut* file has two rows of teeth crossing each other. The angle of the first row is, for general work, from 40 to 45 degrees, and the second row, from 70 to 80 degrees. The angle of the first cut for double-cut finishing files is about 30 degrees and the second cut, from 80 to 87 degrees. The double-cut gives a broken tooth, the surface of the file having a large number of small teeth inclining toward the point and resembling, in shape, the end of a diamond-pointed cold chisel. The second or "up-cut" is usually a little finer and not as deep as the first or "over-cut." The difference between a single and double-cut is shown in Fig. 4, the "hand" file at the top of the illustration being double-cut and the "flat" file, single-cut.

How is the spacing of file teeth designated?

Single- and double-cut files are further classified according to the spacing of the teeth. The names commonly used to designate the different grades of cut are "rough," "coarse," "bastard," "second-cut," "smooth," "dead-smooth," or "super-smooth." "Rough" files are usually single-cut, and the "dead-smooth," double-cut. The other grades are made in both double- and single-cuts; therefore, file teeth may be classified as follows:

Single-cut: Rough, coarse, bastard, second-cut, and smooth.

Double-cut: Coarse, bastard, second-cut, smooth and dead-smooth.

These degrees of coarseness are only comparable when files of the same length are considered, the number of teeth per inch of length decreasing as the length or size of the file increases. Some makers use a series of numbers to designate the cut or coarseness instead of names. The number of teeth per inch varies considerably for different sizes and shapes and on files of different makes.

What are the distinguishing features of a "hand file"?

Hand files are parallel in thickness from the heel (end of file body next to tang or handle) to the middle, and are tapered, as to thickness, from the middle to the point, the latter being about one-half the thickness of the stock (see Fig. 4). The edges of the file are usually parallel throughout the entire length but are sometimes drawn in slightly at the point. The hand file is ordinarily preferred by machinists for finishing flat surfaces. The teeth are usually double-cut, bastard, although many files of this type have teeth of second-cut, smooth, or dead-smooth.

How does a "flat file" differ from a "mill file"?

Flat files are parallel in both longitudinal sections, from the heel to the middle, and tapered in both sections from the middle to the point, the thickness of the point being about two-thirds and the width about one-half that of the stock from which the file is made (see Fig. 4). The flat file is one of the most common files in use and is not confined to any specific class of work, but is employed for a great variety of purposes. Ordinarily, the teeth are double-cut and either bastard, second-cut, or smooth. A single-cut flat

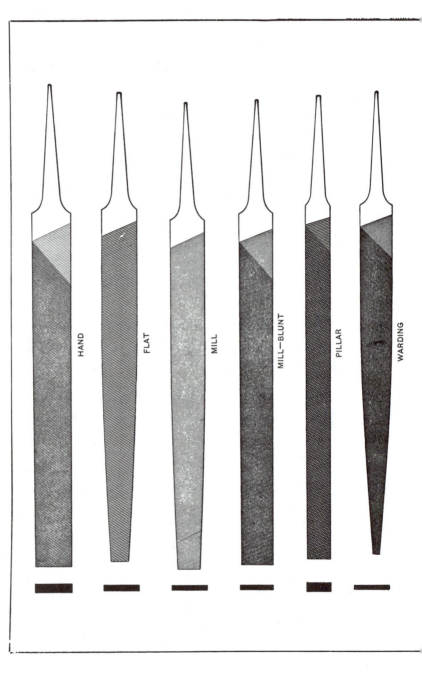

Fig. 4. Standard Shapes of Files

HAND

FLAT

MILL

MILL—BLUNT

PILLAR

WARDING

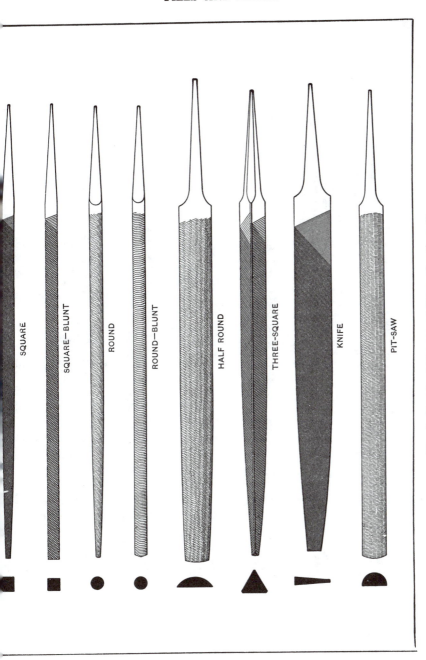

SQUARE

SQUARE—BLUNT

ROUND

ROUND—BLUNT

HALF ROUND

THREE-SQUARE

KNIFE

PIT-SAW

Fig. 5. Standard Shapes of Files (Continued)

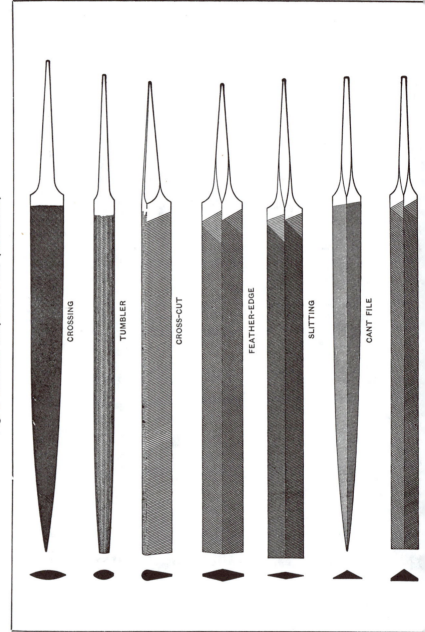

Fig. 6. Standard Shapes of Files (Continued)

CROSSING

TUMBLER

CROSS-CUT

FEATHER-EDGE

SLITTING

CANT FILE

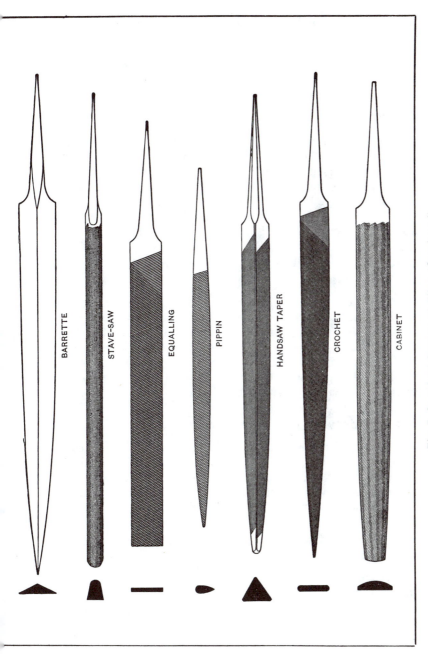

Fig. 7. Standard Shapes of Files (Continued)

file, like the one illustrated, is preferred for some classes of work.

Mill files are parallel in thickness from the heel to the point and usually tapered so that the width at the end equals about three-fourths the width of the stock. The mill file is also made "blunt" or of equal width and thickness throughout its length. Both the tapered and blunt forms are shown in Fig. 4. Quite a number of mill files are used having one round edge, and some are provided with two round edges. The teeth are ordinarily single-cut, bastard. This file is used in machine shops for lathe work, draw filing, and, to some extent, for filing brass and bronze. It is also employed for sharpening metal saws, etc. The mill files of the round edge type are used for filing the gullet or space between saw teeth.

How does a "pillar file" differ from a "hand file"?

Pillar files are parallel as to width, but taper somewhat in thickness toward the point. The cross-section of a pillar file is similar to that of a hand file, except that it is thicker in proportion to the width; these files are made in narrow and extra narrow patterns. They are double-cut and are applicable to general machine shop work, especially in connection with erecting and fitting.

Why are "warding files" so named?

Warding files are considerably tapered in width from the heel to the point (see Fig. 4) and are of uniform thickness. Files of this class are used considerably by machinists, but more especially by locksmiths for filing the ward notches in keys; hence, the name "warding" file.

Are square files tapering or uniform in cross-section?

Square files either taper from the middle toward the point or are made of uniform cross-section throughout. The taper square file has double-cut, bastard teeth, and is extensively used in machine shops generally, principally for enlarging apertures of a square or rectangular shape. The blunt form also has double-cut bastard teeth and is employed by engine builders and in the shops of railroads, ship-yards, etc., for the rougher work in finishing or enlarging mortises, key-ways, or splines, especially when of considerable length.

What types of files are of circular cross-section?

Round files (also called "rat-tail") are made both in the taper and blunt forms, and the cut is mostly bastard. Round files are used either for enlarging round holes or shaping internal surfaces for which quadrangular sections would be unsuitable. The blunt shape is ordinarily used for the heavier classes of work. Both square and round files are made in "slim" forms, which are of regular length but of smaller cross-section.

Gulleting files are also made of round section in the blunt shape. These are single-cut and are used principally for extending the gullets of the teeth of what are known as the "gullet-tooth" and "briar-tooth" saws. There is little demand for files of this type.

Half-round files are not a complete semi-circle, as the name implies, the arc being about one-third of the circle. Files of this class are double-cut and mostly bastard, although many are either second-cut, smooth, or dead smooth, the latter being used to a limited extent. Those having teeth finer than bastard are cut single on the convex side. This type is extensively used in machine shops, especially on curved surfaces.

Pit-saw files are a full half circle in section and some-times referred to as frame-saws. The form is blunt and the teeth single-cut, second-cut. These files are used for filing the teeth of what are known as pit-saws and frame-saws (see Fig. 5).

What types are triangular in shape?

Three-square files are made in taper, slim, and blunt forms. They are double-cut, mostly bastard, and used quite extensively for filing angular surfaces, and for many other purposes. The three sides are of equal width, the angles between them being 60 degrees (see Fig. 5).

Cant-files and *cant-saw files* differ in cross-section as to their angles; the cant-file has 30, 30, and 120 degree angles and the cant-saw, 35, 35, and 110 degree angles. The cant-saw shape was formerly known as "lightening." It is used principally for filing cross-cut saws having M-shaped teeth (Fig. 6).

Handsaw files (Fig. 7) have the same section as a "three-square" type but differ in that the edges are given the proper bluntness to insure durability; the three-square files have comparatively sharp edges so that they are entirely unfit for filing saws. While the term "taper" is commonly used to denote a file which tapers in a lengthwise direction toward the point, custom has also established the term "taper" as a short name for the three-square handsaw file. One class of handsaw file is tapered to a small point and the teeth are single-cut, second-cut. These files are very extensively used for sharpening handsaws. Some saw files are double-cut, second-cut, these being preferred by some for filing fine-tooth saws.

Slim handsaw taper or *slim taper files* are like the ordinary handsaw files, but considerably lighter. They have largely superseded the regular handsaw files, the principal advantage being the greater sweep or stroke obtainable from the same section. There is also the *extra-slim taper* which is of lighter stock than the slim taper. Handsaw files are sometimes made in a blunt shape.

The wedge or knife shape is applied to which type?

Knife files (Fig. 5) derive their name from the fact that they resemble somewhat the blade of a knife. The section is tapering toward one edge and in a lengthwise direction toward the point. The teeth are double-cut, mostly bastard; this type is used quite generally for many purposes to which the knife shape is adapted.

Ginsaw files are also of knife shape and single-cut. This type has been supplanted, to a considerable extent, by the three-square ginsaw file, which is made either tapering or blunt of hand-saw slim steel, and is used for filing cotton ginsaws.

Cross-cut files (Fig. 6) have one round edge and the sides are tapered toward the opposite edge. They are single-cut, the same as mill bastard files of the same size.

Featheredge files taper in cross-section from the center toward each edge. They are of blunt form, double-cut, bastard, second-cut, or smooth. This shape is seldom called for, as the knife file is generally used instead.

Slitting files are similar to the featheredge but the taper is less abrupt and the edges are sharper.

In general practice what grades and shapes of files are used?

The files most commonly used in general practice are 12- and 14-inch, flat and half-round double- and single-cut files in bastard, second-cut, and smooth grades. The coarse and bastard cuts are generally used on coarser grades of work, and the second-cut and smooth, on comparatively fine work. The coarse and dead-smooth cuts are not often used in ordinary practice, although a rough single-cut is sometimes needed for soft material. The dead-smooth, double-cut file is occasionally required for producing very fine surfaces. Single-cut mill files are commonly employed for lathe work. The mill-bastard is adapted to a large variety of lathe filing, and a mill second-cut is used on finer classes of work. The double-cut hand file with one safe-edge is used for finishing flat surfaces. The grade of cut is mostly bastard, although many second-cut and smooth are employed. The double-cut square-bastard in both taper and blunt forms is widely used for enlarging or truing rectangular slots, etc. Flat files are mostly double-cut bastard types, second-cut and smooth being used less frequently. Pillar files with one or both edges safe are useful on narrow work. Special files are sometimes used, but the ordinary standard shapes and cuts will usually meet all requirements.

Sharper files are required for cast irons, brass, and copper than for steel and wrought iron. Broad surfaces also require sharper files than narrow ones. On very thin work, the teeth of a double-cut file bite so freely that the danger of breaking them is great. For work of this character, the long tooth of the single-cut is best adapted, as its form gives it greater strength, and the shear of the cut is smoother, one tooth commencing to cut as another leaves. On broad surfaces, however, the teeth of the double-cut have the advantage.

Does the nominal length of a file include the tang?

The *length* of a file means the distance from the point to the heel and does not include the tang or pointed end which enters the handle. The *heel* is that end of the file body adjacent to the handle.

What is the meaning of "safe-edge" as applied to a file?

Safe-edge means that the edge or side is smooth and without teeth, and may be presented to a surface that does not require filing.

How should a file be held?

When moving a file endwise across the work (which is commonly known as *cross-filing*), the handle should be grasped in such a way that its end fits into and against the fleshy part of the palm opposite the joint of the little finger, with the thumb lying along the top of the handle in a lengthwise direction. The ends of the fingers should point upwards or nearly in the direction of the workman's face. The point of the file should be grasped by the thumb and first two fingers, the hand being in such a position that the ball of the thumb presses upon the top of the file, the thumb being approximately in line with the handle when heavy strokes are required. For lighter strokes, and when less pressure is necessary, the position of the thumb and fingers may be changed so that the thumb lies about at right angles to the file. Those not accustomed to filing will find that the operation will be less difficult if the file is held properly and not grasped at random, as is often done.

What variations in hand pressure are required in filing a flat surface?

To file a surface flat is difficult, especially if it is narrow. While one becomes proficient in work of this kind only as the result of practice, still a hint as to the proper method of procedure should be helpful. When the file begins its stroke, the downward pressure exerted by the left hand holding the outer end should be maximum while a minimum pressure is given by the right hand. As the file advances, the pressure from the left hand decreases while that from the other, increases. This variation in pressure is necessary, because at the beginning of the stroke most of the file extends beyond the surface on one side, whereas at the end of the stroke it overhangs on the other side. After considerable practice, it is possible unconsciously to regulate the pressure on each end of the file so that any excessive rocking motion is prevented. The most natural movement of the hands and arms when filing is to carry the file in circular lines.

When filing oval surfaces or irregular forms, the movements, while not considered so difficult as when filing flat surfaces, nevertheless require considerable practice, in order to blend the strokes of the file upon the round or curved surfaces so as to obtain the required form. Whatever the nature of the work, the pressure on the file should be relieved during the back or return stroke. The teeth are so formed that they can only cut in one direction and, if dragged back over the surface, the sharp points will be considerably damaged.

When a large amount of metal is to be removed quickly, the file may be used at different angles, as shown in Fig. 8. This decreases the area of cut and increases the bite of the file. A new file should never be used for this purpose, as the keen edge of the teeth would be broken off. All surfaces to be filed, especially if narrow, should be held as near the top of the vise jaws as possible, thus preventing vibration.

Are half-round files always used on concave surfaces?

For filing a concave surface, an ordinary half-round file will do for roughing; but a square file (as shown at *A*, Fig. 9) is much to be preferred, as it will "lay to the surface," and make it possible to get the surface straight crosswise with much less effort than with the half-round file. The file should be held straight across, and swept round the curve while it is being advanced, sweeping it both ways so as to insure a smooth curve. The file cuts only on its corners while being used in this way, but it is surprising how much work these corners will do, how little pressure

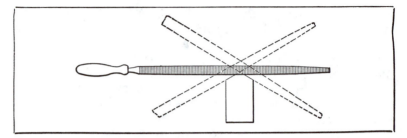

Fig. 8. Method of Removing Metal Quickly by Filing

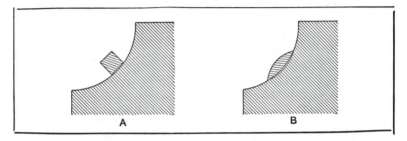

Fig. 9. Filing Curved Surfaces

is required to make the file cut, and how easy it is to keep it from rocking. The crossing file is preferred by many mechanics to the square file for this kind of work, and is generally considered the proper form to use. It should be used with the flattest side next to the surface, or, in other words, the radius of curvature of the file should be *greater* than that of the surface being swept out, as indicated at *B*. The difference between the crossing file and the square file, when applied as shown in Fig. 9, is that the former has more bearing surface on the work. The essential point is that the file has a bearing along its two edges, and *not in the center*. After the piece has been brought to shape and dimensions by cross-filing, it can be finished by draw-filing, using a half-round or crossing file with a radius of curvature *less* than that of the surface.

What is draw-filing?

When a file is held at each end and the motion is sidewise rather than in a lengthwise direction of the file, this is known as *draw-filing*. With this method of filing, the metal is removed more slowly than by cross-filing, provided the same kind of file is used in each case. The surface is left smoother, however, if the draw-filing is properly done, as the scratches are closer, owing to the shearing or shaving cut taken by the file teeth. Files as ordinarily made are intended to cut when used with a forward lengthwise stroke, and the same file cannot work smoothly and to the best advantage when moved sidewise, unless care is taken that the faces of the teeth during the forward movement of the file are presented to the work at a sufficient angle to cut the metal instead of merely scratching it. In order to secure this result, the angle at which the file is held with respect to its line of movement must vary with different files, depending upon the angle at which the last or up-cut is made. In draw-filing, the pressure should be relieved during the back stroke the same as in ordinary filing.

When file teeth become clogged with particles of metal, how are they cleaned?

Sometimes particles of metal collect between the teeth of a file and make deep scratches as the file is passed across the work. This forming of tiny particles between the teeth is commonly known as "pinning" and can sometimes be avoided partially, if not entirely, by rubbing chalk on the file. When these tiny particles of metal are lodged too firmly between the teeth and cannot be removed by tapping the edge of the file against the bench, they should be removed either by using a wire brush or file card (which is drawn across the file in the direction of the teeth) or some other form of cleaning device. When the particles are not lodged too tightly, they may be removed by pushing the end of a stick along through the tooth grooves. In many cases, particles cling so tightly that they must be removed by other means. A piece of No. 1 or No. 2 copper wire having a flattened end is often used. Sheet fiber is also employed for this purpose. If a disk of this fiber, about 6 inches in diameter, is mounted upon an ordinary grinding wheel spindle, files may be cleaned very rapidly by holding them against the revolving disk. Fiber about $\frac{1}{4}$ inch thick is preferable. Clogging of the file teeth is especially marked

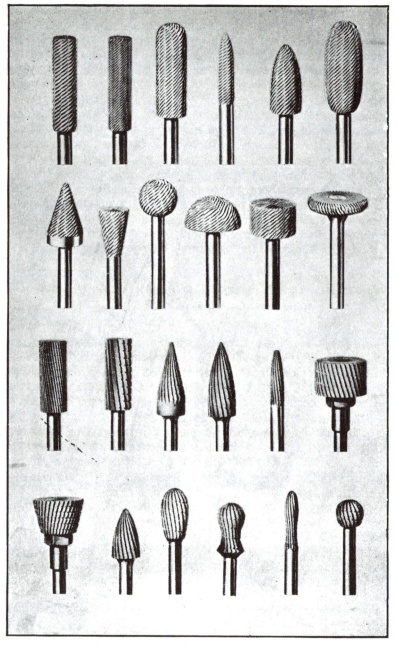

Fig. 10. Typical Shapes for Rotary Files

when the file is used upon wood, horn, or similar materials. When the teeth are clogged with substances of this kind which have become baked and dry, thus preventing the teeth from penetrating the work, the file may be cleaned by holding it in boiling water a few minutes in order to loosen the material prior to carding out the teeth with the file brush. If this operation is quickly performed, the moisture remaining on the file will be evaporated by the heat retained in the file. Files should be cleaned not only at intervals while being used, but before being laid aside. The file card is preferable for all-around use.

At what height should work be held for filing?

For filing in a vise, the vise jaws should be level with the workman's elbows. This height varies from 40 to 44 inches from the floor; hence, 42 inches is a good average height for a vise that is fixed permanently. With the work in this position, the workman is able to get a full swing of his arms from the shoulder. The independent movement of the wrist and elbows should be avoided as much as possible. When parts to be filed are small and delicate, requiring simply a movement of the arms, the vise should be somewhat higher, so that the workman may stand erect and see the work to better advantage. If the parts to be filed are heavy, thus requiring considerable effort, the surface should preferably be below the elbow joint, so that the operator may stand farther away from the work and in a slightly stooping posture. For this class of work, it is desirable to apply some of the weight of the body to the file in order to make it cut; moreover, the arms should be quite rigid, the file being moved largely by the momentum of the body. When filing broad surfaces, the work should be placed quite low, thus enabling the operator to readily file all parts of the surface and put the required pressure on the file.

For what classes of work are power-driven rotary files used?

Power-driven files may be used in making dies, templets, etc. Files and burrs of the rotary type (see Fig. 10), are made in either cylindrical, conical, spherical, concave, or special shapes for finishing the edges or surfaces of punches, dies, metal patterns, and various other classes of work. The file or burr may be rotated by inserting it in a drilling machine spindle, as for finishing the edges of punches or dies, or by using a flexible-shaft or any other drive permitting hand manipulation for varying the position of the file. The twelve rotary files shown in the upper half of Fig. 10 are hand cut and the twelve in the lower half are cut by milling. Most rotary files are made of high-speed steel.

When are power-driven files given a reciprocating motion?

For certain classes of die and tool work, a power-driven reciprocating file is preferable to a rotary file. One hand manipulated type of air-driven filer imparts to the file a very rapid reciprocating motion, the speeds ranging from 500 to 5000 strokes per minute, depending upon the amount of air admitted by the trigger valve. Different shapes of files are used, and they can be quickly inserted or removed to suit the work. The operation of the tool is similar to that of the air chipper shown in Fig. 3. This filer has been found suitable not only for filing work, but also for honing and other operations where rapid strokes approximately ¼ inch long are to be made.

Vertical filing machines are used by diemakers for operations such as filing the openings in blanking dies. One design of die-filing machine has a table which is adjustable about two axes at right angles to each other so that it can be tilted for filing clearance in dies, as well as for other angular work. A screw feed, operated by hand, is provided for feeding the work against the reciprocating file. The stroke of the file is adjustable from zero to 4 inches and it is arranged to clear the work on the return stroke. The cutting stroke can be either on the upward or downward movement, this change being effected by simply shifting the crankpin to the opposite side of the crank-arm. One advantage of the die-filing machine, as compared with hand filing, is that straight or flat surfaces can be filed without difficulty because the file is mechanically guided and moves in a straight line, whereas, when filing by hand, it is difficult to do the work accurately.

Why are some machine parts finished by scraping?

The object in finishing a machined surface by a hand scraping operation may be either to obtain an ornamental effect or to correct slight errors, as, for example, on a planed, milled, or bored surface. Modern machine tools produce such accurate surfaces that scraping to correct errors is seldom required. Assume, however, that surfaces are to be scraped in order to obtain a more perfect fit or greater area of contact between these surfaces. In correcting the slight errors or high spots, the part to be scraped ordinarily is applied to whatever surface it is being fitted; the bearing marks or "high spots" are then noted and removed by scraping. By repeatedly obtaining these bearing marks and then removing them, a more evenly distributed bearing is secured. In this way, bearing boxes are often fitted to their shafts after having been bored. When flat surfaces are scraped to make them more accurate, first apply the work to a standard surface plate, note the bearing marks and, if there is unevenness, correct the error by scraping.

In fitting two flat parts together, it is common practice to first scrape one member to secure as true a surface as possible, and then use it as a standard while fitting the other part. In order to make the bearing marks show clearly, some kind of red or black marking material is generally used. A thin coating is applied to the bearing shaft, surface plate, or whatever surface the work is to be scraped to fit. The work is then rubbed over this surface and the marking material shows just where the high spots are. It is important to keep the marking material in a covered box in order to exclude all grit or chips. The scraper should be made "glass hard" and be given a fine edge by the use of an oilstone.

What forms of scrapers are used?

The different forms of scrapers commonly used are shown in Fig. 11. The flat scraper A is almost invariably used for plane surfaces. For ordinary purposes, the scraper blade is about 3/16 inch thick, from 1 to 1¼ inches wide, and is drawn out at the point to a thickness of about 1/16 inch. The cutting end is as hard as possible and is rounded

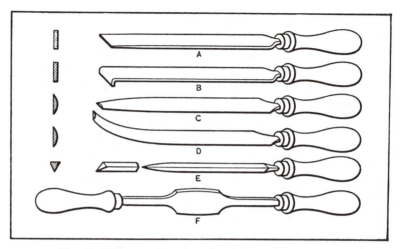

Fig. 11. Different Forms of Scrapers Commonly Used

slightly, in grinding, so that the outer corners will not score the surface being scraped. The grinding should be done, preferably, on a wet grindstone, the edge being finished with an oilstone. When using a scraper of this form, it is generally held at an angle of about 30 degrees. The hook scraper *B* is also used on flat surfaces. It is preferred by some workmen for obtaining a fine, smooth surface and can be used, occasionally, in narrow spaces where there would not be room enough for a straight, flat scraper. Straight and curved scrapers of the "half-round" type are shown at *C* and *D*. These are used for scraping bearings, etc., the sides forming the cutting edges. The curved type *D* is more convenient to use on large half-bearings, as it is held at an angle and the scraping is done by the curved edge. The "three-cornered" or "three-square" scraper shown at *E* is also used to some extent on curved surfaces. When the end is beveled, as shown in the detail view to the left, this form of scraper is convenient for producing sharp corners or for "relieving" them slightly.

The two-handled scraper shown at *F* is an excellent form for scraping bearing boxes and all curved surfaces which are so located that this type can be used. This style of scraper is much superior to the forms shown at *C* and *D*, especially for large work. The straight or curved half-round type works very well on soft bearing metals such as babbitt metal, but on brass or bronze, it cuts slowly and, as soon as the edge is slightly dulled, considerable downward pressure is necessary. The type F requires very much less effort on the part of the workman, and it will cut rapidly. As there are two handles instead of a single handle at one end, the blade can be pressed against the work with little exertion. This form of scraper is largely used for heavy scraping on large bearing boxes. The sides are sometimes ground slightly concave to give the cutting edges "rake," by holding them against the face of the grinding wheel.

In scraping, what kinds of marking materials are used to show the high areas?

The materials commonly used to show the bearing marks in scraping are oil mixed with lampblack, Prussian blue, and red lead. Lampblack mixed with lard oil is good for marking metal surfaces for rough-scraping. Prussian blue paste in tubes is smooth and convenient to use, but a small amount of lard oil mixed with the paste improves it. Red lead mixed with mineral machine oil to the consistency of putty is a good mixture for general scraping operations. Lard oil can be used instead of mineral machine oil when the mixture is applied to the surface plate, but, on account of its smeary nature, it is undesirable to apply it to work on which a dull red-lead surface is required which shows the high spots black and shiny when rubbed with the surface plate.

Mixing Red Lead: The mixing of red lead requires patience, as the lead and oil do not unite readily, and considerable pounding, stirring, and kneading are required. The mixture should be worked until it is smooth. When the compound is kept in a receptacle, it may be allowed to dry out and then be used like shoe or stove polish, mixing in as much machine oil as is required at the time.

Application of Red Lead: For rough-scraping, apply red lead generously to the surface with a rag, using machine oil to assist in spreading. When an even coating has been spread, place the surface plate on the surface to be scraped, rub a little, and remove. The high spots will be clearly defined by the red markings. This is the best way to show the bearing spots. When the spotting shows fairly uniform all over, the process should be reversed, applying the red lead and oil to the work in a somewhat drier form, with the object of attaining a dull red coating which, when rubbed with a clean surface plate, will make a good background for the polished black high spots. The amount of red lead should be gradually reduced and a little gasoline should be mixed with it to act as a drier, finally using just enough red lead to dull the surface without coloring it. This procedure will show the true high spots. They will be small, but no false bearing will appear, which is likely to happen if the surface plate is kept colored. Venetian red is finer than red lead and is sometimes preferred for very accurate work.

How are well-distributed bearing areas obtained by scraping?

In correcting errors on flat or curved surfaces by hand scraping, it is desirable to obtain an evenly spotted bearing with as little scraping as possible. When the part to be scraped is first applied to the surface plate, or to a journal in the case of a bearing, three or four "high" spots may be indicated by the marking material. The time required to reduce these high spots and obtain a bearing that is distributed over the entire surface depends largely upon the way the scraping is started. If the first bearing marks indicate a decided rise in the surface, much time can be saved by scraping larger areas than are covered by the actual bearing marks; this is especially true of large shaft and engine bearings, etc. An experienced workman will not only remove the heavy marks, but also reduce a larger area; then, when the bearing is again tested, the marks will generally be distributed somewhat. If the heavy marks which usually appear at first are simply removed by light scraping, these "point bearings" are gradually enlarged, but a much longer time will be required to distribute them.

The number of times the bearing must be applied to the journal for testing is important, especially when the box or bearing is large and not easily handled. The time required to distribute the bearing marks evenly depends largely upon one's judgment in "reading" these marks. In the early stages of the scraping operation, the marks should be used partly as a guide for showing the high areas, and, instead of merely scraping the marked spot, the surface surrounding it should also be reduced, unless it is evident that the unevenness is local. The idea should be to obtain first a few large but generally distributed marks; then an evenly and finely spotted surface can be produced quite easily.

In scraping to obtain an ornamental effect, what is the general method?

Scraping to obtain ornamental surfaces is commonly known either as "frosting," "snow-flaking" or under the general name of "spotting," and consists in using the scraper in such a way as to obtain a fairly symmetrical series of spots upon the surface to be ornamented. These ornamental surfaces are of two kinds: one is formed of small square spots arranged in checkerboard fashion, although somewhat irregular, whereas the other has a series of crescent-shaped marks arranged in parallel rows. The former marking is usually termed "snow-flaking" and the latter, "frosting." To produce the snow-flaking effect, an end-cutting scraping tool is pushed straight ahead to produce square-shaped marks at even intervals and in parallel rows. Similar spots are then made to fill the intervening spaces by pushing the scraper at right angles to the first direction of movement. Frosting is accomplished by giving the scraping tool a peculiar twist as the cut is made. Surfaces that are ornamented in this manner may often be very carefully hand-scraped in order to secure an accurate plane surface prior to the ornamental scraping, or they may be left as finished by the machine. Ornamenting a surface that has been hand-scraped is done to give a regularity to the mark left by the tool. When a surface is ornamented without previous scraping to correct errors, it may be done wholly as an ornamentation or, as is sometimes the case, to deceive those who are likely to consider such a finish a proof of careful fitting.

For what classes of work are small portable grinders used?

Small portable grinders of the hand type are used for finish grinding miscellaneous parts or surfaces which,

Fig. 12. Finishing Cavity by Grinding with a Kipp Air-driven Grinder

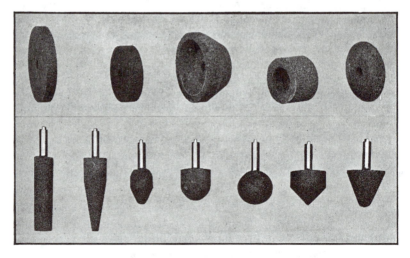

Fig. 13. Wheels and Mounted Points for Grinding Dies
and Finishing Metal Patterns and Castings

because of their shape or location, require a small grinding wheel and hand manipulation. Such grinders are used extensively in diemaking for grinding out cavities or other surfaces. They are applied to various dies used in forming sheet metals and also in finishing certain cavities or passages in die-casting dies, metal patterns, etc. There are several types of these small portable grinders. Some are driven by electricity, others by air, and there is also the flexible-shaft type. Fig. 12 illustrates the use of an air-driven grinder. In this particular instance, the grinder is being used to finish a cavity in a die-casting die. The grinding spindle is driven at a speed of about 40,000 R.P.M. by an air turbine which is at the enlarged end and is rotated by compressed air entering through the connecting hose shown in the illustration. Grinding wheel manufacturers make special shapes of small wheels for use in these portable grinders. These wheels are made in the form of disks, cones, and in cylindrical, spherical and other shapes (See Fig. 13). Portable grinders greatly simplify the machining of many cavities or other surfaces, especially if the part has been hardened.

Tool Steels and Other Metal-Cutting Materials

The term "tool steel" is applied to a wide range of steels differing as to quality and application. The tool steels used for metal-cutting tools were formerly made from so-called "carbon steels" but at the present time "alloy steels" are used very extensively and also other cutting materials which are not classed as steels.

What is the difference between "carbon steel" and "alloy steel"?

The expression "carbon steel" is commonly applied to tool steel containing no alloying metals, the term being used to distinguish such steel from alloy steels which contain tungsten, nickel, chromium, vanadium or other metals. These alloy steels also contain carbon, and some alloy steels have as much carbon as the so-called "carbon steels." The carbon content in steel is generally expressed by giving the percentage of carbon as, for example, 0.90 per cent carbon steel. This is also often expressed as "90-point" carbon steel.

Carbon "Points" in Steel.—The point system used in specifying the carbon content of steel is based on the division of one per cent into one hundred parts; hence, "10 points carbon" means one-tenth of one per cent carbon and not 10 per cent. To express the carbon content in percentage in the case, say, of 50-point carbon steel, the expression should be 0.50 per cent carbon or "one-half per cent" carbon. The term "points" probably originated in an inversion of the reading of the decimal of one per cent; the decimal 0.40, for instance, was read "40-point" instead of "point 40."

Why is "high-speed steel" so named?

The expression "high-speed steel" is derived from the fact that such steel is capable of cutting metal at a much higher rate of speed than ordinary carbon tool steels. The reason why high-speed steel can be used at higher speeds is that it has a special property known as "red hardness," or, in other words, this steel is able to retain its hardness even when heated to a dull red; hence, when cutting at a high rate of speed, the steel, although it becomes heated to a degree which would make an ordinary tool steel useless, retains its cutting qualities. A high-speed steel is not necessarily one conforming to any given analysis but it is some kind of alloy steel. Most high-speed steels contain tungsten as the chief alloying element, but other elements, such as molybdenum, confer the red-hardness characteristic. A high-speed steel should continue to cut when the point of the tool becomes heated to a dull red temperature because of the red-hardness characteristic conferred upon it by tungsten, molybdenum or other alloys. The reason why high-speed steels in general can be heated considerably as the result of high cutting speeds and excessive friction is that some element (or combination of elements), such, for example, as tungsten, so changes the characteristics of the steel that the increase of temperature does not affect it, the same as with ordinary carbon steel.

While high-speed steel is valuable for metal-cutting tools because it will retain a cutting edge even at high temperatures, it is also used for many purposes where temperature is not a factor. This is true, for example, in the case of blanking dies, broaches, certain types of shear blades, etc. High-speed steel in the hardened condition has from five to eight times the wear resistance of a 1 per cent hardened carbon tool steel at ordinary room temperature.

When high-speed steel is designated as "18-4-1," what do these figures represent?

Most high-speed steels contain approximately 18 per cent tungsten, 4 per cent chromium, and 1 per cent vanadium; hence these percentage figures are often combined to designate this high-tungsten steel which has been used much more than any other. The exact composition of these

18-4-1 tungsten steels varies somewhat. For example, the tungsten content may range from 17 to 21, the chromium from 3 to 4½, and the vanadium from 0.7 to 1½. High-speed steels of this general type are easier to harden than some of the other compositions, and they have proved very satisfactory for cutting various materials under normal conditions. The 18-4-1 steel is not only used extensively for forged lathe and planer tools, but for milling cutters, drills, reamers, taps, threading dies, punches, and sheet metal dies, etc.

Another general class of tungsten high-speed steel is known as the 14-4-2 type. This is sometimes preferred for heavy roughing cuts, but is not used as much today as formerly. Because of the lower tungsten content of 14 per cent, this steel is more sensitive to heat-treatment. The carbon content of the various classes of tungsten high-speed steels ranges from about 0.60 to 0.80 per cent. The usual carbon content is from 0.67 to 0.72, as this range gives the best combination of hardness, toughness, and cutting capacity. For a given tungsten and chromium content, the hardness and toughness varies in direct proportion to the carbon content.

Some turning tools are made from an 18-4-2 or "double-vanadium" type of high-speed steel. When applied to broaching, the 18-4-2 steel has proved superior to the 18-4-1 type.

Manufacturers of twist drills generally use practically the same analysis of high-speed steel. This analysis is approximately as follows: Carbon, 0.70; tungsten, 18; chromium, 4; and vanadium, 1 per cent. Steels with somewhat over 0.70 per cent carbon are generally used for small drills, while slightly less than 0.70 per cent carbon is used for the larger sizes. The 14 per cent tungsten high-speed steel is no longer used for drills. High-speed steel containing no tungsten, but instead approximately 7 per cent of molybdenum, has given very good results.

Drill steels generally known as "finishing steels" are a low-tungsten type of tool steel. This is an intermediate type of steel sold as either carbon, super-carbon, or alloy tool steel. The carbon content usually ranges from about 1 to 1.25; the tungsten, from 0.2 to 2.7; and chromium, from 0.5 to 1.2. Some of these steels also have 0.2 or 0.3 per cent of vanadium.

For heavy roughing cuts, what kind of steel has proved efficient?

The high-cobalt or cobalt-tungsten steel is adapted to heavy roughing cuts. These cobalt steels are similar to the 18-4-1 tungsten steel with a certain amount of cobalt added. The high-cobalt tungsten steels contain usually from 7.5 to 12 per cent of cobalt. Tools made of this steel should not be forced to their maximum cutting capacity until the temperature has been raised by the cutting action. With the possible exception of small tools, high-cobalt steel should not be forged because it is more difficult to forge than the high-tungsten steel. As cobalt steels are more expensive than ordinary high-speed steel, it is common practice to weld or braze cobalt-steel tips to a cheaper grade of steel which is used for the shank.

Cobalt high-speed steel drills find wide application in drilling hard metals which are beyond the capacity of ordinary high-speed drills. In resisting the action of abrasion, the cobalt high-speed steel drills, with their higher carbon and alloy content, are superior to those made from ordinary high-speed steel, but they cannot be compared with tungsten-carbide tipped tools. The addition of cobalt to high-speed steel increases the "red hardness." In other words, cobalt high-speed steel drills can be subjected to higher cutting temperatures without destroying the edges.

What kind of tool steel is adapted to finishing cuts?

A low-cobalt high-speed steel has proved very satisfactory for finishing tools requiring tough hard edges. The cobalt type is superior to high-tungsten steel in withstanding relatively high temperatures and maintaining a sharp cutting edge when taking long finishing cuts. This is one reason why cobalt steel is specially adapted for tools used on automatic screw machines or wherever tool replacement involves some difficulty and long tool life is particularly important. A low-cobalt steel may contain from $4\frac{1}{2}$ to 5 per cent of cobalt, 17 to 18 per cent tungsten, and 0.90 to 1.10 per cent vanadium.

Why has molybdenum replaced tungsten in many high-speed steels?

High-speed steels made by substituting molybdenum for tungsten, either partially or entirely, have been used on an increasing scale in recent years. This change was due primarily to war-time conditions which made it necessary to utilize molybdenum which is plentiful in the United States. The molybdenum content in high-speed steels may vary from $3\frac{1}{2}$ to $9\frac{1}{2}$ per cent, tungsten from 0.00 to 0.6 per cent, chromium from 3 to 5 per cent, vanadium from 0.90 to 2.25 per cent, carbon 0.70 to 0.90 per cent, and cobalt up to 9 per cent. See information on heat-treating molybdenum steels in section which follows.

What is stellite?

Haynes Stellite is an alloy of cobalt, chromium and tungsten and is non-ferrous or without iron in its composition. The hardness of this alloy is not materially affected by heat up to 1500 degrees F. and it is actually tougher at red heat than when cold. This important characteristic explains its wide application as a cutting tool material. Haynes Stellite works best when operated at high speed and with a comparatively light feed. The resistance of Stellite to shocks adapts it to interrupted cuts.

J-Metal.—The cutting tool material known as J-Metal is an improved grade of Haynes Stellite. The use of J-Metal results either in higher cutting speeds or in greater production between tool grindings. J-Metal is adapted to various classes of machining operations on practically all kinds of machinable materials, excepting chilled cast iron and manganese steel. The hardness of J-Metal at room temperature is 600 Brinell or Rockwell C, 60-62. It is important to note that the hardness of J-Metal is practically unaffected at red heat and this red hardness is considerably greater than that of high-speed steel.

Haynes Stellite—2400.—This is another cobalt-chromium-tungsten alloy. Cutting tools made of this material have greater edge strength and longer economic tool life at even higher speeds than tools made of J-Metal, without reduction of feed or depth of cut. In fact, the speeds and feeds recommended are from 10 to 50 per cent greater than those for J-Metal. This alloy may be used for roughing or finishing cast and forged steels, cast and malleable irons, nitrided, stainless and other alloy steels.

What are the chief characteristics of carbide cutting tools?

Metal-cutting tools made of cemented or sintered carbides are used in various branches of the machine-building industry either because of the exceptionally high cutting speeds that are possible with these tools, or because of their durability and adaptability in machining either very hard materials or compositions that are destructive to other tools. Cemented carbide is the hardest metallic tool material known. The harder grades are about three times as hard as hardened tool steel.

Sintered carbides are made in a number of different grades (with different combinations of hardness and toughness) to adapt them to the cutting of a large variety of materials. For example, tungsten carbides are especially adapted to cutting cast iron and there are other grades that are more effective as applied to steel. These steel-cutting carbides contain tantalum or titanium carbide, thus changing the properties of the carbide so as to resist the seizing action of the steel chips and prevent formation of a chip cavity back of the cutting edge. When ordinary cemented tungsten-carbide tools suitable for cast iron are used for steel, the formation of these chip cavities shortens the life of the tool and increases the time required for grinding. Information about the particular grade or type of carbide tool for cutting a given material should be obtained directly from the manufacturers. These different grades are designated by various trade names and symbols.

In general, carbide tools are not only very effective in machining all classes of iron and steel, but they are also applicable to non-ferrous alloys such as brass, bronze, and aluminum; to hard rubber, fiber, non-metallic materials for gears, etc.; slate, marble, and other materials which are either too hard for steel or which would quickly dull steel tools by their abrasive action. Sintered or cemented carbides are now applied to various types of tools for turning, milling, reaming, drilling, etc.

Heat-Treatment of Steels Used for Metal-Cutting Tools

The term "heat-treatment" as applied to steel is an operation or combination of operations involving heating and cooling with the object of obtaining certain desirable conditions or properties. When tool steels for metal cutting are heat-treated, the purpose is to obtain the right combination of hardness and toughness, or as much hardness as possible without brittleness that would result in tool breakage. Heat-treatments are not only applied to tool steels, but to various classes of structural steels and to certain other metals used in the manufacture of mechanical products. The object in heat-treating structural steels is to secure greater strength or otherwise improve the physical properties. For example, the strength of many automobile parts is increased greatly by suitable heat-treatments so that lighter parts provide the necessary strength. The general subject of heat-treatment is a very big one and includes a great many classes of equipment and also numerous heat-treating processes. This section, however, deals only with the heat-treatment of tool steels used in metal cutting, and it is restricted to certain fundamental principles and methods likely to prove useful in machine-building practice.

What changes occur when carbon tool steel is heated for hardening?

Carbon steel which has been fully annealed consists chiefly of two constituents. One is the element iron or *ferrite* and the other is a carbide of iron known as *cementite.* A mechanical mixture of a certain proportion of these two elements is called *pearlite* because under the microscope it frequently has the appearance of mother of pearl. A fully annealed steel containing 0.85% carbon would consist entirely of pearlite. If the carbon content is less than 0.85%, it would consist of both pearlite and ferrite whereas if the carbon content is greater than 0.85% it would consist of both pearlite and cementite.

Now when carbon steel in the fully annealed state is heated to a temperature usually varying between 1335 and 1355 degrees F. (depending upon the carbon content) the alternate bands or layers of ferrite and cementite which form the pearlite, begin to merge into each other. The temperature at which this occurs is known as the *lower critical point*. The merging process continues until the pearlite is thoroughly "dissolved," forming what is known as *austenite*. If the temperature of the steel continues to rise and there is present, in addition to the pearlite, any excess ferrite or cementite, this also will begin to dissolve into the austenite until finally only austenite will be present. The temperature at which the excess ferrite or cementite is completely dissolved in the austenite is called the *upper critical point*. This temperature varies with the carbon content of the steel much more widely than the lower critical point. The critical or transformation point at which pearlite is changed into austenite as the steel is being heated is also called the *decalescence point*. If the steel is cooled slowly the critical or transformation point at which the austenite is changed back into pearlite is called the *recalescence point*. These critical points have a direct relation to the hardening of steel because hardening will not occur unless the temperature reaches the decalescence point and hardening also necessitates cooling the steel suddenly before the temperature is reduced to the recalescence point. When the steel is suddenly cooled by plunging it into a bath of water or oil, a new structure is formed as the austenite is largely transformed into *martensite*. The changes indicated by these metallurgical terms are very important in connection with the heat-treating operations.

What temperatures are required in hardening carbon tool steel?

The temperature for hardening carbon tool steel must reach the decalescence point, as previously mentioned, in order to change the pearlite into austenite. When steel so heated is cooled suddenly and before the temperature reaches the recalescence point, the austenite is prevented from changing back into pearlite. The best hardening temperatures for any given tool steel are dependent upon the type of tool and the intended class of service. Wherever possible, the specific recommendations of the tool steel manufacturer should be followed. General recommendations for hardening temperatures of carbon tool steels based on carbon content are as follows: For steel of 0.65 to 0.80 per cent carbon content, 1450 to 1550 degrees F.; for steel of 0.80 to 0.95 per cent carbon content, 1410 to 1460 degrees F.; for steel of 0.95 to 1.10 per cent content, 1390 to 1430 degrees F.; and for steels of 1.10 per cent and over carbon content, 1380 to 1420 degrees F.

How is the hardening temperature checked?

Temperatures for heat-treating operations should be checked preferably by using some type of pyrometer (information about different types of pyrometers will be given later). If the pyrometer shows furnace temperature, care must be taken to allow sufficient time for the work to reach the furnace temperature after the pyrometer indicates that the required hardening temperature has been attained. If the pyrometer indicates work temperature, then, where the work-piece is large, time must be allowed for the interior of the work to reach the temperature of the surface which is the temperature indicated by the pyrometer. Where the hardening temperature is specified as a given temperature rise above the critical point of the steel, a pyrometer which indicates the temperature of the work should be used. The critical point, as well as the given temperature rise, can be more accurately determined with this type of pyrometer.

Why is uniform heating essential?

In the actual heating of a piece of steel, several requirements are essential to good hardening: First, all parts should be heated at the same rate, and second, all parts should be heated to the same temperature. These conditions are facilitated by slow heating, especially when the heated piece is large. A uniform heat, as low in temperature as will give the required hardness, produces the best product. Lack of uniformity in heating causes irregular grain and internal strains, and may even produce surface cracks.

What kinds of quenching baths are commonly used?

To obtain the different rates of cooling required by different classes of work, baths of various kinds are used. These include plain or fresh water, brine, caustic soda solutions, oils of various classes, oil-water emulsions, baths of molten salt or lead for high-speed steels and air cooling for some high-speed steel tools when a slow rate of cooling is required.

Oil Quenching Baths.—Oil is used very extensively as a quenching medium as it results in a good proportion of hardness, toughness, and freedom from warpage when used with standard steels. Oil baths are used extensively for alloy steels. Various kinds of oils are employed such as prepared mineral oils and vegetable, animal and fish oils, either singly or in combination. Prepared mineral quenching oils are widely used because they have good quenching characteristics, are chemically stable, do not have an objectionable odor, and are relatively inexpensive.

Quenching in Water.—Many carbon tool steels are hardened by immersing them in a bath of fresh water, but water is not an ideal quenching medium. In order to secure more even cooling and reduce danger of cracking, either rock salt (8 or 9 per cent) or caustic soda (3 to 5 per cent) may be added to the bath in order to eliminate or prevent the formation of a vapor film or gas pockets, thus promoting rapid early cooling. Brine is commonly used and ¾ pound of rock salt per gallon of water is equivalent to about 8 per cent of salt. Brine is not inherently a more severe or drastic quenching medium than plain water, although it may seem to be because the brine makes better contact with the heated steel and, consequently, cooling is more effective. In still bath quenching, a slow up-and-down movement of the tool is preferable to a violent swishing around.

Why is steel tempered or drawn?

The object of tempering, or "drawing," is to reduce the brittleness in hardened steel, and to remove the internal strains caused by the sudden cooling in the quenching bath. The tempering process consists in heating the piece of work, by various means, to a certain temperature, and cooling it.

The temperature to which the tool is heated determines the degree of toughness and also the degree of softness. Hardened steel is tempered in order to make it less brittle, but unfortunately the tempering process also softens the steel, to some extent. Tempering temperatures for carbon steels may vary from 300 to 600 degrees F. depending upon the type of tool and its use.

In heating steels for tempering, either a furnace or some kind of bath such as oil, various salt mixtures or lead may be used. Different kinds of baths will be referred to later.

How are carbon steel tools tempered by the color method?

When steel is tempered by the color method the temper is gaged by the colors formed on the surface as the heat increases. First the surface is brightened to reveal the color changes, and then the steel is heated either by placing it upon a piece of red-hot metal, a gas-heated plate or in any other available way. As the temper increases, various colors appear on the brightened surface. First there is a faint yellow which blends into straw, then light brown, dark brown, purple, blue and dark blue, with various intermediate shades. When only one end of a chisel or other tool is to be hardened and tempered, this is done by first heating the cutting end to a cherry-red, and then quenching the part to be hardened. When the tool is removed from the bath, the heat remaining in the unquenched part raises the temperature of the cooled cutting end until the desired color (which will show on a brightened surface) is obtained, after which the entire tool is quenched. The foregoing methods are convenient, especially when only a few tools are to be treated, but the color method of gaging temperatures is not dependable, as the color is affected, to some extent, by the composition of the metal.

Why is steel annealed?

Annealing involves reheating and cooling and it usually implies relatively slow cooling. The purpose of annealing may be either (1) to remove stresses, (2) to soften a metal as for machining, (3) to change the ductility, toughness, electrical, magnetic or other physical properties; or (4) to refine the crystalline structure. The annealing temperature and rate of cooling depend upon the material and purpose of the treatment.

A common method of annealing steel is to pack it in a cast-iron box containing some material, such as powdered charcoal, charred bone, charred leather, slaked lime, sand, fireclay, etc. The box and its contents are then heated in a furnace to the proper temperature, for a length of time depending upon the size of the steel. After heating, the box and its contents should be allowed to cool at a rate slow enough to prevent any hardening. It is essential, when annealing, to exclude the air as completely as possible while the steel is hot, to prevent the outside of the steel from becoming oxidized.

The temperature required for annealing should be slightly above the critical point, which varies for different steels. Low-carbon steel should be annealed at about 1650 degrees F., and high-carbon steel at between 1400 and 1500 degrees F. This temperature should be maintained just long enough to heat the entire piece evenly throughout. Care should be taken not to heat the steel much above the decalescence or hardening point. When steel is heated above this temperature, the grain assumes a definite size for that particular temperature, the coarseness increasing with an increase of temperature. Moreover, if steel that has been heated above the critical point is cooled slowly, the coarseness of the grain corresponds to the coarseness at the maximum temperature; hence, the grain of annealed steel is coarser, the higher the temperature to which it is heated above the critical point. If only a small piece of steel or a single tool is to be annealed, this can be done by building up a firebrick box in an ordinary blacksmith's fire, placing the tool in it, covering over the top, then heating the whole, covering with coke and leaving it to cool over night. Another method is to heat the steel to a red heat, bury it in dry sand, sawdust, lime, or hot ashes, and allow it to cool.

What is "water annealing"?

Rapid annealing can be partially effected by what is known as "water annealing." The steel is slowly heated to a cherry red, and is then removed from the furnace. A piece of soft wood is used to test the heat of the piece of steel as it is decreasing, the heat being tested by touching the steel with the end of the stick. When the piece of steel has cooled so that the wood ceases to char, the steel is plunged quickly into soapy water. Very often a piece of steel annealed in this manner will be found to be much softer than if annealed in the regular way by being packed in charcoal and allowed to cool over night.

What general types of furnaces are used for heating steel in hardening?

The furnaces used for the hardening or tempering of steel are heated either by gas, oil, electricity or solid fuel. Furnaces are made in many different styles and sizes to suit various classes of work. When oxidation or the formation of scale is particularly objectionable, furnaces of the muffle type are often used. These furnaces contain a refractory retort in which the steel is placed, thus excluding the products of combustion. These muffles must be replaced quite frequently and more fuel is required than for an oven type of furnace. Some modern furnaces are equipped with special means for producing a protective atmosphere in the heating chamber. Electrically-heated furnaces are generally considered very satisfactory for the heat-treatment of high-grade work. This type of furnace gives a uniform heat and is adapted to accurate regulation. Electrically-heated furnaces are also used in conjunction with heating baths, the current being transmitted through a bath of metallic salts. The solid-fuel type of furnace is inferior to other types, for most purposes, because it is almost impossible to maintain a uniform temperature, and the gases of combustion are liable to injure the steel.

Why are molten baths often used for heating steels prior to hardening?

When steel is heated for hardening by immersing it in a molten bath having the required hardening temperature, the principal advantages are as follows: No part of the work can be heated to a temperature above that of the bath; the temperature can be easily maintained at whatever degree has proved, in practice, to give the best results; the sub-merged steel can be heated uniformly, and the finished surfaces are protected against oxidation. The liquid baths commonly used for heating steel tools preparatory to hardening are molten lead, sodium cyanide, barium chloride, a mixture of barium and potassium chloride and other metallic salts. The molten substance is retained in a crucible or pot and the heat required may be obtained from gas, oil, or electricity.

Lead Bath.—The lead bath is extensively used, but is not adapted to the high temperatures required for hardening high-speed steel, as it begins to vaporize at about 1190 degrees F., and, if heated much above that point, rapidly volatilizes and gives off poisonous vapors; hence, lead furnaces should be equipped with hoods to carry away the fumes. Lead baths are especially adapted for heating small pieces which must be hardened in quantities. It is important to use pure lead that is free from sulphur. The work should be preheated before plunging it into the molten lead.

Salt Baths.—Molten baths of various salt mixtures or compounds are used extensively for heat-treating operations such as hardening and tempering; they are also utilized for annealing ferrous and non-ferrous metals. Commercial salt-bath mixtures are available which meet a wide range of temperature and other metallurgical requirements. For example, there are neutral baths for heating tool and die steels without carburizing the surfaces; baths for carburizing the surfaces of low-carbon steel parts; baths adapted for the usual tempering temperatures of, say, 300 to 1100 degrees F.; and baths which may be heated to temperatures up to approximately 2400 degrees F. for hardening high-speed steels. Salt baths are also adapted for local or selective hardening, the type of bath being selected to suit the requirements. Various proportions of salts provide baths of different properties. Sodium cyanide is often used and costs less than potassium cyanide.

When tools are heated in baths for tempering, what are the advantages?

A common method of tempering, especially in quantity, is to heat the hardened parts to the required temperature in a bath of molten lead, heated oil or other liquids; the parts are then removed from the bath and quenched. The bath method makes it possible to heat the work uniformly, and to a given temperature within close limits.

Oil Bath.—Oil baths are extensively used for tempering tools (especially in quantity), the work being immersed in oil heated to the required temperature, which is indicated by a thermometer. It is important that the oil have a uniform temperature throughout and that the work be immersed long enough to acquire this temperature. Cold steel should not be plunged into a bath heated for tempering, owing to the danger of cracking it. The steel should either be preheated to about 300 degrees F., before placing it in the bath, or the latter should be at a comparatively low temperature before immersing the steel, and then be heated to the required degree. A temperature of from 650 to 700 degrees F. can be obtained with heavy tempering oils; for higher temperatures, either a bath of nitrate salts or a lead bath is generally used. In tempering, the best method is to immerse the pieces to be tempered in the oil before starting to heat the latter. They are then heated with the oil. After the pieces tempered are taken out of the oil bath, they should immediately be dipped in a tank of caustic soda, and after that in a tank of hot water. This will remove all oil which might adhere to the tools.

Lead Bath.—The lead bath is commonly used for heating steel in connection with tempering, as well as for hardening. The bath is first heated to the temperature at which the steel should be tempered; the preheated work is then placed in the bath long enough to acquire this temperature, after which it is removed and cooled. As the melting temperature of pure lead is about 620 degrees F., tin is commonly added to it to lower the temperature sufficiently for tempering. Reductions in temperature can be obtained by varying the proportions of lead and tin.

Why are pyrometers commonly used in heat-treating steel?

Pyrometers are of great value in connection with the heat-treatment of steel, as they make it possible to determine high temperatures accurately; moreover, the temperature, when heating for hardening, can be regulated to conform with the temperature that has given the best results in practice. The most commonly used pyrometers are of the *thermo-electric type.* In this type, temperature variations are determined by the measurement of an electric current generated by the action of heat on the junction of two dissimilar metals. The thermo-couple, consisting of two pieces of dissimilar metals, is placed at some point within the furnace and is connected by wires with a meter which may be close to the furnace or in some other part of the plant.

Many of the pyrometers used in heat-treating plants may be designated as the *indicating type,* since the temperature variations are shown by the position of a hand or pointer relative to a graduated scale. The indicating instrument may be located close to the furnace or in some central station or controlling room. When it is by the furnace, the furnace operator controls the temperature either according to his experience with similar work, or possibly by reference to data previously recorded. This is a common method in small plants, but where a large heat-treating department is installed, a centralized system of control is quite general.

A *recording pyrometer* is provided with some kind of marking device which traces either a continuous or a dotted line upon a chart graduated with reference to temperature and time. By referring to one of these charts, the temperature at any period within the range of the chart is shown graphically. Where a heat-treating plant contains two or more furnaces, a pyrometer may be installed that will record automatically on a chart temperature variations in each furnace to which it is connected. This type of pyrometer is generally used when the heat-treating process requires a half hour or more for its completion. When four, six, or eight records are needed, these may be printed on the chart in different colors to avoid confusion.

The pyrometer that automatically controls furnace temperatures is so arranged that the moving element of the instrument not only indicates the temperature by its position relative to a scale, but by combined mechanical and electrical apparatus, controls the temperature, within certain limits, by regulating the heat supply. The pyrometer can be set for any temperature desired within certain maximum and minimum limits and may be applied to furnaces heated either by gas, oil, or electricity.

In hardening high-speed steel, what is the general procedure when a furnace is used?

The heat-treatment of high-speed steel requires several steps or separate operations. The first consists in preheating the steel at a temperature which may be about 800 degrees below the hardening temperature. This is done in a preheating furnace. The second step is to transfer the steel to a furnace which raises the temperature to that required for hardening, when the steel is quenched or otherwise cooled to harden it. The next step is to temper or draw the steel. All steel should be in the annealed condition before heat-treatment. This annealing prior to hardening may be done by heating the steel to the forging heat and then burying it in lime or ashes so that it will cool gradually. Forged tools should always be annealed prior to heat-treatment.

To what temperature is high-speed steel preheated to avoid internal strains?

The preheating furnace should have a temperature of 1500 to 1600 degrees F. This preliminary heating should be done very gradually, particularly if the tool consists of thin and thick sections which would result in unequal expansion and enormous strains. Preliminary warming by placing the steel on top of the furnace is desirable. When the steel is inserted in the furnace, place it near the opening and turn it frequently to insure uniform heating. In this preliminary heating, the tool is allowed to "soak" or remain in the furnace until it is heated evenly throughout all sections. Many tools are ruined because they are not heated gradually.

To what temperature is high-speed steel finally heated for hardening?

The hardening temperature varies somewhat for steels of different compositions. The hardening furnace temperatures usually are between 2350 and 2450 degrees F. Cobalt steels require somewhat higher temperatures than the ordinary "18-4-1" steels which are described in the section on tool steels. It is important to quench the steel while the temperature is rising and to insure this some prefer to maintain the furnace temperature at 50 to 100 degrees above the actual hardening temperature for the steel. The preliminary heating of the steel, as mentioned, must be gradual and considerable time is required to obtain slow uniform heating; but when the tool is transferred to the hardening furnace, it quickly reaches the hardening temperature (sometimes in a few seconds for small tools) and it should then be removed and quenched as quickly as possible, as otherwise the steel will be injured.

How is it possible to determine when high-speed steel has reached the hardening temperature?

There are two general methods. One is to adjust the furnace until the temperature (as shown by a pyrometer) conforms to whatever hardening temperature is required for the steel. The tool then is allowed to remain in the furnace until experience or judgment indicates that it has been heated uniformly to the furnace temperature, and then it is quenched. Many experienced heat-treaters prefer to rely upon the appearance of the steel in determining the exact moment when the tool should be removed and quenched. The furnace temperature may be 50 to 100 degrees above the desired hardening temperature.

In watching the steel as it approaches the hardening heat, either colored glasses or a colored glass screen in front of the furnace opening should always be used to protect the eyes from the intense glare of the heating chamber. As the steel approaches the hardening temperature, small round globules will appear on its surface. These globules increase rapidly in size, and, finally, they spread or run together. In hardening lathe and planer tools, quench the tool as soon as this spreading begins. In hardening more delicate tools, such as cutters with thin teeth and thick bodies, quenching should be done as soon as the flux bubbles first appear. The actual temperature at which the flux bubbles are formed varies with high-speed steels of different grades and compositions. The steel should never be allowed to soak at the quenching temperature, even for a short time, as this tends to coarsen the structure and injure the steel.

Why is oil used in preference to cold water for cooling high-speed steel?

If it were possible to cool, instantly and uniformly, the entire body of a steel tool, the greatest degree of hardness would be obtained. Uniform cooling, however, is impracticable and it is necessary to delay the cooling somewhat to avoid sudden and excessive stresses which are caused by very rapid, uneven cooling. For example, if steel heated to a temperature of, say, 2400 degrees F. were suddenly plunged into very cold water, the sudden lowering of the temperature of some parts, while others were still very hot, might result in cracking the steel. To avoid such troubles, oil or a dry air blast is used for cooling at a slower rate. A quenching tank of oil is generally used, and this should be very close to the hardening furnace so that the temperature of the steel will still be rising as it is plunged deeply into the oil. As the steel is immersed, it should be moved about quickly for a few moments to insure quick and more even cooling. If one tool after another is being hardened, the quenching oil temperature will, of course, rise unless provision is made for keeping it cool. It is preferable to use a tank provided with a jacket and circulating cooling water so that the oil never becomes hot. After quenching, the tool may be rolled in sawdust to remove the oil.

In tempering high-speed steel, how is the temperature established?

The tempering or drawing temperature for high-speed steel tools usually varies from 900 to 1200 degrees F. This temperature is higher for turning and planing tools than for such tools as milling cutters, forming tools, etc. If the temperature is below 800 degrees F., the tool is likely to be too brittle. The general idea is to temper tools at the highest temperature likely to occur in actual service. Since this temperature ordinarily would not be known, the general practice is to temper at whatever temperature experience with that particular steel and tool has proved to be the best. The furnace used for tempering usually is kept at a temperature of from 1000 to 1100 degrees F. for ordinary high-speed steels and from 1200 to 1300 degrees F. for steels of the cobalt type. These furnace temperatures apply to tools of the class used on lathes and planers. Such tools, in service, frequently heat to the point of visible redness. Milling cutters, forming tools, or any other tools for lighter duty, may be tempered as low as 850 or 900 degrees F. When the tool has reached the temperature of the furnace and has been heated evenly throughout, it should be allowed to cool gradually in the air and in a place that is dry and free from air drafts. In tempering, the tool should not be quenched, as this tends to produce strains which may result later in cracks. If a tool for any reason must be rehardened, it should always be annealed first, as otherwise a coarse poor structure will be obtained.

Double Tempering.—In tempering high-speed steel tools, it is common practice to repeat the tempering operation or "double temper" the steel. This is done by heating the steel to the tempering temperature (say 1050 degrees F.) and holding it at that temperature for two hours. It is then cooled to room temperature, re-heated to 1050 degrees F. for another two-hour period, and again cooled to room temperature. After the first tempering operation, some untempered martensite remains in the steel. This martensite is not only tempered by a second tempering operation but is relieved of internal stresses, thus improving the steel for service conditions.

How are salt baths applied in heat-treating high-speed steels?

Salt baths are used for preheating high-speed steels, for heating to the hardening temperature, and also for quenching and tempering. They are adapted for heat-treating the 18-4-1 steels, the molybdenum steels, and the cobalt types. A well-known type of electric salt bath furnace is equipped with electrodes (usually a pair) which are immersed in the molten salt. The latter serves as an electrical conductor. An electromagnetic stirring action throughout the entire bath maintains a uniform temperature which is automatically controlled by a pyrometer set for whatever temperature is required. The salt bath furnace is conducive to uniform heating and hardening, prevention of decarburization, and serves to protect finished surfaces or sharp edges. In heat-treating high-speed steel, three baths may be employed. The temperature of the first or preheating bath is usually about 1500 to 1600 degrees F. After preheating, the tool is transferred to another bath having a hardening temperature ordinarily between 2200 and 2400 degrees F.

Quenching in Salt Bath.—The object in using a liquid salt bath for quenching (instead of an oil bath) is to obtain maximum hardness with minimum cooling stresses and distortion which might result in cracking expensive tools, especially if there are irregular sections. The temperature of the quenching bath may be around 1100 or 1200 degrees F. Quenching is followed by cooling to room temperature and then the tool is tempered or drawn in a bath having a temperature range of 950 to 1100 degrees F. In many cases the tempering temperature is about 1050 degrees F.

Salt-Bath Method of Heat-treating Molybdenum High-speed Steels.—Molybdenum high-speed steels may be divided into three general groups. The molybdenum-tungsten group contains from 1.25 to 2.30 per cent tungsten, and 8 to 9.5 per cent molybdenum; the molybdenum-vanadium group contains no tungsten, but from 7.5 to 9.5 per cent molybdenum; the tungsten-molybdenum group contains from 5 to 6 per cent tungsten, 3.5 to 5.5 per cent molybdenum. These steels also contain varying amounts of vanadium, chromium, carbon and cobalt. The composition of the steel may have an important influence on the heat-treatment, and it is advisable to consult with the steel manufacturers. Difficulty may be experienced due to decarburization, especially if attempts are made to heat-treat the molybdenum-tungsten and molybdenum-vanadium types in a conventional type of furnace. The salt-bath furnace is preferable.

Molybdenum high-speed steels, when hardened in salt baths, are entirely surrounded by neutral molten salt. A salt film is retained on the tool throughout the hardening procedure, thus preventing decarburization or scaling. The salt-bath hardening method provides uniform heating, and this generally results in less distortion. All sections of intricately shaped tools are uniformly heated by this method. The salt bath permits selective hardening. When a salt bath is properly selected and maintained, there is no chemical attack on the molybdenum high-speed steels. The original surface of the steel is retained. Molybdenum high-speed steels will take all the special surface treatments, including nitriding when immersed in molten cyanide, that are applied to tungsten high-speed steels for certain applications.

In general, immersed-electrode furnaces are being used when there is sufficient production to keep the furnaces operating at a reasonable capacity. The immersed electrodes generate heat directly in the molten salt bath by the electrical resistance of the bath material, and produce a positive circulation of the bath, due to the internal stirring action caused by the electrical flow between the electrodes. This stirring action increases the speed of heating and eliminates local overheating, thus aiding close temperature control, which is always advantageous. A properly selected and maintained salt bath prevents scaling or oxidation of the work, as well as surface decarburization.

Procedure for Salt-Bath Hardening.—Briefly, the procedure for salt-bath hardening is as follows:

1. Clean work to free it from scale, rust, oil, grease, and moisture. Use either a solvent degreaser or a suitable alkaline cleaner, followed by a clean hot water rinse, and thorough drying. Every precaution must be taken to prevent moisture on tools that are going into the salt bath, as wet tools may cause a steam explosion and burn the operator.

2. Immerse in preheat salt bath having a temperature of 1500 to 1550 degrees F. Allow sufficient time for the work to reach the temperature of the bath.

3. Transfer to high-heat salt bath having a temperature of 2150 to 2250 degrees F. Allow sufficient time for work to reach temperature of bath plus proper soaking time at temperature.

4. Transfer to quench bath having a temperature of 1100 to 1200 degrees F. Allow sufficient time to cool to bath temperature. Two to five minutes will suffice, depending upon size. Remove from salt bath and cool in air or oil. Quenching in oil is not recommended for work of intricate design or where distortion is likely to occur.

5. After the work has cooled to room temperature, wash off all salts in a hot alkaline cleaner. In cases where a furnace (atmospheric) is to be used for tempering, the work should be shot- or sand-blasted, or cleaned by other methods, to insure the removal of all salts. If this is not done, the salt will attack the work during the tempering. Where a salt bath is to be used for tempering, the work need only be cleaned in a hot alkaline solution or hot water.

6. For tempering, reheat slowly and uniformly to 950 to 1100 degrees F. For general work, 1050 degrees F. is most common. Hold at this temperature at least one hour. Two hours is a safer minimum, and four hours is maximum. The time and temperature depend on the hardness and toughness required.

Nitriding High-Speed Steel Tools.—Nitriding as applied to high-speed steel is for the purpose of increasing tool life by producing a very hard skin or case, the thickness of which ordinarily is from 0.001 to 0.002 inch. This nitriding is done after the tool has been fully heat-treated and finish-ground. (The process differs entirely from that which is applied to certain alloy steels in order to surface harden them by heating in an atmosphere of nitrogen or ammonia gas.) The temperature of the high-speed steel nitriding bath, which is a mixture of sodium and potassium cyanides, is equal to or slightly lower than the tempering temperature. For ordinary tools, this temperature usually varies from about 1025 to 1050 degrees F.; but if the tools are exceptionally fragile, the range may be reduced to 950 or 1000 degrees F. Accurate temperature control is essential to prevent exceeding the final tempering temperature. The nitriding time varies considerably. In some cases, a period of 10 or 15 minutes is sufficient, whereas in others the time may be extended to 30 minutes or longer, as determined by experiment. The shorter periods are applied to tools for iron or steel or any shock-resisting tools, and the longer periods are for tools used in machining non-ferrous metals and plastics. This nitriding process is applied to tools such as hobs, reamers, taps, box tools, form tools, milling cutters, etc. Nitriding may increase tool life 50 to 200 per cent or more, but it should always be preceded by correct heat-treatment.

Numerical Control   by James J. Childs*

Numerical control is a form of automation which, **for** certain types of work, is capable of producing machine shop type parts faster, more accurately, and at substantially less cost than when produced by conventional machine tools. The result is to create more products per man hour thus reducing the cost and consequently increasing the opportunity for more people to buy manufactured goods. Because of this, machine tool operators may be temporarily displaced due to the need for different skills and the higher production rates obtainable with numerically controlled equipment. However, new talents are required in order to operate and maintain numerically controlled equipment properly. For example, there have been many more job openings for qualified persons who can program numerically controlled equipment than there were qualified people. This is one of the major factors which has held back the extended use of numerical control and, although suppliers of this equipment offer training programs, these are usually limited to the requirements peculiar to their product. The basic knowledge of blueprint reading, mathematics, and a detailed understanding of machining operations which is also required, must be obtained through formal technical school and apprentice programs.

What is numerical control?

Numerical control, commonly referred to as N/C, is a semi-automated method of machining wherein the motions of the machine are controlled generally by an electronic system. Machine instructions such as: The distance to be moved, spindle speeds, feedrates, and the automatic turning of the coolant on and off are "described" on a punched tape which is "read" by the electronic control system. This "description" generally takes the form of a series of holes

*President, James J. Childs Associates, Inc., Alexandria, Virginia

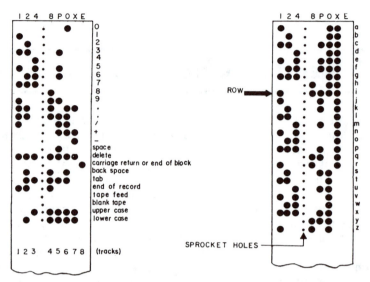

Fig. 1. Electronic Industries Association Standard 1-Inch Punched Tape—Configuration of Punches and Their Significance

conforming to a coded numerical pattern which are punched in rows on the 1-inch wide tape. Any digit, 0 thru 9, or a letter of the alphabet may be punched on one line across the tape, as shown in Fig. 1. By combining the numerals, it is possible to express whole numbers. Thus if it is required to move a machine table 2.2341 inches, the instruction would be expressed by inserting successively the punched codes for numerals 2 2 3 4 and 1. The decimal points are usually ignored. This coded instruction expressed on tape would appear as shown in Fig. 2. Unlike the

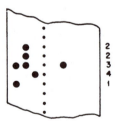

Fig. 2. Configuration of Holes Expressing the Dimension 2.2341 Inches

usual form of automation as practiced in the automotive industry where thousands of identical parts such as engine blocks progress along a transfer line with successive machining units performing a complete series of machining operations without manual aid or control, numerical control has been restricted to general-purpose types of machine tools which are utilized for machining relatively small quantities of different kinds of parts.

By adapting an electronic control system to a machine tool such as a drill press or lathe it is transformed from a manually operated piece of equipment wherein the operator is required to direct each movement, to an automatic machine obtaining its instructions from a previously prepared tape. While there is perhaps less technical demand on the operator, a new and more exacting requirement has been developed—that of planning the detailed step-by-step machining instructions required for preparing the tape.

What are the two types of numerical control?

There are two types of numerical control: "point-to-point," sometimes referred to as "positioning," and "contouring," also known as "continuous path." In point-to-point machining the object is to move a cutting tool to a specific point and once at the required location perform an operation such as the drilling of a hole. The path that the

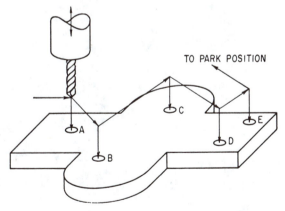

Fig. 3. Point-to-Point Controlled Motions for Drilling Holes with a Simple Type Numerically Controlled Drill Press

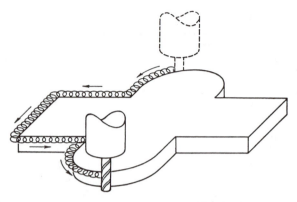

Fig. 4. Path of a Milling Cutter Tracing the Perimeter or Profile of a Part Under Continuous Path Numerical Control

Fig. 5. Simple N/C Point-to-Point Drill Press

drill takes in moving to the required location is immaterial, providing it doesn't take too long in getting to its destination, since the drill is operating in air and not cutting metal when moving from one hole location to another. The illustration in Fig. 3 indicates a sequence of moves that a drill would automatically follow in machining the holes A, B, C, D, and E as shown.

N/C contouring differs from point-to-point in that the entire path that the cutting tool moves along must be carefully controlled to within close machining accuracies since it is continuously cutting the work as it moves along its

Fig. 6. N/C Contouring Milling Machine

path. N/C contouring has generally been applied to profile type milling machines and engine lathes. An N/C contouring example is described in Fig. 4 which shows a flat end milling cutter moving along the perimeter of a part.

A popular type point-to-point drill press is shown in Fig. 5. The cabinet at the right contains the necessary electronic components which position the table under the drill head. When the table reaches its assigned position the drill automatically descends into the work-piece. Coded commands are read from the one-inch wide punched tape which moves through the tape reader in an intermittent motion. The period between tape movements, depends on the length of time in moving from one point to the next, plus the length of time required to perform the drilling operation.

An example of a contouring arrangement is shown in Fig. 6. The machine tool in this instance is utilized for relatively light milling cuts on jobs which require that the two horizontal motions (side to side and front to back)

Fig. 7. Examples of Complex Parts Produced on
Milling Machine of Type Shown in Fig. 6

of the table be coordinated with the vertical motion of the spindle. With all three motions being obtained simultaneously it is possible to machine complex parts such as are shown in Fig. 7.

The electronic control system required for contouring is much more complex and costly than that required for point-to-point machines since the various traversing motions must be precisely coordinated with cutting operations such as milling or turning.

How does a numerical control machine work?

Practically all numerical control systems operate on the "closed loop" or feedback principle as shown schematically in Fig. 8. Essentially this involves comparing the difference between the position to which the cutting tool or machine table is to move and its present position. A feedback device which transmits signals back to the control system indicates the position of the table or head. When commands are inserted into the control system an unbalanced condition is created due to the difference between where the table should

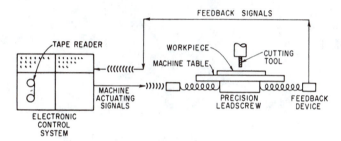

Fig. 8. Feedback Principle of an N/C Machine

be and where it is. This difference is accurately measured by the electronic control system which then transmits electrical signals to the drive mechanism causing the table or cutting tool to move, thus reducing the difference. When there is no variation between the feedback position and the input command the moving element has reached its pre-scribed position and stops.

Machine elements may be driven by either hydraulic or electrical motors, the latter usually being reserved for the smaller, lighter machine tools. Feedback devices may be geared to the lead screw or fastened directly to the table or head of the machine tool.

What types of machine tools have been adapted to N/C?

Theoretically N/C can be adapted to any machine requiring controlled motion. In addition to machines such as the standard type drill presses and mills as shown in Figs. 5

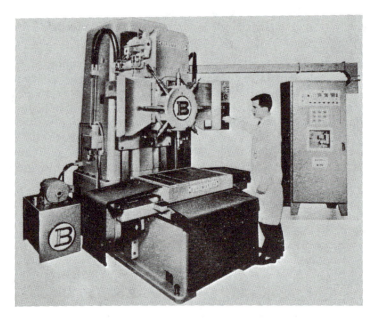

Fig. 9. N/C Turret Drill with Automatic Indexing

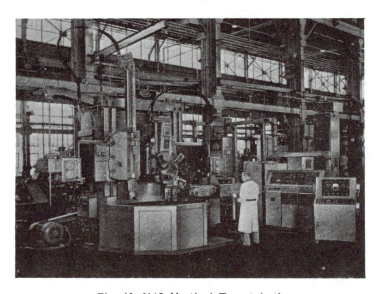

Fig. 10. N/C Vertical Turret Lathe

and 6 N/C has been successfully applied to engine lathes, turret drill presses, horizontal boring mills, vertical turret lathes, radial drills, surface and other type grinders, large multi-spindle profilers, and complex machines that mill, drill, bore, tap and ream at any angle to the work-piece in addition to having the capability of automatically changing cutting tools.

Figure 9 shows a point-to-point turret type drill press. The main difference between this type machine and that shown in Fig. 5 is in the turret which automatically rotates to bring the proper tool into cutting position in accordance with the tape instructions thus saving the delay time required for manually changing the cutting tools.

Figure 10 shows a large vertical turret lathe (sometimes referred to as a vertical boring mill) having one vertical moving ram, a five-sided turret which also moves in a vertical direction and a third cutting mechanism which moves in and out horizontally from the side. This machine may be equipped with either a point-to-point or a contouring system depending on the type of work to be performed.

Figure 11 shows a large numerically controlled multi-spindle vertical profiler capable of contour milling three parts simultaneously.

The machining center shown in Fig. 12 has the capability of performing numerous operations such as milling, boring, turning, drilling, tapping, and reaming without removing the work-piece from the table. A rotary table and head also make it possible to machine any side of a part, except that resting on the table, without removing the part from the machine. Any one of sixty cutting tools may be selected by noting its associated code on the tape. Also the part itself when fixed to a specially designed pallet can be automatically positioned and removed from the work-table. Theoretically the machine is capable of cutting material over 95 per cent of the time which typifies the high efficiency of N/C equipment.

How is the tape prepared?

The tape may be prepared either manually by the use of a special typewriter which is designed to punch the proper holes or automatically by means of a large scale general

Fig. 11. N/C Multiple Spindle Milling Machine

Fig. 12. N/C Machining Center with Automatic
Tool and Pallet Changer

purpose electronic computer. The tape for most point-to-point machining and for very simple contouring can be prepared efficiently by hand. However, with the more complicated point-to-point pieces as well as the great majority of contour parts it is more practical to utilize a computer because of the lengthy mathematical calculations involved. Whether the tape is prepared manually or by the use of a computer the final arrangement of the whole pattern on the tape is identical.

A flow diagram showing the steps required for preparing a tape manually is outlined in Fig. 13. The key individual in the flow pattern is the part programmer, whose respon-

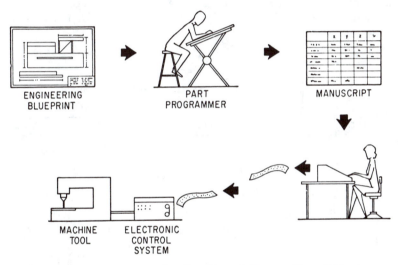

ENGINEERING
BLUEPRINT

PART
PROGRAMMER

MANUSCRIPT

MACHINE
TOOL

ELECTRONIC
CONTROL
SYSTEM

Fig. 13. Flow Pattern for the Manual Preparation of Tape

sibility it is to detail the step-by-step machining instructions from data supplied from the engineering blueprint. The "manuscript," which is prepared by the part programmer, lists the machine tool movements together with the proper speeds, feeds and any other automatic functions required of the machine tool. The manuscript is then forwarded to a typist who prepares the tape which may then be inserted directly into the electronic control system. Since the machine's operations depend entirely on the coded tape instructions which are prepared directly from the manuscript, the manuscript must be precise and entirely free from error.

Although more steps are involved when utilizing a computer for assistance in preparing tapes, the overall length of time required is generally far less since computers have the capability of performing calculations in a minute fraction of the time required by humans. The manuscript that the part programmer prepares for the computer is less detailed and contains far fewer calculations than that required by the typist who must prepare the tape directly for the machine tool control system. Tapes that would take several weeks to prepare manually can be prepared in a few hours with the assistance of a computer. The flow pattern required in preparing tapes utilizing a computer is shown in Fig. 14. The description of the part to be machined

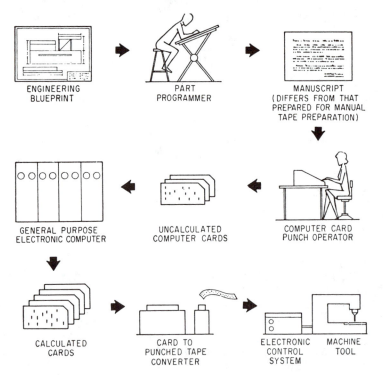

Fig. 14. Flow Pattern for Preparing Tapes with the Assistance of a Computer

plus generalized instructions noted on the manuscript, are typed on cards which are then processed by the computer. Actual computer processing time may require from a few minutes to a half hour, depending on the complexity of the part. Calculated cards, which are the computers output are then processed through a converter which automatically punches the tape.

How is a point-to-point part programmed manually?

Let us assume that it is required to drill several holes in the flat stock shown positioned on the table in Fig. 15. Pro-

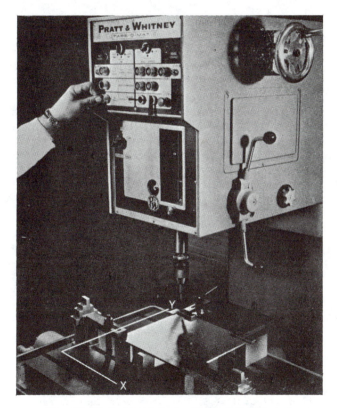

Fig. 15. Part About to be Drilled Positioned on an N/C Drill Press

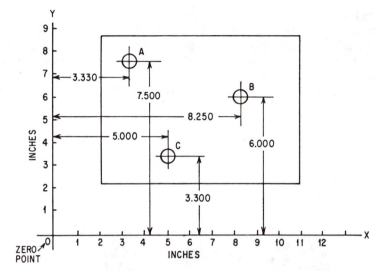

Fig. 16. Positions of Holes are Described as Distances
from the X and Y Axes

gramming the part would first entail positioning it with
respect to two lines or axes that intersect at right angles,
the horizontal line being designated as the X axis and the
vertical line as the Y axis as shown in Fig. 16. The dis-

MANUSCRIPT			
POINT	x	y	COMMENT
A	3.330	7.500	Use 1/2" Drill
B	8.250	6.000	
C	5.000	3.300	

Fig. 17. Simplified Manuscript Form—Special Instructions Peculiar to
the Machine and Control System have been Omitted for Simplicity
Although these Special Notations are Seldom Complex and
Are Easily Learned

tances between hole centers are described on the "manuscript" as x and y distances from these axes which is consistent with the standard mathematical coordinate system. If hole A is to be drilled first the coordinates, or distances from the X and Y axes are noted first on the manuscript as shown in Fig. 16. The x dimension for point A, which is the distance from the Y axis, parallel to the X axis, is 3.330 inches and this is entered in the x column, as shown in Fig. 17. The y dimension which is the distance up from the X axis and parallel to the Y axis is described as 7.500 inches on the manuscript in the y column. Next hole B is to be drilled and the distances from the X and Y axes are noted on the manuscript. The x dimension for hole B would be 8.250 inches and the y dimension or distance up from the X axis is 6.000 inches. Point C is then described in the same manner. After the manuscript is completed, the x and y dimensions are typed on tape together with any special codes peculiar to the machine tool. In setting up the job it is necessary that the drill be positioned accurately with respect to the part. In the example shown, the operator would "target" the drill over a point on the table which is consistent with the "0" point described in Fig. 16. With some control systems it is required that the part be accurately positioned to specific points on the table. After the tool is positioned with respect to the work-piece, the next step is to push the start button. The drill will automatically position over the center of hole A; perform its drilling operation at a speed and feed that may either be set manually on the machine or controlled directly from tape instructions. After the drill has retracted from hole A it will then move to the B location, perform its operation and then move on to C. After completing hole C, the drill may be either stopped above hole C or moved to any other convenient point, such as back to the "0" target point.

What is the difference between a part program and a computer program?

A part program is a detailed description of a particular part which is in a form that can be inserted into a computer or used directly in manually preparing a tape. It is required that the part programming information which is to be used for manually preparing the tape be far more detailed than a part program that is to be inserted into a computer since the computer performs the bulk of the calculations. The part programming information is listed on a form known as a manuscript. A part programmer is a person who prepares part programs. His background generally consists of well rounded machine shop experience together with a thorough knowledge of blueprint reading and shop mathematics including trigonometry. An understanding of analytic geometry has been found to be very helpful for contour programming although it is generally not necessary with point-to-point programming. It is not at all necessary that the part programmer understand the functioning of a computer or the higher mathematics involved. An ideal part programmer is usually a person that has a good grasp of high school mathematics, including trigonometry, and has had in addition training in either a two year technical school or in an apprenticeship program. Machine shop experience coupled with the above training is naturally extremely desirable. It goes without saying that a knowledge of cutters, cutting practices, and the machine tool is a must.

Courses are being established at various technical institutes throughout the country which cover both the operation of numerical control equipment and the programming function.

A computer program differs from a part program in that it consists of a generalized mathematical solution with associated computer routines which is inserted into a computer prior to the computer's acceptance of the part program. A computer program is therefore used to solve the part program. The preparation of a computer program generally involves a relatively high degree of complex mathematics in addition to computer knowledge. Numerical control computer programmers generally possess advanced mathematical degrees in addition to general computer operation experience. However, a computer programmer's knowledge of machine tool practice need not be too extensive.

How is a point-to-point manuscript prepared for computer programming?

Although programming for most parts requiring point-to-point machining can be successfully and economically performed manually, certain configurations of hole patterns lend themselves nicely to computer programming. Consider for example, the bolt hole pattern shown in Fig. 18. Calculating the x and y coordinates of all twelve hole centers would be time consuming, requiring trigonometric tables in

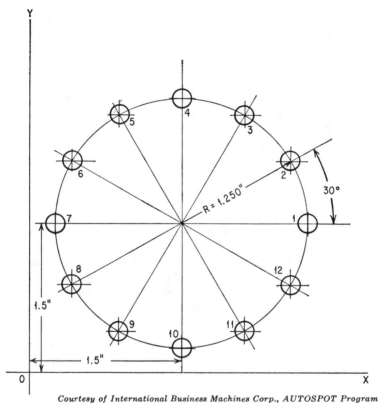

Courtesy of International Business Machines Corp., AUTOSPOT Program

Fig. 18. This Bolt Hole Pattern can be Programmed within Minutes with the Aid of a Computer

order to determine the sines and cosines of the numerous triangles about the center of the circular pattern.

The computer program consists of one shorthand statement describing the particulars of the pattern. The program which would be typed on input computer cards (one punch per letter or number) is as follows:

Drill, 0101/DAA, AT (1.500, 1.500) R (1.250)
SA (0.0) 1A (30.0) NH (12) $

Explanation for the above statements is as follows:

Drill = Denotes the operation which is to be performed

0101 = The assigned number of the cutting tool describing the point angle as well as the diameter of the drill

DAA, AT (1.500, 1.500) = Describes the center of the hole pattern as being 1.5 inches in the x direction and 1.5 inches in the y direction

R (1.250) = Denotes that the radius of the hole pattern is 1.250 inches

SA (0.0) = SA stands for starting angle and describes the angle that hole No. 1 makes with the horizontal, or x axis. In this instance the center of the hole lies on a line parallel to the x axis, consequently the angle is 0.0 degrees

IA 30 = IA stands for incremental angle and specifies the angle between the holes. In this instance it is 30°

NH 12 = NH stands for the number of holes. In this case there are 12

$ = Denotes the end of the instruction. In this case the program

The computer, after accepting the above instructions, calculates all of the necessary x and y coordinates of the hole centers in addition to specifying special machine notations on the output computer cards. As described in Fig. 14 the calculated cards are then processed through a card-to-tape converter which automatically prepares the finished tape.

It should be pointed out that, prior to inserting the program for a specific part such as described above, it is necessary that the computer be primed by first inserting information and generalized mathematical data which will enable the computer to accept and calculate the information describing the particular part. Quite a number of generalized computer programs are available, both for point-to-point and contour programming.

How is a contour part program prepared for a computer?

Just as with point-to-point computer programs there are a number of contour computer programs available. One such program that has been accepted far more than any other is known as the APT program. APT stands for Automatically Programmed Tool. In place of countless numerical values the APT program substitutes an English-like abbreviated language. As with the point-to-point computer program described earlier each abbreviated English-like command is interpreted by the computer and considered along with the calculating functions performed by the computer.

Figure 18 shows a part, the perimeter of which is to be machined. The part may be considered to be a flat cam. As will be observed there are few dimensions noted which describe the part. Instead certain abbreviated characteristics such as LINE 1, PT 1, and CIR are noted.

The "Part Description" shown below the illustration defines the part characteristics. The "SET PT" which is the starting point of the cutting tool is noted as having coordinates $x = 1.0''$ and $y = 1.0''$. "PT 1" is defined as a point having coordinates of $x = 1.0''$ and $y = 3.0''$. "Line 1" is a line through "PT 1" and at an angle of zero degrees with the horizontal X axis. The circular arc which is noted as "CIR 1" is defined as a circle whose center is "PT 2" and having a radius of 2.0''. Under "Machining Instructions" are noted the 1'' diameter of the cutting tool; the tolerance required in inches; and the feed rate in inches per minute. The next instruction specifies turning on the spindle and coolant. The fifth and sixth lines instruct the cutter to go from the "SET PT," in the direction of point 1, to "LINE 1." Line eight instructs the cutter to make a right turn and go along "LINE 1." "TL RGT" means that the cutting tool is to keep to the right of the part when looking in the direction of travel. The "GØ FWD/CIR 1" and "GØ LFT/LINE 2" mean that the cutter is to continue around the arc and make a left turn along "LINE 2." "GØ PAST/LINE 1" and "GØ/SET PT" instructs the cutter to continue along "LINE 2," past the extension of "LINE 1," and then go to the starting point which is noted as "SET PT." "ØFF SPIN" and "ØFF KUL" stop the spindle rota-

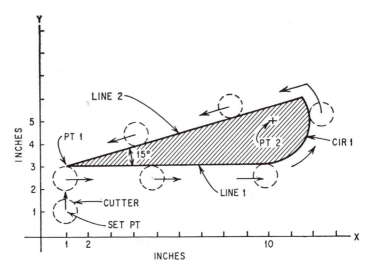

Fig. 19. Example of a Part Program in the APT Language

tion and coolant flow. The notation END stops the calculation of the program in the computer and FINI designates that the part is completed.

Both the Part Description and Machining Instructions as shown in Fig. 19 are typed on computer cards exactly as shown and inserted into the computer. The computer interprets the APT language noted on the cards and transforms the information into mathematical terms which are then calculated and transformed into the coded language required for operating the machine tool. These instructions are then transferred to punched tape which is inserted into the reading head of the machine control unit.

Definitions of Shop Terms in General Use

Definitions of terms that are frequently used in the mechanical industries will be found in this section. Many of these terms are used in this treatise on machine-building practice. This section is not intended as a complete mechanical dictionary, but is restricted chiefly to terms which are often used in the machine shop and, in some instances, in connection with machine design.

ABRASIVE.—An abrasive is the material used in making grinding wheels or abrasive cloth and paper and it may either be natural or artificial. The natural abrasives, such as emery and corundum, have been replaced largely by artificial abrasives, of which there are two general classes. One class is known as *silicon carbide* abrasives and the other as *aluminous* abrasives. The raw materials used in making the silicon carbide abrasives are pure glass or silica sand and carbon supplied by coke of various grades. The aluminous abrasives are made from bauxite, which is mined in the southern part of the United States and in various other parts of the world.

ADAPTER.—The term "adapter" is commonly applied to any device for holding a milling cutter or arbor, which, without an adapter, would not fit into the spindle hole or onto the spindle "nose", as the case may be. For example, the standard taper for milling machine spindles is 3½ inches per foot, and the largest diameter of a No. 50 taper is 2¾ inches. If the comparatively small shank of an end-mill, for example, is to be held in this spindle, an adapter must be used. The outside of this adapter fits into the machine spindle, and a hole in the center has the same taper as the shank of the end mill. This term adapter may also be applied to work-holding or other devices which serve as an intermediate supporting member.

ADDENDUM.—Height of gear tooth above pitch circle or the radial distance between the pitch circle and the top of the tooth.

AERO-THREAD.—A screw thread system especially applicable to high-strength studs and cap-screws inserted in holes tapped in aluminum and magnesium alloys. The screw does not bear directly in the tapped hole but against a coiled hard bronze insert. This insert fits into a tapped 60-degree V-thread in the hole and also in a shallow circular groove formed around the screw. The main object in using the insert is to utilize fully the strength of steel screws when inserted in relatively soft light alloys.

ALLOWANCE.—The term "allowance," as applied to the fitting of machine parts, means a difference in dimensions prescribed in order to secure classes of fits: For instance, if the hole in a crank disk is 3 inches in diameter and the shaft is made 3.005 inches in diameter in order to secure a forced fit, the 0.005 inch would represent the *allowance* for that part. According to American Standard terminology, allowance is the minimum clearance (positive allowance) or maximum interference (negative allowance) between such parts.

ALLOY.—An alloy is an intimate mixture of two or more metals melted together. Mixtures of this kind are generally mechanical in their nature, but are homogeneous; in some cases, they may form chemical compounds. As a rule, when two metals are melted together to form an alloy, the substance formed is, for all practical purposes, a new metal. Brass, bronze, and German silver are examples of well-known alloys.

Non-Ferrous Alloys: Alloys may be divided into ferrous and non-ferrous; the former contain iron as their chief component, while the latter do not. The most important of the ferrous alloys are the alloy steels and castings. Of non-ferrous alloys, the bronzes, brasses, aluminum, and magnesium alloys are the most important.

ALLOY STEEL.—A steel containing some metallic element other than iron and carbon, such as nickel, chromium, tungsten, vanadium, etc., is generally known as an "alloy" or "special" steel. These various metals when added to steel in certain (generally small) percentages, give it distinct properties.

AMERICAN WIRE GAGE.—A gage used for bare and insulated wire of aluminum and copper; for all bare wire made of brass, phosphor-bronze, German silver, or zinc; for resistance wire of German silver or other alloys; for rods of brass, copper, and aluminum; for sheets of brass, phosphor-bronze, aluminum, and German silver. The American Wire Gage is also known as the Brown & Sharpe.

AMERICAN ZINC GAGE.—A gage used in the United States for sheet zinc. The Continental Zinc Gage is used in Belgium, France, and Germany for zinc sheets, and has metric sizes and weights.

ANGLE DIAMETER.—The pitch diameter of a screw thread is sometimes called the "angle diameter," because it is measured in the angle of the thread either by using a special type of micrometer or by means of the well-known three-wire method. The pitch or angle diameter is the diameter measured halfway between the theoretical top and bottom of a screw thread, and, therefore, equals the theoretical outside diameter minus the thread depth. The term "pitch diameter" is recommended.

ANGLE OF THREAD.—The angle included between the sides of the thread measured in an axial plane.

ANGLE PLATE.—A cast iron or forged piece having two surfaces at an angle to each other, usually a right angle, and used for holding work to be machined. One face is clamped to the machine face-plate or table and the work is supported by the other face.

ANNEALING.—Annealing is a heating and cooling operation of a material in the solid state. *Note a.*—Annealing usually implies a relatively slow cooling. *Note b.*—Annealing is a comprehensive term. The purpose of such a heat-treatment may be: (1) To remove stresses; (2) To induce softness; (3) To alter ductility, toughness, electrical, magnetic or other physical properties; (4) To refine the crystalline structure; (5) To remove gases.

The temperature for the annealing operation and the rate of cooling depend upon the material being heat-treated and the purpose of the treatment. Certain specific heat-treatments such as "normalizing" and "spheroidizing" come under the comprehensive term "annealing."

APRON.—In machine construction, a protecting cover which encloses a mechanism is sometimes called an apron, this term being applied particularly to the apron of a lathe which covers the mechanism employed for transmitting feed motion to the cross-slide and for engaging and disengaging the feed motion of either the cross-slide or the carriage.

ARBOR.—The name *arbor* is often used to designate a shaft or spindle that is employed for holding bored parts while turning the outside surfaces in a lathe, but tools of this class are known as "mandrels" by most small-tool manufacturers, whereas the spindles or supports for milling cutters, saws, etc., are called "arbors." In many machine shops and tool-rooms, the term arbor is commonly used to indicate a tool or shaft for holding parts while turning, although some forms of work-holding devices are known as mandrels even by those who ordinarily use the name arbor. The presses used to force mandrels into bushings, etc., preparatory to turning are usually known as *arbor* presses.

ARC OF ACTION.—Arc of the pitch circle through which a gear tooth travels from the first point of contact with the mating tooth to the point where contact ceases.

ARC OF APPROACH.—Arc of the pitch circle through which a gear tooth travels from the first point of contact with the mating tooth to the pitch point.

ARC OF RECESSION.—Arc of the pitch circle through which a gear tooth travels from its contact with the mating tooth at the pitch point to the point where its contact ceases.

AUTOMATIC SCREW MACHINE.—A type of machine tool adapted to a large variety of operations, not only on parts which are turned from bar stock, but on separate castings or forgings that are automatically fed to the machine by a special feeding mechanism. Characteristic features of screw machines in general are means for automatically locating successive tools in the correct working position, the automatic changing of feeds and speeds to secure economical operation, and the presenting of new stock to the tools for a similar series of operations. These various movements, which are entirely automatic, are obtained principally from cams which are rotated at predetermined speeds, and are so formed and set relative to one another that the parts of the machine which they control all operate at the proper time, and at suitable speeds.

AXLE LATHE.—An axle lathe is a type that is equipped with two tool carriages, so that both ends of a car axle may be turned at the same time. On most lathes of this class, the axle is revolved by a special driving head, which is located in the center of the lathe bed. The axle is gripped in the middle by clamps on the head, and the ends are supported by tailstocks. With this arrangement, the work is rotated on "dead centers" (non-rotating centers), which is desirable, and the ends are accessible for the turning operations.

BABBITT METAL.—Babbitt is the name given to a large variety of white metal alloys used as linings for bearings. The name is derived from that of the inventor, Isaac Babbitt, who, in 1839, obtained a patent for a special type of bearing enclosing a soft metal alloy. The exact composition of the original babbitt metal is not known, but the ingredients were copper, tin, and antimony, in approximately the following proportions: 89.3 per cent tin; 3.6 per cent copper; and 7.1 per cent antimony.

BACK-GEARS.—Back-gears are applied to various types of belt-driven machine tools such as lathes, boring mills, drill presses, etc., in order to increase the range of speeds obtainable with a cone-pulley drive. The fastest speed with the back-gears in mesh is somewhat slower than the slowest speed when driving with the back-gears out of mesh.

BACKLASH.—The play between mating gear teeth or the shortest distance between the non-driving surfaces of adjacent teeth.

BACK PITCH OF RIVETED JOINT.—The distance between the center-lines of any two adjacent rows of rivets is sometimes called the "back pitch." The distance, which is measured at right angles to the direction of the joint, should be at least twice the diameter of the rivets for boiler work.

BACK-REST.—Any support employed in machine tools for supporting revolving work, and specifically applied to rear supports for long, slender shafts or similar work while being turned or ground.

BASIC DIMENSION.—The basic size of a screw thread or machine part is the theoretical or nominal standard size from which variations are made. For example, a shaft may have a *basic* diameter of 2 inches, but a maximum variation of minus 0.010 inch may be permitted. The minimum hole should be of basic size in all cases where the use of standard tools represents the greatest economy. The maximum shaft should be of basic size in all cases where the use of standard purchased material, without further machining, represents the greatest economy, even though special tools are required to machine the mating part.

BELL CENTER PUNCH.—A center punch which is mounted inside of a cone-shaped bell-mouthed casing. By placing the bell-mouthed casing over the end of a bar, the point of the punch is automatically located at the center of a bar.

"BELL MOUTHED" HOLE.—A hole that is enlarged slightly at the ends, instead of being truly cylindrical throughout its length, is said to be "bell mouthed." This may be caused, in internal grinding, by traversing the wheel entirely out of the hole. This may result in a slight deflection, especially if there is any play in the spindle bearings.

BELT CREEP.—A term indicating lateral shifting movement of the belt. If the driving and the driven shafts are not parallel and the pulleys are cylindrical, the belt will creep or move toward the "low side" of the pulley or toward the side where the shafts are closer. This creeping movement is due to the fact that any given point on the edge of the belt adjacent to the low side comes into contact with the pulley before a corresponding point on the opposite side. The result is that the belt is gradually shifted over toward the low side of the pulley.

BENDING DIE.—Dies of this class are designed for bending sheet metal or wire parts into various shapes which are usually irregular and are produced either by pushing the stock into cavities or depressions of corresponding shape in the die or by the action of auxiliary attachments such as slides, etc., which are operated as the punch descends.

BIRMINGHAM GAGE (B.G.).—The British legal standard for iron and steel sheets and hoops. The initial letters B.G. should be used in specifications to distinguish this iron and steel sheet and hoop gage from other gages. This 1914 Birmingham gage differs from the older Birmingham or Stub's iron wire gage.

BIRMINGHAM OR STUB'S IRON WIRE GAGE.—A gage used for wall thickness of seamless tubing of brass, copper, bronze, steel, and aluminum; strip steel, spring sheet steel, saw steel; sometimes applied to flat rolled steel; generally used for iron and steel telephone and telegraph wire. This gage is being superseded by American wire gage. (Stub's iron wire gage differs from Stub's steel wire gage.)

BLACK DIAMOND. — An inferior variety of diamond used in the industries for truing hard grinding wheels. It is more expensive than *bort*, but is more economical to use for hard wheels.

BLANCHARD LATHE.—The Blanchard type of lathe is named after the inventor, Thomas Blanchard, who built the first lathe of this kind in 1822. This machine is designed especially for turning wooden parts of irregular shape, and has been extensively used for turning the stocks of guns and rifles. There is a former or model which corresponds to the shape required.

BLANKING DIE.—Dies of the "blanking class" are used for cutting blanks usually from flat sheets or strips of stock; such blanks may or may not be drawn, formed or bent, either by other parts combined with the blanking members, or by means of separate dies. If the chief or only function of the die is to cut blanks, it is a blanking die; if the blanking operation is followed by a more important operation in the same die such as drawing, then the term drawing die would be applied, the blanking part being considered a secondary feature of the design.

BLIND HOLE.—In machine construction, a hole which does not pass through a part but has a closed inner end is commonly known as a "blind hole."

BLOCK INDEXING.—With the multiple or "block" system of indexing, which is sometimes used in gear-cutting, a number of teeth are indexed at one time instead of cutting the teeth consecutively, and the gear is revolved several times before the teeth are all finished.

BOLSTER.—Dies are usually held in position on the bed of a punch-press by means of a *bolster* or *diebed*, although large dies are often attached directly to the press bed. The principal functions of a bolster are: 1. That of supplying an adequate support for the die, and means of holding the die in its proper position relative to the punch. 2. To furnish a means of attachment to the press.

BOLT CUTTER.—The machines used for cutting the threads on bolts are known as "bolt cutters." A typical design is called a *single* bolt cutter because it has one spindle. Some bolt cutters have two, three, or four spindles and are known as *double, triple,* and *quadruple* bolt cutters, respectively. The thread is cut by means of a die-head attached to and revolved by the spindle of the machine.

BORT.—Bort is an inferior variety of diamond which is used in the industries for truing soft grinding wheels and for making diamond dies for wire drawing and similar purposes. It is not as hard as the variety of diamond known as the *carbon* or *black diamond,* and is considerably lower in price; but it is not as economical to use as the black diamond for truing hard grinding wheels. Bort is a semi-transparent stone known as an "imperfect brilliant," and, therefore, is useless as a precious stone.

BOX-JIG.—A jig which is made somewhat in the form of a box, and encloses the work, thus permitting guide bushings to be placed on all sides and also making it comparatively easy to locate and securely clamp the part in the proper position for drilling. As a rule the piece to be drilled can be inserted only after one or more covers or leaves have been swung out of the way.

BOX-TOOL.—A type which is equipped with some form of back-rest opposite the turning tool for supporting the work; it usually encloses or surrounds to some extent the part being turned; hence the name a *box-tool*. Tools of this type are extensively used on turret lathes and screw machines for turning parts from bar stock. There are many different designs.

BRAKE FOR BENDING.—A bending brake is a form of press used in sheet-metal work for forming strips and plates. Brakes are made in both hand-operated and power-operated types. As compared with other presses for forming sheet metals, brakes are wide between the housings and are designed for holding long, narrow forming edges or dies for giving the flat stock whatever shape is required. Brakes are used extensively in the manufacture of various kinds of metal furniture, and for miscellaneous sheet metal bending and forming operations.

BRASS.—Brass is an alloy composed mainly of copper and zinc as a general rule. Brass castings usually contain 65 per cent of copper and 35 per cent of zinc. So-called "low" brasses, which are especially suitable for hot-rolling, contain from 37 to 45 per cent of zinc. The "high" brasses, which are used for cold-rolling and drawing, contain from 30 to 40 per cent of zinc. If lead is present to an amount exceeding 0.1 per cent, the ductility of brass is decreased, and sheet brass intended for drawing should be as free from lead as possible.

BRASS LATHE.—A "brass lathe" is so called because it is designed especially for operating on brass parts of various kinds. A typical machine is equipped with a screw-chasing attachment similar to the type found on Fox or monitor lathes, and the tailstock is mounted on a cross-slide. There is no tool carriage, but a T-rest for supporting hand-manipulated tools. When a turret is added to a lathe of this general design, it may be known either as a *turret lathe*, *Fox lathe*, or *monitor lathe*.

BRAZING.—Brazing is a method of joining metal parts together by means of an alloy known as *spelter solder*, or simply as *spelter*, which is melted into the joint and unites with the metals. Brazing is practically the same as *hard soldering*, but, according to the commonly accepted meaning of the two terms, there is the following distinction: Brazing means the joining of metals by a film of brass (a copper-zinc alloy); hard soldering is the term ordinarily applied when silver solder is used, the latter being an alloy of silver, copper, and zinc.

BRIGGS PIPE THREAD.—The Briggs pipe thread which is used for pipe joints and is the standard for this purpose in the United States, derives its name from Robert Briggs, and is now known as the American Standard.

BRINELL HARDNESS TEST.—The Brinell test for determining the hardness of metallic materials consists in applying a known load to the surface of the material to be tested through a hardened steel ball of known diameter. The diameter (or depth) of the resulting permanent impression in the metal is measured. The Brinell hardness number is taken as the quotient of the applied load divided by the area of the surface of the impression, which is assumed to be spherical.

BRANCH PIPE.—A very general term used to signify a pipe either cast or wrought, that is equipped with one or more branches. Such pipes are used so frequently that they have acquired common names such as tees, crosses, side or back outlet elbows, manifolds, double-branch elbows, etc. The term branch pipe is generally restricted to such as do not conform to usual dimensions.

BROACH.—A broach is a metal-cutting tool having a series of teeth which increase slightly in size from one end of the tool to the other, so that, when the broach is forced through a hole or is traversed by the work, as in external broaching, the teeth successively cut the hole or surface to the required form. Broaches may be *pull broaches* or *push broaches*, depending upon whether they are to be pulled or pushed through the hole in the work.

BRONZE.—Bronze is an alloy composed mainly of copper and tin in variable proportions, and sometimes containing small percentages of zinc, antimony, lead, aluminum, phosphorus, or manganese.

BROWN & SHARPE TAPER.—A standard taper used for taper shanks on tools such as end mills and reamers, the taper being approximately ½ inch per foot for all sizes except for taper No. 10, where the taper is 0.5161 inch per foot. Brown & Sharpe taper sockets are used for many arbors, collets, and machine tool spindles, especially milling machines and grinding machines.

BROWN & SHARPE WIRE GAGE.—The Brown & Sharpe wire gage, also known as the American wire gage, is the gage universally recognized in the United States as the standard gage for copper wires and wires of metals other than steel. The diameters of the wires of successive numbers increase according to a geometrical ratio. The diameter of each succeeding number can be found by multiplying the diameter of the preceding number by 1.123, this being the ratio of the geometrical progression. The Brown & Sharpe Wire Gage is generally used in the United States for all bare wire of brass, copper (except bare copper telephone wire) phosphor-bronze, German silver, aluminum, and zinc; for resistance wire of German silver and other alloys; for insulated wire of aluminum and copper; for rods of non-ferrous metals, such as brass, copper and aluminum; and for sheets of brass, phosphor-bronze, aluminum and German silver.

BUFFING.—Buffing is the process of obtaining a very fine surface, having a "grainless" finish, on metal objects, by means of soft wheels of felt to which a fine polishing material is applied, or by wheels formed of layers of cotton cloth. The term "buffing" is often used interchangeably with "polishing." The operation is performed with any wheel to the face of which the abrasive is loosely applied.

BULLDOZER.—A machine especially adapted for bending operations. It is closely allied to the forging machine, and many operations can only be done successfully on forging machines when the bulldozer is used for performing a preliminary operation.

BULL WHEEL.—The gear in a planer drive which meshes with the rack beneath the platen and through which the motion of the platen is obtained.

BURNISHING.—The burnishing of metals is a method of securing smooth finished surfaces by compressing the outer layer of the metal, either by the application of highly polished tools, or by the use of steel balls which, by rolling contact, produce smooth surfaces.

BUSHING, PIPE.—A pipe fitting for the purpose of connecting a pipe with a fitting of larger size, being a hollow plug with internal and external threads to suit the different diameters.

CABINET FILE.—A cabinet file is flat on one side and rounded on the other, and it is wider and thinner than a regular half-round file. It is double-cut, with coarse, bastard teeth. This type is made for cabinet makers and wood-workers generally.

CALIBRATION.—Calibration, in its mechanical sense, denotes an accurate comparison of any measuring instrument with a standard, and more particularly the determination of the errors of a scale used in a measuring device. The term "calibration" refers not only to measurements of length, but to measurements of all other engineering units; thus, for example, ammeters, voltmeters, pyrometers, dynamometers, and all other measuring instruments, are calibrated by comparison with a standard.

CALKING.—The riveted joints of steam boilers and other vessels which subjected to pressure are made tight by "upsetting" and compressing the metal along the edges of the joint, which operation is known as *calking*. The calking tool by means of which the material is compressed is either operated by a pneumatic hammer or, if such a tool is not available, it is struck repeatedly by a hand hammer.

CAMS.—Many machine parts require either an intermittent or an irregular motion. The most common method of obtaining an irregular motion is by means of cams which have grooves or surfaces of such shape or form that the required motion is imparted to the driven member when the cam is in motion. The exact movement derived from any cam depends upon the shape of its operating groove or edge which may be designed according to the motion required.

CANT-FILE.—Cant-files and cant-saw files are files of triangular cross-section, and differ in cross-section as to their angles; the cant-file has 30, 30, and 120 degree angles and the cant-saw, 35, 35, and 110 degree angles. The cant-saw shape was formerly known as "lightening." It is used principally for filing cross-cut saws having N-shaped teeth.

CAPE CHISEL.—A form of chisel for metal cutting, having a narrow blade for the cutting of grooves or keyways.

CAP-SCREW.—A screw having either a hexagonal or a slotted head, and (according to the American Standard) ranging in diameter from ¼ inch up to certain maximum sizes which vary from ¾ inch for flat-head and button-head screws to 1¼ inches for the hexagonal form. The screw thread ordinarily conforms to the Coarse-thread Series of the American Standard.

CAPSTAN LATHE. — In England, turret lathes are often called *capstan lathes.* The terms *"capstan"* and *"turret,"* however, are sometimes used interchangeably, although many firms observe a sharp distinction in their application, in that they apply the name "capstan" only to those machines which have a slide moving in a saddle that is bolted down to the bed, whereas the name "turret" is used when the turret-slide is mounted directly on the bed.

CARBIDE TOOLS.—Metal-cutting tools made of cemented or sintered carbides are used in various branches of the machine-building industry either because of the exceptionally high cutting speeds that are possible with these tools, or because of their durability and adaptability in machining either very hard materials or compositions that are destructive to other tools.

Sintered carbides are made in a number of different grades (with different combinations of hardness and toughness) to adapt them to the cutting of a large variety of materials. For example, some carbides are especially adapted to cutting cast iron (as well as a variety of other materials) and there are other grades that are more effective as applied to steel. These steel-cutting carbides include tantalum or titanium in the binder, thus changing the properties of the carbide so as to resist the seizing action of the steel chips and prevent formation of a chip cavity back of the cutting edge.

CARBON STEEL.—The expression "carbon steel" is applied to tool steel containing no alloying metals, the term being used to distinguish such steel from alloy steels which contain, in addition to carbon, tungsten, nickel, chromium, or other metals. The carbon content in steel is generally expressed by giving the percentage of carbon as, for example, 0.90 per cent carbon steel.

CARBORUNDUM.—Carborundum is an abrasive which is produced artificially in the electric furnace; it is a chemical combination of the two elements, carbon and silicon, *carborundum* being a trade name.

CARBURIZING.—Adding carbon to iron base alloys by heating the metal below its melting-point in contact with carbonaceous material. *Note.*—The term "carbonizing" used in this sense is undesirable and its use should be discouraged.

CASEHARDENING. — Carburizing and subsequently hardening by suitable heat-treatment, all or part of the surface portions of a piece of iron base alloy.

Case: That portion of a carburized iron base alloy article in which the carbon content has been substantially increased.

Core: That portion of a carburized iron base alloy article in which the carbon content has not been substantially increased.

CAST IRON.—According to the specifications adopted by the International Association for Testing Materials, *cast iron* is defined as iron containing so much carbon that it is not malleable at any temperature. To conform to this definition, iron containing more than 2.2 per cent of carbon is classified as cast iron. Generally, commercial cast iron, however, has a carbon content of between 3 and 4 per cent. This carbon may be present as graphite, in which case the iron is known as *gray cast iron*, or it may be present in the form of cementite or combined carbon, in which case the iron is known as *white cast iron.*

CAST STEEL.—The term "cast steel" is sometimes used to designate what is known as tool steel or crucible steel, but this usage is becoming more and more obsolete and should be discontinued, as it is confusing. Steel castings made by pouring molten steel into suitable molds are sometimes referred to as cast steel, but the latter term should not be applied to the high-carbon steel which is made by the electric or crucible processes and is suitable for cutting tools.

CHANGE-GEARS. — A term indicating that gearing on a machine tool is changeable for obtaining whatever ratio may be required. Prior to the introduction of the quick change gear mechanism on lathes, an assortment of these gears was provided with every screw-cutting lathe, different sizes being employed for cutting threads of various pitches; hence, the name "change-gears." The gears of a milling machine, used to drive a dividing or index head from the table feed screw for such work as cutting spirals or helices, are also known as change-gears, and this term may be applied to various other changeable gears

CHASER.—A chaser is a form of threading tool having a number of teeth instead of a single point like the threading tools commonly used in connection with lathe work.

CHATTERING.—The term "chattering," as applied by machinists, means the formation of slight ridges or nicks upon a part that is either being turned, planed, milled, or ground. Chattering is caused by vibration.

CHILLED CASTING.—A chilled casting is one which has been cooled suddenly by casting it in contact with some material which will rapidly conduct heat away from the surface of the casting. The effect is to produce a surface of great hardness which will withstand considerable wear.

CHIP BREAKER.—The term "chip breaker" indicates a method of grinding turning tools, that will break up the chips into short pieces.

thus preventing the formation of long or continuous chips which would occupy considerable space and be difficult to handle. The chip-breaking form of cutting end is especially useful in turning with carbide-tipped steel turning tools because the cutting speeds are high and the chip formation rapid. Some tools have attached chip breakers.

CHUCKING MACHINE.—A term applied to turret lathes which are not arranged for turning parts from bar stock, but are designed exclusively for machining castings or forgings which must be held in a chuck.

CHUCKING REAMER.—Reamers of this class are so named because they are used largely for reaming parts held in the chuck of some machine such as an engine lathe or turret lathe. The *fluted* type is used for enlarging drilled holes and finishing them true to size; the *rose* type is used for enlarging cored or drilled holes and is so constructed that a considerable amount of metal can be removed by it.

CIRCULAR PITCH.—Length of the arc of the pitch circle between the centers or other corresponding points of adjacent gear teeth.
Normal Circular Pitch: The shortest distance on the pitch surface between the centers or any other corresponding points of adjacent teeth—applied to helical gearing.

CIRCULAR THICKNESS.—The thickness of the gear tooth on the pitch circle—also known as arc thickness.

CLEARANCE IN GEARING.—Radial distance between the top of a gear tooth and the bottom of the mating tooth space.

CLUSTER GEARS.—The term "cluster gears" is applied when two or more gears are formed on one solid piece. Cluster spur gears may be used in geared speed-changing mechanisms because they are stronger and more compact than single gears fastened together.

COLD-ROLLING.—The cold-rolling of shafting or bar stock consists in passing the shaft or bar through burnishing rolls which leave a smooth dense surface. Most shafts and bars, however, which are designated as "cold-rolled," have been finished by a cold-drawing process which involves pulling the stock through dies.

COLLET CHUCK.—The collet type of chuck consists of a split sleeve or collet which has a tapering or conical end that fits into a seat of corresponding taper so that a lengthwise movement of the collet causes a contraction or expansion of the gripping surfaces. The collet type of chuck is used for holding rods or bar stock and it is applied to toolroom lathes, bench lathes, turret lathes (when operating on bar stock), and on turning machines of the automatic screw machine class.

COMPARATOR.—This term often is applied to various types of instruments which are used to check the sizes of parts by comparison with some standard instead of by actual measurement; thus, the operation is one of *checking* rather than *measuring*. To illustrate the principle, a comparator may have an indicating hand which points to zero when the size of the work conforms to the known size of a standard, such, for example, as a plug gage or possibly a combination of precision gage blocks.

COMPOUND REST.—The compound rest of a lathe consists of an upper slide mounted on the lower or main cross-slide. The upper slide can be turned to any angular position so that the tool, which ordinarily is moved either lengthwise or crosswise of the bed, can be fed at an angle. The base of the compound rest is graduated in degrees and the position of these graduations shows to what angle the upper slide is set. The term *compound slide* is sometimes used.

COUNTERBORING.—This operation is for the purpose of enlarging some part of a cylindrical bore or hole. For example, if a machine screw hole is enlarged at one end to receive a fillister-head screw, this is counterboring and the tool used is known as a *counterbore*. Counterboring is also done in connection with lathe and boring-mill work, as, for example, when the bore of a cylinder is enlarged at the ends to form a clearance space or counterbore, as it is called.

COUNTERSINKING.—This operation, which is performed by a conical form of cutting or reaming tool, is for the purpose of tapering or beveling the end of a hole, as, for example, when the end must be made conical to form the seat for a "flat head" or countersink type of machine screw. When a hole is enlarged at one end by counterboring, the enlarged part is cylindrical in form; but when the enlargement is tapering or conical, this is known as countersinking.

COUPLING, PIPE.—A threaded sleeve used to connect two pipes. Commercial couplings are threaded inside to suit the exterior thread of the pipe. The term coupling is occasionally used to mean any jointing device and may be applied to either straight or reducing sizes.

CRITICAL TEMPERATURES.—The temperatures at which certain changes in the chemical condition of tool steel take place, during both heating and cooling, are referred to as the decalescence and recalescence or critical points. The critical points vary for different kinds of steel and must be determined by tests in each case. It is the variation in the critical points that makes it necessary to heat different steels to different temperatures, for hardening.

CROSS-CUT FILE.—A type having one round edge with sides tapered toward the opposite edge. A cross-cut file is single-cut, the same as a mill bastard file of the same size.

CROSSING FILE.—Crossing or cross files have a double oval section, one side being shaped like a half-round file and the other like a cabinet file. The cut is either bastard, second-cut, or smooth.

CROWN GEAR.—In bevel gearing, when the pitch-cone angle of one of the gears is 90 degrees, this gear is called a *crown gear*. In this case, there is, properly speaking, no pitch cone, but rather a pitch plane. The crown gear of bevel gearing is equivalent to the rack of spur gearing.

CUTTING-OFF MACHINES.—In general, any machine which is designed exclusively for cutting either bar stock or structural steel. Machines which utilize a revolving saw for severing the material are commonly listed as *cold-saw cutting-off machines*, or simply as cold saws. The term "cold" is used in this connection to indicate that the machine is intended for cutting unheated stock. Among other machines used for cutting off stock, which may properly be inserted under the general classification of cutting-off machines, are hacksaw machines, metal-cutting band saws, abrasive wheel cutting-off machines, and the friction saw.

CUTTING SPEED.—The rate at which metal is cut by a tool as in turning, milling, planing, or other metal-cutting operations. Cutting speed is expressed in feet per minute. This speed may represent the movement of the part being cut as in turning or planing, or the speed of the metal-cutting tool as in drilling or milling.

CYANIDING.—Surface hardening of an iron base alloy article or portion of it by heating at a suitable temperature in contact with a cyanide salt, followed by quenching.

DECALESCENCE POINT. — The decalescence point indicates the temperature at which a decided change in the internal condition of steel takes place, and above which steel must be heated in order that it may be properly hardened by quenching. Generally speaking, the decalescence point of any carbon steel marks the correct quenching temperature of that particular steel. The decalescence point is not the same for all steels, but occurs for most carbon steels at temperatures between 1350 and 1450 degrees F.

DEDENDUM. — Depth of gear tooth space below pitch circle or radial dimension between the pitch circle and the bottom of a tooth space.

DIAMETRAL PITCH. — Ratio of the number of gear teeth to the number of inches of pitch diameter — equals number of gear teeth to each inch of pitch diameter. The diametral pitch system is applied to most of the cut gearing produced in the United States. If gear teeth are larger than about one diametral pitch, it is common practice to use the circular pitch system.
Normal Diametral Pitch: The diametral pitch corresponding to the normal circular pitch and equal to number of teeth divided by product of pitch diameter and cosine of helix angle; also equals diametral pitch in plane of rotation, divided by cosine of helix angle.

DIAMOND LAP. — Very small holes in precision work are often finished after drilling, by using a rotary diamond-charged lap. Laps of this kind are made of mild steel, and the slightly enlarged working end is charged with diamond dust, thus converting it into an efficient grinding wheel for small holes.

DIE-CASTING.—The term "die-casting" generally refers to a casting that has been made in a metallic mold or die, into which molten metal has been forced under the influence of either mechanical or pneumatic pressure. Die-castings do not include so-called "hot-pressed" or "die-pressed" forgings, because in this process, the metal placed in the dies is not molten, but merely in a plastic or semi-plastic condition. The definition also excludes castings that are poured by gravity into metallic molds, the latter generally being known as "permanent-mold" castings.

DIE FOR PRESS WORK.—The term "die" is often applied to an entire press tool including both upper and lower members, while the names "punch" and "die" are used to designate parts or sections of a complete die. These main sections ordinarily are classified with reference to shape, rather than by location, notwithstanding the fact that the punch is usually but not invariably the upper member. When the name "die" is applied to part of a press tool, it refers to the member that has an opening or cavity to receive a punch, for blanking, drawing, or otherwise forming whatever stock or part is confined between the punch and die members.

DIE-SINKING.—Die-sinking is the process of forming an impression in a die for drop forging or other forming operations. It is done by means of a die-sinking type of machine or attachment, often in conjunction with hand chipping, filing, and scraping.

DIVIDING HEAD.—The dividing or indexing head is an attachment used principally on the milling machine for dividing circular parts into equal spaces or divisions, as when cutting gears, fluting milling cutters, etc. It is also used for imparting a rotary motion to cylindrical work (while the latter is fed axially by the longitudinal feeding movement of the table) for milling helical or spiral grooves. The dividing or indexing head is sometimes known as a *spiral head.*

DOUBLE-CUT FILE.—A *double-cut* file has two rows of teeth crossing each other. The angle of the first row is, for general work, from 40 to 45 degrees, and the second row, from 70 to 80 degrees.

DOWEL PINS. — Dowel pins are used either to retain parts in a fixed position or to preserve alignment. For parts which have to be taken apart frequently, and where driving out of the dowels would tend to wear the holes and thus loosen the dowel, and also for very accurately constructed tools and gages which have to be taken apart, or which require to be kept in absolute alignment, the taper dowel is preferable.

DRAW-FILING.—When a file is held at each end and the motion of the file is lateral instead of in a lengthwise direction, this is known as *draw-filing.* With this method of filing, the metal is removed more slowly than by cross-filing, provided the same kind of file is used in each case. The surface is left smoother, however, if the draw-filing is properly done, as the scratches are closer, owing to the shearing or shaving cut taken by the file teeth.

DRAWING DIES.—Drawing dies are used for drawing parts from flat stock into cylindrical and various other shapes. There are several different classes of drawing dies, including plain drawing dies, combination dies, double-action dies, and triple-action dies.

DRAWING STEEL.—Steel is "drawn" or tempered by reheating it after hardening to some temperature below the critical temperature range and then cooling the steel. This heat-treatment is often referred to as drawing, but the term tempering is preferable.

DRILL PRESS.—This term is often applied to metal drilling machines in general, evidently because the drill is pressed or forced through the metal as it revolves.

DRILL ROD.—High-carbon tool-steel rods are generally referred to as *drill rod*. Drill rod is either polished or unpolished. Carbon-steel drill rod is kept in stock by steel manufacturers in various sizes, including the standard "letter" and "number" sizes for drills.

DRIVING FIT.—When a plug or a shaft is made slightly larger than the hole into which it is to be inserted and the allowance is such that the parts are assembled by driving, this is known as a *driving fit*. Such fits are employed when the parts are to remain in a fixed position relative to each other.

DROP FORGING.—Forgings produced in dies by a falling hammer which is lifted by mechanical means and known as a drop-hammer, are called *drop-forgings*. The shape of the forging is cut out or "sunk" into dies, so that often a single blow of the hammer on the dies shapes the heated iron bar to the desired form.

ELBOW (ELL).—A fitting that makes an angle between adjacent pipes. The angle is always 90 degrees, unless another angle is stated.
Service Ell: An elbow having an outside thread on one end. Also known as street ell.
Drop Elbow: A small sized ell that is frequently used where gas is put into a building. These fittings have wings cast on each side. The wings have small countersunk holes so that they may be fastened by wood screws to a ceiling or wall or framing timbers.

EMERY.—Emery is an abrasive which, in the past, has been very extensively used. At one time, practically all grinding wheels were made from it, but artificial abrasives possessing superior cutting qualities are now employed for machine grinding.

EXTRUSION OF METALS.—The extrusion process is a method by means of which shapes of fairly plastic metals are produced by forcing the metal, which is usually heated, under high pressure through an aperture of the shape to be produced. In this manner, a continuous bar or pipe of the cross-section of the aperture or die is produced.

FACE MILLING.—This term as generally used, means the production of a plane surface by the teeth of a milling cutter which operate in a plane that is at right angles to the axis of the cutter.

FACE OF GEAR TOOTH.—That surface of the gear tooth which is between the pitch circle and the top of the tooth.

FACE WIDTH OF GEAR TOOTH.—Width of the pitch surface. The *active face width* is the width which actually makes contact with a mating gear. When herringbone gears have a central clearance groove, the width of this groove is not included in the active face width.

FEED MECHANISM. — The term "feed mechanism" or "feeding mechanism" as applied to machine tools, usually relates to some form of mechanism (1) for feeding either a cutting tool or the work as in turning, planing, drilling or milling, and usually with means of varying the rate of tool or work movement per revolution or stroke; (2) a mechanism for feeding raw material or parts from some source of supply to the working or operating position.

FEED OF CUTTING TOOL.—In metal-cutting operations, such as turning, milling, drilling, planing, etc., the relatively slow advancing movement of the cutting tool, as in turning, or of the work, as in milling, that is required to obtain a progressive cutting action, is known as a feeding or feed movement. For example, in turning the tool feeds a certain distance per revolution of the work; in planing, the tool feeds laterally after each cutting stroke; and in milling, the work feeds past the milling cutter a certain amount per cutter revolution.

FERROUS ALLOY.—Ferrous alloys differ from non-ferrous alloys in that they contain iron. Steel and cast iron are outstanding examples of ferrous alloys, whereas in the non-ferrous group there are the various brass, bronze, aluminum and other alloys.

FILE DEFINITIONS.—The length of a file means the distance from the point to the heel and does not include the tang. The *heel* is that end of the file body adjacent to the handle. A *blunt* file is one having the same sectional shape from the point to the tang. An *equaling* file is similar in appearance to the blunt form, but has a very slight curvature from the point to the tang. The *back* is the convex side of the half-round, cabinet, pit-saw and similar forms. *Bellied* is the term used to describe a file having a fullness in the center. The coarse grades of single-cut files are sometimes called *floats*. *Safe-edge* means that the edge or side is smooth and without teeth, and may be presented to a surface that does not require filing. *Over-cut* is a term used to describe the first series of teeth on a double-cut file. *Up-cut* means the series of teeth superimposed on the over-cut series of a double-cut file. The term *superfine* (or *super*) cut is used by Lancashire file makers to designate the grade of cut known in this country as "dead-smooth." *Taper* is used to distinguish a file having tapering sides from one that is blunt or straight. Custom has also established this term as a short name for "three-square" or triangular hand-saw files.

FITTINGS.—The term "fittings" as applied to pipe work, includes the various parts used in pipe lines for connecting different pipes, *viz.*, ells or elbows, tees, and crosses, as well as pipe flanges. They are made from cast iron, wrought iron, malleable iron, or composition metal.

FIXTURE. — A special work-holding device used on some type of machine tool, especially in the manufacture of duplicate parts of machines. The terms "jig" and "fixture" are frequently used interchangeably, but, as a general rule, a jig is a tool which, while it holds the work, at the same time also contains guides for the respective tools to be used (for example, a *drill jig*), while a fixture only holds the work while the cutting tools are performing the operation upon the piece, without containing any special arrangements for guiding the tools.

FLANK OF GEAR TOOTH.—That surface of the tooth which is between the pitch circle and the bottom land. The flank includes the fillet.

FLAT FILE. — A flat file is parallel in both longitudinal sections, from the heel to the middle, and tapered in both sections from the middle to the point, the thickness of the point being about two-thirds and the width about one-half that of the stock from which the file is made. The flat file is one of the most common files in use and is not confined to any specific class of work, but is employed for a great variety of purposes. Ordinarily, the teeth are double-cut, and either bastard, second-cut, or smooth. A single-cut flat file is preferred for some classes of work.

FLAT TURRET LATHE.—The flat turret lathe is so named because the turret is a flat circular plate mounted on a low carriage to secure direct and rigid support for the tools, from the lathe bed. The tools, instead of being held by shanks inserted in holes in the turret, are clamped firmly onto the low circular turret plate so that they do not overhang, but have an unyielding support directly below the cutting tools. This type of turret lathe was designed by James Hartness. Lathes of the flat-turret class are sometimes referred to as *turntable lathes.*

FLOATING TOOL-HOLDER.—A tool is said to "float" when it is not held rigidly but is free to move within certain limits. The floating or free movement may be in one direction only or in any direction. Many die- and tap-holders are of the floating type to allow the die or tap to move in the direction of its axis, so that it is free to follow its own lead in case the forward movement is retarded by the backward pull or drag of the turret or tool-slide to which it may be attached. When a tool-holder is arranged to allow a die, tap, reamer, or other tool to move laterally or possibly in any direction, this is to permit the tool to align itself in case a hole is slightly off center. When the work to be operated upon is placed in a chuck either by hand or automatically from a magazine, a lateral or universal floating movement for the tools is especially desirable because of the difficulty of chucking parts in perfect alignment.

FLY-CUTTER.—The fly-cutter is a simple type of formed milling cutter that is often used for operations that will not warrant the expense of a regular formed cutter. The milling is done by a single tool or cutting edge which has the required outline and is held in an arbor. The advantage of the fly cutter is that a single tool can be formed to the desired shape, at a comparatively small expense.

FOLLOWER.—The name "follower" is often applied to the driven member of a gear train or other mechanism having a part that receives motion from another member and follows it; usually the "follower" in a train of mechanism is the last driven member.

FOLLOW-REST.—For turning long slender parts, such as shafts, etc., a follow-rest is often used for supporting the work. A follow-rest differs from a steadyrest in that it is attached to and travels with the lathe carriage.

FORCED FIT.—"Forced" or "pressed fit" is the term used when a pin, shaft, or other cylindrical part is forced into a hole of slightly smaller diameter, ordinarily by the use of a hydraulic press or some other type of press capable of exerting considerable pressure. A forced fit has a larger allowance than a driving fit, and therefore requires greater pressure for assembling.

FORGING MACHINES.—Forging machines are made in a variety of designs, some being intended especially for bolt and rivet heading, and others for more general work. The form or shape into which a part is forged is governed by dies of the required shape and also by a heading tool or plunger which bends or upsets the heated bar of metal and forces it into the die impression. The die may have a single impression, or two or three impressions may be required in order to forge the part by successive operations.

FORMED CUTTERS.—When pieces having an irregular outline are to be milled, it is necessary to use a cutter having edges which conform to the profile of the work. This is called *form milling*, and the cutter a *form* or *formed* cutter. There is a distinction between a *form* cutter and a *formed* cutter, which, according to the common use of these terms, is as follows: A formed cutter has teeth which are so relieved or "backed off" that they can be sharpened by grinding, without changing the tooth outline, whereas the term "form cutter" may be applied to any cutter for form milling, regardless of the manner in which the teeth are relieved.

FORM GRINDING.—The grinding of machine parts by using a broad wheel which is shaped to conform to the shape required, and without traversing either wheel or work laterally, is known as *form grinding*. The wheel is wide enough to cover the surface to be ground, and, for round work, is fed straight in, thus grinding the entire surface at the same time, without a traversing movement such as is common to ordinary cylindrical grinding.

FORMING DIES.—Forming dies are a type of dies in which a blank is formed into a hollow shape by simply being pushed into a cavity of the required shape in the die, or a previously drawn cup is given a different shape by compressing it between a punch and die which conform to the shape desired.

FORMING TOOLS.—Tools used on lathes or other turning machines for forming curved or irregular shapes by reproducing the shape of the cutting edge.

FREE-CUTTING STOCK.—The term "free cutting" is applied to stock which may readily be machined and which does not form long tough chips that tend to clog cutting tools. This free-cutting property is especially important in connection with automatic screw machine and turret lathe practice.

GAGE.—Any tool or instrument used for taking measurements might properly be called a "gage," but this term, as used by machinists and toolmakers, is generally understood to mean those classes of tools which conform to a fixed dimension and are used for testing sizes, but are not provided with graduated adjustable members for measuring various lengths or angles. There are exceptions, however, to this general classification.

GAGE TERMS.—The definitions which follow apply to certain terms used in connection with the American Gage Design Standards.

American Gage Design Standard: The caption "American Gage Design Standard" has been adopted to designate gages made to the design specifications promulgated by the American Gage Design Committee.

Anvil: The gaging member of a snap gage when constructed as a fixed nonadjustable block, or as the integral jaw of the gage.

Adjustable Snap Gage: An external caliper gage employed for the size control of external dimensions, comprising an open frame, in both jaws of which gaging members are so held that one or more pairs can be set and locked to any predetermined size within the range of adjustment.

Solid Snap Gage: An external caliper gage employed for the size control of external dimensions, comprising an open frame and jaws, the latter carrying gaging members in the form of fixed, parallel, nonadjustable anvils.

Taper Lock: Term designating that construction in which the gaging member has a taper shank, which is forced into a taper hole in the handle.

Annular Plug Gage: A shell type plug gage in which the gaging member is in the form of a ring, the external surface of which is the gaging section, the central portion of the web being machined away for the purpose of reducing weight, ball handles being provided for convenience in handling. This construction is employed for plain and thread plug gages in the ranges above 8.010 inches.

Plain Cylindrical Plug Gage: A complete unthreaded internal gage of single- or double-ended type for the size control of holes. It consists of handle and gaging member or members, with suitable locking means.

Progressive Cylindrical Plug Gage: A complete unthreaded internal gage consisting of handle and gaging member in which the "go" and "not go" gaging sections are combined in a single unit secured to one end of the handle.

Reversible Plug Gage: A plug gage in which three wedge-shaped *locking prongs* on the handle are forced into corresponding *locking grooves* in the gaging member by means of a single through screw, thus providing a self-centering support with a positive lock.

Thread Plug Gage: A complete internal thread gage of either single- or double-ended type, comprising handle and threaded gaging member or members, with suitable locking means.

Plain Ring Gage: An unthreaded external gage of circular form employed for the size control of external diameters. In the smaller size it consists of a gage body into which is pressed a *bushing* that is accurately finished to size for gaging purposes.

Thread Ring Gage: An external thread gage employed for the size control of threaded work. means of adjustment being provided integral with the gage body.

Thread Ring Gage Locking Device: Means of expanding and contracting the thread ring gage during the manufacturing or resizing processes. It also provides an effectual lock.

GANG MILLING.—Milling with a combination or "gang" of two or more cutters mounted on one arbor. If a plain cylindrical cutter were placed between two side mills a gang cutter would be formed for milling several surfaces. This would not only be a rapid method, but one conducive to uniformity when milling duplicate parts.

GANG PLANING.—When a number of duplicate parts have to be planed, much time can often be saved by arranging the castings in a straight row along the planer table so that they can all be planed at the same time. This method, called gang planing, enables a number of parts to be finished more quickly than would be possible by machining them separately, and it also insures duplicate work.

GIBS.—A gib (also known as an adjusting or takeup strip) may be defined as a wedge or adjusting shoe. the object of which is to insure a proper sliding .fit between two machine parts, and to make possible adjustment to compensate for wear after the proper adjustment has been lost through continued service. Gibs are used extensively on various classes of machine tools

GLAZED GRINDING WHEEL.—A wheel is "glazed" when the cutting particles have become dull or worn down even with the bond, which latter is so hard that the abrasive grains are not dislodged when too dull to cut effectively. Glazing may indicate either that the wheel is too hard for the work, or that the wheel speed is too high. The remedy, then, for glazing is to decrease the speed or use a softer wheel.

GULLETING FILE.—This file is made of round section in the blunt shape. It is single-cut and used principally for extending the gullets of the teeth of what are known as the "gullet-tooth" and "briar-tooth" saws.

HALF-NUTS.—The term "half-nuts" is applied to the two-part nut used on engine lathes for engaging and disengaging with the lead-screw. The half-nuts are opened or closed by a lever and are used to control the movement of the lathe carriage while cutting screw threads.

HALF-ROUND FILES. — Such files do not form a complete semi-circle, as the name implies, the arc being about one-third of the circle. Files of this class are double-cut and mostly bastard, although many are either second-cut, smooth, or dead smooth, the latter being used to a limited extent. Those having teeth finer than bastard are cut single on the convex side.

HAND FILE.—This type of file is parallel in thickness from the heel (end of file body next to tang or handle) to the middle, and is tapered, as to thickness, from the middle to the point, the latter being about one-half the thickness of the stock. The edges of the fire are usually parallel throughout the entire length but are sometimes drawn in slightly at the point. The hand file is ordinarily preferred by machinists for finishing flat surfaces. The teeth are usually double-cut, bastard, although many files of this type have teeth of second-cut, smooth, or dead-smooth.

HAND REAMER.—A "hand" reamer is used by hand for producing holes that are to be smooth and true to size. The reamer consists of a cutting portion, a shank, and a square by which it is turned when in use. Between the cutting part and the shank, there is a short neck, the purpose of which is to provide clearance for the grinding wheel when grinding the cutting edges and the shank of the reamer.

HAND TAP.—Hand taps usually are made in sets of three, termed "taper," "plug," and "bottoming" taps. When all three taps are employed for tapping a hole, they are used in the order named, thus distributing the work of cutting a full thread.

HARD SOLDERS.—Hard solders, such as are used for silver soldering, and known as *silver solders*, are composed of silver, copper and zinc or brass; whereas hard solders which are used for brazing are alloys formed of copper and zinc. The hard solder used for brazing is commonly known as *spelter* or *spelter-solder*. The composition of silver solders varies considerably according to the nature of the work. A silver solder extensively used by jewelers contains 70 parts of silver and 30 parts of copper.

HEADER.—A large pipe into which one set of boilers is connected by suitable nozzles or tees, or similar large pipes from which a number of smaller ones lead to consuming points. Headers are often used for other purposes—for heaters or in refrigeration work. Headers are essentially branch pipes with many outlets, which are usually parallel. Largely used for tubes of water-tube boilers.

HEADSTOCK.—That part of a lathe, and of certain other machine tools, which contains the main spindle with its driving mechanism, is known as the headstock. Lathes which are commonly referred to as the *geared-head* type, have a headstock which contains a system of gearing so arranged that the drive may be transmitted through different combinations of gearing for varying the spindle speeds. The relative positions of the levers for obtaining a given speed are indicated by an index plate.

HEAT-TREATMENT.—An operation or a combination of operations involving the heating and cooling of a metal or an alloy which is in the solid state, for the purpose of obtaining certain desirable conditions or properties. Heating and cooling for the sole purpose of mechanical working is not classified as heat-treatment. Generally speaking, the heat-treatment of steel includes hardening and tempering of high-carbon steels, casehardening of low-carbon steels, and annealing of steel.

HELIX.—The curvature of a screw thread represents a helix or helical curve. Every point along a helix is equidistant from the axis and the curve advances at a uniform rate around a cylindrical area. The teeth of helical gears are another example of helical curves. The term "spiral" is often applied to such gears, although the teeth do not have a spiral curvature. The spiral has a constantly increasing radius of curvature and a watch spring represents roughly the general form.

HELIX ANGLE.—The term "helix angle" indicates the inclination of a helical curve relative to its axis. If the circumference of a helical gear is divided by its lead (or the distance that a tooth would advance axially in a complete turn), the quotient equals the tangent of the helix angle as related to the axis. See also Lead Angle.

HIGH-SPEED STEEL.—A name applied to alloy tool steels which are capable of cutting metal at a much higher rate of speed than ordinary carbon tool steels. A high-speed steel is not necessarily one conforming to any given analysis, nor is tungsten a necessary element. Most high-speed steels contain tungsten, but other elements, such as molybdenum, confer the "red-hardness" characteristic.

IDLER GEAR.—An idler or intermediate gear simply transmits motion from one gear to another but it has no effect on the speed ratio, or the number of revolutions made by a driven shaft in a given time. This would also hold true if there were several intermediate gears. An idler, however, does change the *direction* in which the driven gear revolves. When driving and driven gears are located on fixed centers and when their sizes must be varied to obtain different speed ratios, an adjustable idler may be used as an intermediate member.

IMPERIAL STANDARD WIRE GAGE (S.W.G.).—The standard gage in Great Britain and Canada for all wires; also customary standard for non-ferrous sheets; sometimes used for iron and steel sheets.

INDEXING.—The process of dividing a circular or straight part into equal spaces by means of an indexing- or dividing-head or any other type of mechanical dividing mechanism, is known as indexing. There are three systems of indexing known as the plain or simple system, the compound system, and the differential system.

INDEXING HEAD.—This is an attachment which forms a part of the equipment of all milling machines of the universal type, and of many plain machines. By means of the indexing or dividing-head, the circumference of a cylindrical part can be divided into almost any number of equal spaces, as, for example, when it is necessary to cut a certain number of teeth in a gear. It is also used for imparting a rotary motion to work (in addition to the longitudinal feeding movement of the table) for milling helical or spiral grooves, and is sometimes called a "spiral head."

INTERCHANGEABLE MANUFACTURE. — There are several degrees of interchangeability in machinery manufacture. Strictly speaking, interchangeability consists in making the different parts of a mechanism so uniform in size and contour that each part of a certain model will fit any mating part of the same model, regardless of the lot to which it belongs or when it was made. However, as often defined, interchangeability consists in making each part fit any mating part in a certain series; that is, the interchangeability exists only in the same series. The strict definition of interchangeability does not imply that the parts must always be assembled without hand work, although that is usually considered desirable. It does mean, however, that when the mating parts are finished, by whatever process, they must assemble and function properly, without fitting individual parts one to the other.

JARNO TAPER.—The taper per foot of all Jarno taper sizes is 0.600 inch on the diameter. The diameter at the large end is as many eighths, the diameter at the small end is as many tenths, and the length as many half inches as are indicated by the number of the taper.

KNURLING.—The forming of a series of fine ridges upon the periphery of a circular part, such as a screw-head, handle, or knob, is known as *knurling* or *nurling*. The purpose of this checked or milled surface is usually to increase the grip of the hand and thus facilitate rotating the knurled part, although knurling is also done in many cases merely to produce an ornamental effect. The handles of gages and other tools are often knurled, and the round thumb-screws used on instruments, etc., usually have knurled edges.

LAND OF GEAR TOOTH.—The *top land* is the top surface of a tooth, and the *bottom land* is the surface of the gear between the flanks of adjacent teeth.

"LAND" OF TAP OR CUTTER. — The term "land" as applied to metal-cutting tools such as taps, reamers, milling cutters, etc., refers to the top surface of a tooth. In the case of a tap, the land is the surface between two flutes, the land width being measured along the outer circumference from the front face of the tooth to the heel. The land of a reamer or milling cutter tooth is the top or clearance surface back of the cutting edge, but does not include the steeper slope at the rear which forms the back of the tooth and part of the flute or chip clearance space.

LAPPING. — Lapping is a refined abrading process generally employed for correcting errors in hardened steel parts and securing a smooth, accurate surface, or for reducing the size a very small amount.

LATHE SIZE.—The size of an engine lathe, according to the practice followed in the United States, is based upon the "swing" or the maximum diameter that can be rotated over the ways or shears of the bed. The nominal sizes listed by lathe manufacturers, however, ordinarily do not represent the maximum swing, but a diameter which is somewhat less. For instance, a lathe which is listed as a 24-inch size may actually swing 24½ or 25 inches.

LEAD.—The lead of a screw thread equals the distance that a nut would advance in one turn or revolution; hence, lead is the distance measured along the axis that a thread advances in one turn. The lead of a helical gear is the distance along the axis that a tooth would advance if it made one complete turn. See Helix and Helix Angle.

LEAD ANGLE.—The helix angle of a screw thread, according to customary practice, is not measured relative to the axis but from a plane perpendicular to the axis, and it is known as the "lead angle." The helix angle of a helical gear is measured from the axis. The helix angle in each case and for any given diameter of screw thread or gear, depends upon the lead of the thread or gear tooth. The term "lead angle," however, is applied only to screw threads, worms, etc., to indicate that the angle is measured from a plane perpendicular to the axis. This angle is more useful in connection with screw threads, and the angle relative to the axis is more useful in designing helical gears.

Lead Angle of Turning Tool: The term "lead angle" is sometimes applied to the angle of the side or leading cutting edge of a turning tool. The angle thus designated is the same as the one known as "side cutting-edge angle" in the American Standard for single-point tools. (See Volume I, Fig. 1, of the section on Single-Point Tool Forms and Tool Grinding.)

LEAD BATH.—The lead bath is extensively used in connection with the heat-treatment of steel, but is not adapted to the high temperatures required for hardening high-speed steel, as it begins to vaporize at about 1190 degrees F., and, if heated much above that point, rapidly volatilizes and gives off poisonous vapors. Lead baths are especially adapted for heating small pieces that must be hardened in quantities. It is important to use pure lead that is free from sulphur.

LEAD JOINT.—(1) Generally used to signify the connection between pipes which is made by pouring molten lead into the annular space between a bell and spigot, and then making the lead tight by calking. (2) Rarely used to mean the joint made by pressing the lead between adjacent pieces, as when a lead gasket is used between flanges. Lead wool may be used in place of molten lead for making pipe joints. It is lead fiber, about as coarse as fine excelsior, and when made in a strand, it can be calked into the joints, making them very solid.

LEAD OF SCREW THREAD.—The distance a screw thread advances axially in one turn. On a single-thread screw, the lead and pitch are identical; on a double-thread screw, the lead is twice the pitch; on a triple-thread screw, the lead is three times the pitch, etc.

LEAD-SCREW.—The lead-screw of an engine lathe is used for feeding the carriage when cutting threads. The carriage is engaged with this screw by means of two half-nuts that are free to slide vertically and are closed around the screw by operating a lever. Any screw which performs a similar function on other machine tools may properly be classed as a lead-screw.

LIMIT.—With the modern system of interchangeable manufacture, machine parts are made to a definite size within certain limits which are varied according to the accuracy required, which, in turn, depends upon the nature of the work. The established limits are the maximum and minimum dimensions specified on the drawings. The words *limit* and *tolerance* are often used interchangeably but tolerance represents the difference between the minimum and maximum limits.

LIMIT GAGE. — In order to insure having all machine parts of a given size or class within the prescribed limit, so that they can readily be assembled without extra and unnecessary fitting, what are known as *limit gages* are used. One form of limit gage for external measurement is double-ended and has a "go" end and a "not go" end; that is, when the work is reduced to the correct size, one end of the gage will pass over it, but not the other end. Limit gages are very generally used for the final inspection of machine parts, as well as for testing sizes during the machining process.

LINE OF ACTION IN GEARING.—The term "line of action" as applied to gearing means the line that would be described by the point of contact of two gear teeth from the time they come into contact until they separate.

LOADED GRINDING WHEEL.—A grinding wheel is "loaded" when the pores or interstices between the cutting particles are partly or entirely clogged with the material being ground. Loading prevents the wheel from cutting and causes excessive heat to be generated. If a wheel becomes loaded, the bond may be too hard or the speed too slow. The remedy for loading is to increase the speed or use a softer wheel.

MACHINABILITY.—A word coined in the mechanical industries to indicate the cutting resistance of a metal or the effect of a metal-cutting operation upon the cutting capacity of a given tool. A metal which may not be machinable with one type of machine or cutting tool, may be machinable under other conditions. The term machinability ordinarily is applied to metal cutting that is on a practical basis. To illustrate, if a metal offers such high resistance to cutting that the best available tools are dulled very quickly, such a metal is not machinable from a practical or commercial point of view.

MACHINE SCREW.—A screw having a head slotted to receive a screwdriver, and (according to the American Standard) ranging in size from No. 1 (0.073 inch) to ½ inch diameter. Most of the machine screw threads conform to the Coarse-thread Series of the American Standard.

MACHINE STEEL.—Machine steel is a black stock and contains from 0.25 to 0.45 per cent carbon. It is commonly used for all machine parts that are not subject to strain or shock. For short shafts, studs, arbors, etc., it will give long service and withstand considerable strain if casehardened.

MACHINING.—This is a general term that is applied to any metal-cutting operation on a machine tool, regardless of the type of machine used. For example, either turning, milling, drilling, planing, grinding, or broaching might be referred to as a machining operation.

MAGNAFLUX.—A trade name applied to a method of inspection that exposes fatigue cracks, grinding checks, seams developed by the rolling mill, etc. The parts to be inspected are magnetized so as to set up a polarity between any cracks or breaks either on the surface of the metal or below it for approximately 2 inches. The part is then immersed in an oil in which finely powdered black magnetic iron oxide is held in suspension or else it is sprayed or flooded with the oil. Particles of the iron oxide will adhere to the surface of the work wherever the polarity of a flaw attracts them, and these particles form a black line showing the defect.

MAJOR DIAMETER.—The largest diameter of a screw thread. The term major diameter applies to both internal and external threads and replaces the term "outside diameter" as applied to the thread of a screw and also the term "full diameter" as applied to the thread of a nut.

MALLEABLIZING.—Malleablizing is a type of annealing operation with slow cooling whereby combined carbon in white cast iron is transformed to temper carbon and in some cases the carbon is entirely removed from the iron. *Note.*—Temper carbon is free carbon in the form of rounded nodules made up of an aggregate of minute crystals.

MANDREL.—A cylindrical rod or shaft having centered ends and used for supporting bushings or other bored parts, while turning the outer surfaces, as in a lathe.

MANIFOLD.—(1) A pipe fitting with numerous branches used to convey fluids between a large pipe and several smaller pipes. (2) A header for a coil.

MIKE. — A micrometer is often called a "mike" in the shops and tool-rooms of the United States.

Mikechecks: This term is frequently applied to the standards used for checking the accuracy of micrometers, especially when there is likely to be some inaccuracy in anvils or screws as a result of long usage. These standards are in the form of disks or short cylinders.

The size of each standard is stamped on it and the diameter of the disk or cylinder conforms to this size within, possibly, 0.000025 inch. A one-inch standard frequently is used for checking purposes; or there may be a range of several sizes so that the accuracy of a micrometer can be checked with reference to a number of different dimensions.

MINOR DIAMETER.—The smallest diameter of a screw thread. The term minor diameter applies to both internal and external threads and replaces the terms "core diameter" and "root diameter" as applied to the thread of a screw and also the term "inside diameter" as applied to the thread of a nut.

MODULE. — Ratio of the pitch diameter to the number of teeth. Ordinarily, module is understood to mean ratio of pitch diameter *in millimeters* to the number of teeth. The English module is a ratio of the pitch diameter in inches to the number of teeth.

MONITOR LATHES.—Turret lathes which are intended principally for brass work are often referred to as *monitor lathes*, the name "monitor" in this connection indicating a revolving turret. This name is not applied to the same design of turret lathe by different manufacturers, although, in general, it indicates a comparatively small turret lathe which, in many cases, is provided with a thread-chasing attachment of the Fox lathe type and is designed principally for turning, boring, and threading parts made of brass. Some lathes which are listed as the monitor type have a stock-feeding mechanism, whereas others do not have this feature.

MULT-AU-MATIC.—This is a trade name applied to a vertical, multiple, automatic machine tool of the station type. This machine has work-holding chucks with tool slides and tools above for performing the necessary operations progressively as the chucks and work index periodically from one station or tool position to the next. Thus the tools of each position operate simultaneously and every indexing movement brings a finished part to the first position, where it is replaced with a rough casting or forging.

MUSIC WIRE.—Music or piano wire is made from a high grade of steel in diameters of from 0.004 to 0.180 inch. There are many different gages to which this class of wire is made, but the piano wire gage designated as the "American Steel & Wire Co.'s Music Wire Gage" is adopted as standard for piano wire in the United States.

NEEDLE VALVE.—A valve provided with a long tapering point in place of the ordinary valve disk. The tapering point permits fine graduation of the opening. At times called a needle point valve.

NIPPLE.—(1) A tubular pipe fitting usually threaded on both ends and under 12 inches in length. Pipe over 12 inches long is regarded as cut pipe.

Close Nipple: One the length of which is about twice the length of a standard pipe thread and is without any shoulder.

Short Nipple: One whose length is a little greater than that of two threaded lengths or somewhat longer than a close nipple. It always has some unthreaded portion between the two threads.

Shoulder Nipple: A nipple of any length, which has a portion of pipe between two pipe threads. As generally used, however, it is a nipple about halfway between the length of a close nipple and a short nipple.

Space Nipple: A nipple with a portion of pipe or shoulder between the two threads. It may be of any length long enough to allow a shoulder.

NORMALIZING.—Normalizing is a special annealing process. It is the heating of iron-base alloys above the critical temperature range followed by cooling to below that range in still air at ordinary temperature. Normalizing is intended to put steel into a uniform unstressed condition, and to obtain the proper grain size and refinement so that the steel will respond properly to further heat-treatment. Normalizing is particularly important in the case of forgings which are to be heat-treated later.

"OLD MAN."—An "old man" is the supporting bracket for the feed screw end of a ratchet drill. It consists of an L-shaped member which is usually bolted, or clamped to the part being drilled, and an adjustable arm against which the pointed end of the ratchet feed-screw bears. The device is also used with pneumatic and electric portable drills to provide "backing" or a resisting support for the drill.

PACK-HARDENING.—Pack-hardening is the process whereby high-carbon steels, or steels that will harden without the aid of carbon added by carburizing, may be protected from furnace gases and air. By packing in some carbonaceous material, the work is protected by the carbon gases given off, and in quenching from the pot, the work is clean and free from scale.

PARALLELS.—Parallels are placed beneath parts to be planed or ground, usually for the purpose of raising them to a suitable height, or to align a finished surface on the under side with the platen, when such a surface cannot be placed in direct contact with the platen. These parallels are made in pairs of different sizes, and opposite sides are parallel to each other.

PILLAR FILE.—A file that is parallel as to width, but tapers somewhat in thickness toward the point. The cross-section of a pillar file is similar to that of a hand file, except that it is thicker in proportion to the width; these files are made in narrow and extra narrow patterns. They are double-cut and are applicable to general machine shop work, especially in connection with erecting and fitting.

PILOT FOR CUTTING TOOLS.—A pilot for metal-cutting tools, such as boring-bars and reamers, is a cylindrical part which extends beyond the cutting end and enters a close-fitting hole in some rigid member, thus supporting the cutting edges and preventing the lateral deflection that might otherwise occur.

PITCH CIRCLE.—A circle the radius of which is equal to the distance from the gear axis to the pitch point.

PITCH DIAMETER OF GEAR.—Diameter of the pitch circle (generally understood to mean the diameter obtained by dividing the number of teeth by the diametral pitch or the diameter of the pitch circle when the center-to-center distance between mating gears is standard).

PITCH DIAMETER OF SCREW.—On a straight screw thread, the "pitch diameter" is equivalent to the diameter of an imaginary cylinder which would pass through the threads at such points as to make the width of the threads and the width of the spaces cut by the surface of the cylinder equal. Thus the pitch diameter is equal to the outside diameter minus one thread depth.

PITCH OF GEAR.—Pitch, as used with reference to gear teeth, defines the sizes of the teeth. Two kinds of pitches are used, *circular pitch* and *diametral pitch*. The circular pitch is the distance along the pitch circle from the center of one tooth to the center of the next. The pitch circle at the larger ends of the bevel teeth is used for determining the circular pitch of a bevel gear. The diametral pitch of a spur or bevel gear is the ratio or quotient obtained by dividing the number of teeth by the pitch diameter.

PITCH OF RIVETS.—The pitch of rivets is the distance from the center to center of adjacent rivets. The pitch should be as large as possible without impairing the tightness of the joint when under pressure.

PITCH OF SCREW THREAD.—The pitch of a screw thread is the distance from the center of one thread to the center of the next thread, measured parallel to the axis. This definition applies whether the screw has a single, double, triple, or quadruple thread. The *lead* of a screw thread is equal to the distance a nut will move forward on the screw, if it is turned around one full revolution.

PITCH POINT.—The point of tangency of the pitch circles or the point where the center-line of mating gears intersects the pitch circles.

PLANER SIZE.—The size of a planer is equivalent to the width and height of the largest part that will pass between the housings and under the cross-rail, when the latter is raised to its highest position. For instance, a 38- by 38-inch planer is one that will plane work approximately 38 inches wide and 38 inches high. Sometimes the maximum length that can be planed is included when designating the planer size. Thus a 36-inch by 36-inch by 8-foot planer means that a piece 36 inches square will pass between the housings, and that a length of 8 feet can be planed.

POLISHING.—In general polishing is an operation performed by using any wheel that has a polishing abrasive glued to its face, regardless of whether the wheel is made from leather, canvas, or some other material. The term polishing embraces everything from the "flexible-grinding" operations performed on axes and picks, and the removal of flash from table knives and forks, to the production of the bright luster, such as is given to surgical instruments, high quality scissors, and other kinds of general hardware.

PRESSURE ANGLE.—The pressure angle of a pair of mating involute gears is the angle between the line of action and a line perpendicular to the center-line of these gears.

Normal pressure angle: Applied to helical gears to indicate pressure angle in a plane normal or perpendicular to the teeth as distinguished from a plane that is perpendicular to the axis of the gear.

PROFILING MACHINE.—A profiling machine or "profiler" is a type of milling machine which is largely used for making parts of guns, pistols, typewriters, sewing machines, and for similar work. Profiling machines are adapted to milling duplicate pieces having an irregular shape or contour, especially in connection with interchangeable manufacture.

PROTRACTOR.—The protractor is used either for locating lines at a given angle or for measuring angles between lines. The type generally used by machinists and toolmakers is known as the *bevel protractor*. It has a straight edge or blade which can be set at any angle with the base or stock; the angle for any position is shown by graduations.

PUNCH.—A punch, as the term is applied in pressed-metal work, is that part of a press tool which enters into an opening or cavity formed in the die section, as in drawing, forming, or blanking. The punch usually is the upper member, being attached to the press slide or ram, but it may be the lower member, as in the case of press tools of inverted design.

PYROMETER.—A pyrometer is used for measuring relatively high temperatures, especially in connection with the heat-treatment of steel. There are a number of different types. The most commonly used pyrometers are of the *thermo-electric* type. Temperature variations are determined by the measurement of an electric current generated by the action of heat on a thermo-couple consisting of two pieces of dissimilar metals. The current is conducted by wires leading to the meter or indicating part of the pyrometer outfit. The instrument may be calibrated or graduated to give readings directly in degrees.

QUARTERING MACHINE.—A quartering machine is a special design of horizontal boring machine that is employed exclusively for boring the crankpin holes in pairs of locomotive driving wheels. The holes in each pair of wheels must be 90 degrees apart and they are bored after the wheels are forced on the axle.

QUARTER-TURN BELT DRIVE.—When two pulleys are mounted on shafts located at right angles to each other and are connected by a belt, this is known as a *quarter-turn drive*. Such drives should be avoided, if possible, because the belt is distorted as it twists around from one pulley to another, and, moreover, the contact between the belt and the pulleys is reduced, owing to the angular position of the belt.

QUENCHING.—Immersing to cool in connection with a heat-treating operation. *Note.*—Immersions may be in liquids, gases, or solids.

QUICK CHANGE-GEARS. — A combination of gears permanently assembled in a machine tool and so arranged that ratio changes are obtained merely by the shifting of levers. For example, on many modern lathes, the changes of feed for turning and screw cutting are obtained by means of a system of gearing which enables the changes to be made rapidly by shifting one or more levers. A table or index plate, attached to the machine, shows what rates of feed or pitches will be obtained for different positions of the levers.

QUILL FOR BENCH LATHE.—A quill is an auxiliary spindle that is used on bench lathes for holding and revolving parts that require extreme accuracy in the location of holes, etc. The spindle revolves in its own bearing or "quill rest," which, in turn, is mounted on the bench-lathe bed in front of the headstock. The work may be held either in a "chuck quill" or be attached to a "faceplate quill"; special fixtures are also attached to the end of the quill spindle. The *quill driver* is a special coupling used for driving a bench-lathe quill so that any jar that may be imparted to the lathe spindle by the belt joint as it passes over the cone pulley is not transmitted to the quill spindle.

RADIAL DRILLING MACHINE.—A radial drilling machine differs from the regular upright machine in that the drilling head is mounted upon a radial arm adjustable vertically and also horizontally by swinging the arm about its supporting column. The drilling head may also be moved along this radial arm, to the required position for drilling, instead of adjusting the work or table each time a new hole is to be drilled. Because of this feature, the radial drilling machine is especially adapted to heavy work.

RAKE OF METAL-CUTTING TOOL.—When a lathe or other metal-cutting tool is so ground that the surface against which the chips bear, while being severed, inclines in such a way as to increase the keenness of the cutting edge, it is said to have "rake." If the inclination is such as to give the tool less keenness than is equivalent to a rake angle of zero, the term *negative rake* is often used.

RATIO OF GEARING. — In referring to the ratio of gearing, it is customary to give the ratio of the number of teeth on the larger gear to the number of teeth on the smaller gear, instead of reversing this order. Thus, if the ratio is 2 or "2 to 1," this usually means that the smaller gear or pinion makes two revolutions to one revolution of the larger gear, or that the larger gear has twice as many teeth as the smaller gear or pinion. To find the ratio of gearing, divide the number of teeth on the larger gear by the number of teeth on the smaller gear (or pinion). The same result will be obtained if the pitch diameters are used instead of the numbers of teeth.

REAMER.—Reamers have two applications: (1) for producing a hole that is smooth and true to size, and (2) for enlarging cored or drilled holes. *Hand reamers* are used by hand for producing holes that are smooth and true to size. *Fluted chucking reamers* are used in machines for enlarging holes slightly and finishing them smooth and true to size. *Rose chucking reamers* are used in machines for enlarging cored or drilled holes and are so constructed that they are able to remove a considerable amount of metal.

RECALESCENCE POINT.—The recalescence point (sometimes designated Ar. 1) is the temperature at which the internal structure of steel which has been heated above the decalescence point and then allowed to cool slowly, changes back to the structural conditions existing before the steel was heated above the decalescence point.

RECESSING TOOL.—A recessing tool is a cutting tool intended for cutting an internal groove or recess in a machine part. Recesses are often cut on the inside of castings and forgings in places which may be rather inaccessible. Special tool-holders and devices, all of which are generally classified as "recessing tools," are sometimes used for this purpose.

RED BRASS.—The alloy known as "red brass," and used for castings, contains 85 per cent of copper, 5 per cent of tin, 5 per cent of lead, and 5 per cent of zinc. This is the recognized standard red brass. There are numerous modifications, for various purposes.

RED HARDNESS.—A term sometimes applied in connection with high-speed steel because of its property of retaining a sufficient hardness for cutting metals even when heated to a temperature high enough to cause dull redness. The property of red hardness is conferred upon the steel by the presence of tungsten and by the heat-treatment to which it is subjected.

REDUCER.—(1) A pipe fitting having a larger size at one end than at the other. Some have tried to establish the term "increaser"—thinking of direction of flow—but this has been due to a misunderstanding of the trade custom of always giving the largest size of run of a fitting first; hence, all fittings having more than one size are reducers. They are always threaded inside, unless specified flanged or for some special joint. (2) Threaded type, made with abrupt reduction. (3) Flanged pattern with taper body. (4) Flanged eccentric pattern with taper body, but flanges at 90 degrees to one side of body. (5) Misapplied at times, to a reducing coupling.

RELIEVING.—Relieving, also known as *backing off*, is the process of removing, by turning, grinding, or milling, some of the metal behind the cutting edge of a cutting tool in order to provide clearance; applied specifically to milling cutters, taps, dies, reamers, and drills. Many milling cutters for gear cutting and form milling are so relieved that the cutting edges retain the same shape or curvature as the front faces of the teeth are ground repeatedly for sharpening.

RELIEVING ATTACHMENT.—A relieving attachment is a device applied to lathes (especially those used in tool-rooms) for imparting a reciprocating motion to the tool-slide and tool, in order to provide relief or clearance for the cutting edges of milling cutters, taps, hobs, etc.

RIVETING MACHINES.—Machines for riveting may be classified according to the method of forming the rivet head, which may be either by (1) compression; (2) by a succession of rapid blows; (3) by rapid blows accompanied by rotary motion of the rivet set; (4) by combined compressive and rolling or spinning action; (5) or by the application of pressure to an electrically-heated rivet. "Riveting machines," according to common usage, differ from "riveters" in that the riveting operation with a machine is effected by a succession of blows or by a compressive rotating action, whereas a riveter merely subjects the rivet to compression.

ROCKWELL HARDNESS TEST.—The Rockwell hardness tester is essentially a machine that measures hardness by determining the depth of penetration of a penetrator into the specimen under certain fixed conditions of test. The penetrator may be either a steel ball or a diamond sphero-conical penetrator. The hardness number is related to the depth of indentation and the number is higher the harder the material.

ROD.—A "rod" as the term is applied in rolling mill practice, is generally understood to be a round bar. The United States Government limits wire rods to sizes larger than No. 6 B. W. G. (0.203 inch) in diameter; all smaller sizes are termed *wires.* (In length measure, one rod = 5.5 yards = 16.5 feet = 25 links.)

ROOT DIAMETER. — The root diameter of a screw thread is the diameter across the bottom or root of the thread groove, measured at right angles to the axis of the screw. According to the American Standard definitions for screw threads, the terms "root diameter" and "core diameter" have been replaced by "minor diameter," which also replaces "inside diameter" as applied to the thread of a nut. The minor diameter is the smallest diameter of an external or internal screw thread.

ROSE REAMER.—This is an end-cutting reamer used for enlarging cored or drilled holes. The cylindrical part of the reamer has no cutting edges, but merely grooves cut for the full length of the reamer body, providing a way for the chips to escape and a channel for lubricant to reach the cutting edges. There is no relief on the cylindrical surface of the body.

ROTARY FILE.—Files of the rotary type are made in either cylindrical, conical, spherical, concave, or special shapes for finishing the edges or surfaces of punches, dies, metal patterns, and various other classes of work. The file may be rotated by inserting it in a drilling machine spindle, as for finishing the edges of punches or dies, or by using a portable drive, as when the position of the file must be varied by hand control.

ROTARY MILLING.—Castings or forgings which are so shaped as to be readily clamped or released from a fixture are sometimes milled by a continuous circular milling operation. This may be done by the use of a circular attachment on a vertical milling machine, or on a special machine equipped with a circular revolving table.

ROTARY PLANER.—The rotary planer or "end milling machine," as it is sometimes called, is especially adapted to the planing or slab milling of flat surfaces on heavy castings or forgings. The distinguishing feature of a rotary planer is the large circular cutter head which carries inserted tools or cutters which successively cut away the metal as the cutter head revolves.

ROUND FILE.—Round files (also called "rat-tail") are made both in the taper and blunt forms, and the cut is mostly bastard. Round files are used either for enlarging round holes or shaping internal surfaces.

ROUTING.—Routing is a name given the operation of milling when the feeding movements of the work are controlled by hand, in order to follow an irregular routine, as when roughing out the impressions in drop-forging dies, etc.

RUN.—(1) A length of pipe that is made of more than one piece of pipe. (2) The portion of any fitting having its ends "in line" or nearly so, in contradistinction to the branch or side opening, as of a tee. The two main openings of an ell also indicate its run, and when there is a third opening on an ell, the fitting is a "side outlet" or "back outlet" elbow, except that when all three openings are in one plane and the back outlet is in line with one of the run openings, the fitting is a "heel outlet elbow" or a "single sweep tee" or sometimes a "branch tee."

RUST JOINT.—Employed to secure rigid connection between pipes. The joint is made by packing an intervening space tightly with a stiff paste which oxidizes the iron, the whole rusting together and hardening into a solid mass. It generally cannot be separated except by destroying some of the pieces. One recipe is 80 pounds cast-iron borings or filings, 1 pound sal-ammoniac, 2 pounds flowers of sulphur, mixed to a paste with water.

SADDLE.—A machine tool saddle is a slide which is mounted upon the ways of a bed, cross-rail, arm, or other guiding surfaces, and the saddle usually supports one or more secondary slides for holding either metal-cutting tools or a work-holding table. On a knee-type milling machine the saddle is that part which slides upon the knee and which supports the work-holding table. The saddle of a planer or boring mill is mounted upon the cross-rail and supports the tool-holding slide. The saddle of a lathe is that part of a carriage which slides directly upon the lathe bed and supports the cross-slide.

S.A.E. STEEL.—This abbreviated term means that the steel is one of the standard compositions approved by the Society of Automotive Engineers, Inc. A system of numbering is used to indicate the general class of steel and the approximate percentages of the chief elements

	Cast and malleable iron (Also for general use of all materials)		Electric windings, electro-magnets, resistance, etc.
	Steel		Concrete
	Bronze, brass, copper, and compositions		Brick and stone masonry
	White metal, zinc, lead, babbitt, and alloys		Marble, slate, glass, porcelain, etc.
	Magnesium, aluminum, and aluminum alloys		Earth
	Rubber, plastic electrical insulation		Rock
	Cork, felt, fabric, leather, fiber		Sand
	Sound insulation		Water and other liquids
	Thermal insulation		Wood— Across grain
	Firebrick and refractory material		Wood— With grain

American Standard Section Lines used on Mechanical Drawings to Indicate General Classes of Structural Materials

SCALE OF A DRAWING.—Many drawings of machines and machine parts are drawn to a reduced size or scale because a full-size drawing would require too much room. Thus 1½, **3**, or **6** inches on the scale may represent one foot. These are the ordinary scales used by mechanical draftsmen.

SCLEROSCOPE.—The scleroscope is an instrument which measures the hardness of the work. A diamond-tipped hammer is allowed to drop from a known height on the metal to be tested. As this hammer strikes the metal, it rebounds, and the harder the metal, the greater the rebound. The extreme height of the rebound is noted or recorded, and an average of a number of readings taken on a single piece will give a good indication of the hardness of the work. The surface smoothness of the work affects the reading.

SCREW MACHINE.—A turret lathe that is designed more particularly for turning comparatively small screws, pins, etc., from steel rods or bar stock, is commonly (although not invariably) known as a *hand screw machine*, or as a *turret screw machine*. According to the practice of some manufacturers, the name screw machine is applied to small turret lathes which have a collet chuck in the spindle and a "wire feed" or a mechanism for feeding a wire rod or bar stock through the spindle. When the machine is intended for either bar or chuck work, or for chuck work exclusively, the name turret lathe is commonly used, and such a machine may or may not have a stock-feeding mechanism which operates in conjunction with the spindle chuck. The foregoing method of distinguishing between the two types, however, is not universal, and there is no general agreement in the use of these names. See also Automatic Screw Machine.

SECTION LINING.—A term indicating a system of lines or symbols used on the cross-sections of mechanical drawings to indicate the general classes of commonly used materials. The accompanying chart shows the lines and symbols which have been adopted as the American Standard.

SELF-ALIGNING.—This term is applied to machine members that are so mounted that they can adjust themselves within certain limits which usually are small. The self-aligning principle, for example, is employed in certain bearings which are so mounted that the bearing can adjust itself to the alignment of the shaft.

SEMI-AUTOMATIC MACHINE.—This term is generally understood to describe a machine which performs a complete cycle of operations automatically, but which requires the attention of an operator each time a part is finished. Many machine tools which actually are semi-automatic are classed as "automatic" by their manufacturers.

SET-SCREW.—The principal difference between a set-screw and a cap-screw is that the former bears on its point, whereas a cap-screw bears on its head. Set-screws are generally used to prevent relative motion between two machine parts, as, for example, when a set-screw passes through a tapped hole in the hub of a pulley and bears against a shaft which drives the pulley. Keys are preferable to set-screws for locking pulleys. gears, etc., to their shafts, and for similar work, although set-screws may serve the purpose when not subjected to heavy loads. Set-screws are not only used for locking parts together, but also as a means of obtaining slight adjustments, either to eliminate unnecessary play by means of gibs, or for changing the location of a tool or other part.

SHAVING TOOL.—When forming work of irregular contour in the automatic screw machine, a shaving tool may be used which is operated tangentially to the work and passes either under it or over it as conditions may require. It is customary to place the shaving tool on the rear cross-slide, so that the shaving operation can be accomplished at the same time as the turret operations, when the spindle is running forward. The chief use of this tool is for finishing work after it has been rough-formed with a circular form or other external cutting tool.

SHELL REAMER.—A shell reamer has a hole through the center by means of which it is mounted on an arbor or detachable shank. By making reamers in this manner, one arbor can be used for a number of sizes.

SHIM.—In mechanical work a shim is a thin sheet of material (usually brass, steel or some other metal) which is sometimes applied between parts to provide convenient means of making adjustment either to compensate for wear or for other reasons. The laminated shim is an improved form consisting of layers of metal which can be peeled off to obtain the desired thickness. This laminated form provides a quick and accurate method of obtaining adjustments by the shim method.

SHRINKAGE FIT.—A cylindrical part which is to be held in position by a shrinkage fit is first turned a few thousandths of an inch larger than the hole in which it is to fit; the diameter of the latter is then increased by heating, and after the part is inserted, the heated outer member is cooled, causing it to grip the pin or shaft with tremendous pressure.

SHRINKAGE RULE.—Except in unusual cases, a pattern-maker does not figure shrinkage by adding it to dimensions measured by the standard rule. but uses a shrinkage rule instead. These rules can be procured in all standard shrinkages; they are oversize the amount of shrinkage per foot.

SILICON CARBIDE.—This is an artificial abrasive produced from a coke and sand mixture and used in making grinding wheels. The electric resistance furnace is used in its production. Wheels made from this abrasive are adapted for grinding materials of low tensile strength, such as soft brasses and bronzes, cast and chilled iron, aluminum, copper, marble, granite, leather, and other non-metallic substances.

SILVER SOLDER. — Silver solders are made in strip, sheet, and granular form, and in a number of different grades of fusibility. The melting points of silver solders vary between 1250 and 1500 degrees F.

One of the best silver solders used is made of 61 per cent silver, 29 per cent copper, and 10 per cent zinc. Many alloys of low silver content are used, in which the silver ranges from 5 to 50 per cent. Silver solder is especially suitable for jointing monel metal, nickel, and stainless steel, since it gives the necessary whiteness to the seam or joint, whereas with ordinary brazing solder, a red or yellow color is noticeable at the joint.

SINE-BAR.—The sine-bar or *sine-protractor* is used either for measuring angles accurately or for locating work to a given angle. It consists of an accurate straightedge to which are attached two hardened and ground plugs of the same diameter. The sine of the required angle is found and this sine is multiplied by the distance between the plug centers, to obtain the vertical distance between the plug centers, for that particular angle. The bar is then adjusted until this vertical distance coincides with the dimension found.

SINGLE-CUT FILE.—A *single-cut* file or "float," as the coarser cuts are sometimes called, has single rows of parallel teeth extending across the face at an angle of from 65 to 85 degrees with the axis of the file. This angle depends upon the form of the file and the nature of the work it is intended for.

SLIDE-REST.—An adjustable tool-holding slide on a machine tool. The *compound type* of slide-rest for lathes has angular adjustment in a horizontal plane and it enables a tool to be fed at right angles to the lathe bed, parallel with it, or at any intermediate angle.

SLITTING SAW.—A slitting saw or cutter is used for cutting off stock or for milling narrow slots, like the screw-driver slots in screwheads, and for similar purposes.

SLOTTER.—The slotting machine or "slotter," as it is commonly called, operates on the same general principle as a shaper, except that the ram which carries the planing tool moves in a vertical direction and at right angles to the work-table. Slotters are used for finishing slots or other enclosed parts which could not be planed by the tool of a horizontal machine like a planer or shaper. The slotter is also used for various other classes of work requiring flat or curved surfaces which can be machined to better advantage by a tool which moves vertically.

SLOTTING ATTACHMENT.—A slotting attachment is sometimes used for adapting a milling machine to slotting. The tool-slide, which has a reciprocating movement like the ram of a slotter, is driven from the main spindle of the machine by an adjustable crank which enables the stroke to be varied.

SMALL TOOLS.—The expression "small tools" is generally used in the metal-working industries and is understood to mean such tools as taps, dies, reamers, milling cutters, drills, and counterbores or any other small tool, especially if used for metal cutting.

SOCKET-HEAD.—The socket-head type of cap-screw has either a hexagonal or square wrench socket formed in the end of a cylindrical head, so that the cap-screw can be turned (by inserting a wrench in the socket) until the head, which enters a counterbored hole, is flush with the outer surface of the part held by the cap-screw. The socket-head cap-screw bears the same relation to ordinary cap-screws having projecting heads as the safety hollow set-screw does to ordinary set-screws.

Fluted-socket: The American Standard fluted socket has six flutes which engage six splines on the wrench.

SOFT HAMMER.—"Soft hammers" are used in machine shops either to prevent marring finished surfaces or to avoid upsetting the ends of arbors, bolts, etc. Soft hammers are made of copper, babbitt metal, rawhide, or brass. While a heavy blow may be struck with one of these soft hammers, a finished surface will not be marred by dents, because of the relative softness of the hammer face.

SPEED LATHE.—The name "speed lathe" is usually applied to light lathes which have neither back-gears nor a tool carriage, and are intended for rotating parts rapidly either for polishing, hand turning, or filing. When turning parts by hand-manipulated tools, the ends of the tools are supported by a T-shaped rest that is clamped to the bed. Lathes of this class are sometimes known as "hand lathes." Many lathes of the "speed" class have either a lever or a combination lever-and-screw feeding motion for the tailstock spindle, the lever being very convenient for feeding drills or reamers which are held in the end of the spindle. The term "speed lathe" is also applied by some manufacturers to a design which has a hand-operated carriage, and one that is without back-gears or a power feeding mechanism.

SPELTER.—Spelter is a name which, in the past, was frequently applied to zinc, but, at the present time, it is used only commercially, and then only when it refers to zinc cast in ingots. Spelter solder, an alloy of copper and zinc, used in hard soldering or brazing, is also frequently spoken of simply as "spelter," but in that case the expression is an abbreviation of the term "spelter solder."

SPHEROIDIZING.—Prolonged heating of iron base alloys at a temperature in the neighborhood of, but generally slightly below, the critical temperature range, usually followed by relatively slow cooling. *Note a.*—In the case of small objects of high carbon steels, the spheroidizing result is achieved more rapidly by prolonged heating to temperatures alternately within and slightly below the critical temperature-range. *Note b.*—The object of this heat-treatment is to produce a globular condition of the carbide.

SPIRAL.—A spiral may be defined mathematically as a curve having a constantly increasing radius of curvature. Spirals are often confused with helices, as for instance, when speaking of "spiral" gears which should properly be termed "helical" gears. Both the spiral and the helix are exemplified in spring design, the spiral being represented, approximately, by a watch spring, while the helix is represented by a coil spring.

SPLINE.—When the hub of a gear or other part must be free to slide axially along its shaft, but rotate with the shaft, a *feather* or *spline* is used. This is simply a form of key, and is either fixed to the shaft or to the hub. For instance, in some cases, the feather is sunk into the shaft like an ordinary key, but it is longer than the hub to allow for axial movement. Multiple splines are also formed on shafts by milling or hobbing a series of equally-spaced grooves along the shaft. These *multiple-spline fittings* may either be sliding or fixed.

SPOT FACING.—This term is generally applied to the operation of truing a comparatively small surface or spot, as, for example, when a true seat is formed on a casting or forging for the under side of a screw head. The spot that is faced true may be a circular raised pad on a casting or merely the surface around a bolt hole. Spot facing for screw heads may be done with a counterbore or counterboring tool, which then is used merely to remove enough metal to form a true seat. If this same type of tool is used to enlarge a hole, as, for example, when the enlarged part is to receive the fillister head of a machine screw, then the operation becomes counterboring.

SPRING BRASS.—This is a brass containing about 66 per cent of copper and 33 per cent of zinc with a small percentage (not exceeding 1.5 per cent) of tin.

STANDARD WIRE GAGE.—The Standard Wire Gage (also known as the Imperial Wire Gage and as the English Legal Standard) is used in England for all wires. The abbreviation S. W. G. is sometimes used for Standard Wire Gage, also the abbreviation N. B. S. for New British Standard Wire Gage. This gage was legalized in Great Britain in 1883. The Birmingham Wire Gage sizes are very generally used for iron and steel telephone and telegraph wires, but the sizes of bare copper telephone wires, usually conform in the United States, to the Standard Wire Gage.

STEADYREST.—Occasionally long slender shafts, rods, etc., which are to be turned, are so flexible that it is necessary to support them at some point between the lathe centers. An attachment for the lathe, known as a "steadyrest," is often used for this purpose. The term "center-rest" is also used.

STEEL.—The word "steel" is applied to many compositions which differ greatly from each other in their chemical as well as physical qualities. The ingredient that exerts the most influence on steel is carbon. High-grade razor steel contains about 1.25 per cent of carbon; spring steel, 1 per cent; steel rails, from 0.50 to 0.75 per cent; soft steel boiler plate 0.060 per cent; and steel used for casehardened parts often has about 0.20 carbon. Steel which is very low in carbon can easily be welded, but it cannot be hardened by merely heating and quenching. Steel with carbon above 0.75 per cent can readily be hardened. In tool steel, other ingredients than carbon are sometimes used to influence its hardness, such as nickel, manganese, chromium, tungsten, etc., the last named playing an important part in so-called "high-speed steels."

STEEL WIRE GAGE.—The Steel Wire Gage (also known as (1) Washburn & Moen, (2) American Steel & Wire Co. (3) Roebling, and (4) National Wire Gage) is used for bare wire of galvanized and annealed steel and iron (except telephone and telegraph), and also for spring steel wire.

STELLITE.—Stellite or Haynes Stellite is an alloy of cobalt, chromium and tungsten and is non-ferrous or without iron in its composition. The hardness of this alloy is not materially affected by heat up to 1500 degrees F. and it is actually tougher at red heat than when cold. This important characteristic explains its wide application as a cutting tool material. Haynes Stellite works best when operated at high speed and with a comparatively light feed.

J-Metal: The cutting tool material known as J-Metal is an improved grade of Haynes Stellite. The use of J-Metal results either in higher cutting speeds or in greater production between tool grindings.

Haynes Stellite — 2400: This is another cobalt-chromium-tungsten alloy. Cutting tools made of this material have greater edge strength and longer economic tool life at even higher speeds than tools made of J-Metal, without reduction of feed or depth of cut. This alloy may be used for roughing or finishing cast and forged steels, cast and malleable irons, nitrided, stainless and other alloy steels.

STOCKING CUTTER. — Roughing cutters which are intended primarily for removing surplus stock are sometimes classed as "stocking cutters". For example, when the pitch of a gear is large enough to warrant taking both roughing and finishing cuts, a "stocking cutter" may be used for removing the bulk of the metal, leaving a small amount on the sides of the teeth for finishing. Several different types of stocking or roughing cutters are in use.

STRADDLE MILLING.—When it is necessary to mill opposite sides of duplicate parts so that the surfaces will be parallel, two cutters can often be used simultaneously. This is referred to as *straddle milling.* The two cutters which form the straddle mill are mounted on one arbor, and they are held the right distance apart by one or more collars and washers. Side-mills, which have teeth on the sides as well as on the periphery, are used for work of this kind.

STUB'S STEEL WIRE GAGE.—This gage is used for cold-drawn tool steel rods and wire and is used by some manufacturers of drill rods. It differs from Stub's iron wire gage which is the same as the Birmingham Wire Gage.

STUD SETTING.—The screwing of studs into the tapped holes of some part such as a steam cylinder, pump cylinder, etc., is known as *stud setting.* Studs may be screwed into place either by hand or power. Special stud-setting chucks are often employed. One type is so arranged that it can be used for either studs or nuts, and the jaws open automatically and release either the stud or nut, as the case may be, just like a die of the self-opening type, and without stopping or reversing the spindle of the machine.

SUB-PRESS.—A sub-press cannot be defined as a special class of die, but merely as a principle which may be applied in constructing different kinds of dies. The sub-press principle is simply that the upper and lower portions of the die are combined into one self-contained unit so arranged as to always hold the upper and lower members in exact alignment with each other.

SUB-SURFACE MILLING.—The term "sub-surface milling" has sometimes been applied when a tilted type of rotary milling machine is used in conjunction with an auxiliary reservoir in which cutters and work can be submerged in a bath of cooling compound while the milling operation is being performed.

SUPERFICIAL HARDENING.—When low-carbon steel is subjected to the cyanide hardening process, a very thin but extremely hard surface is obtained, and this is known as superficial hardening. This hard outer skin may be only a few thousandths of an inch thick, and this is the important difference between superficial hardening and ordinary casehardening.

TAILSTOCK.—A tailstock is that part of a machine tool, such as an engine lathe or cylindrical grinding machine, which is used to support upon a conical center the outer end of a rod, shaft, or other piece which is being turned or ground. The body forming the tailstock may be clamped in different positions along the machine bed for accommodating work of different lengths, and the tailstock spindle containing the supporting center is adjustable to allow for small variations in length.

TAP. — A tap is an internal thread-cutting tool having teeth which conform to the shape of the thread.

Ground Tap: Grinding taps after hardening serves to correct the distortion due to the hardening process and it also leaves the tap with keen cutting edges. The advantages of grinding include accuracy of shape or thread form, accurate lead, as accurate a diameter as is necessary to meet commercial requirements. Most ground taps are made from high-speed steel.

Adjustable Tap: Adjustable taps are made for the purpose of permitting adjustment to a correct standard size. There are various designs.

Collapsible Tap: The collapsing tap is so designed that when the tap has been fed to depth, a gage or trip comes into contact with the end of the work, which causes the chasers to collapse automatically. The tool is then withdrawn. As collapsible taps need not be backed out of the hole at the completion of the thread, this reduces the actual tapping time and naturally increases production.

TAPER ATTACHMENT.—Many modern lathes are equipped with a special device for turning tapers, known as a *taper attachment,* which permits the lathe centers to be kept in alignment for taper work the same as for cylindrical turning, and enables more accurate work to be done.

TAPPING ATTACHMENT.—Some drilling machines are equipped with special gearing which can be utilized for reversing the rotation of the spindle when tapping, so that a special reversing tap chuck is not necessary. This mechanism for reversing the spindle when the tap has reached the required depth is often known as a tapping attachment.

TAPPING CHUCK.—In tapping by power, the tap ordinarily is fed down into the hole to the required depth and its rotation is then reversed for screwing it out of the hole. There are different methods of obtaining this reverse motion. When the tapping is done in an ordinary drilling machine, special tap chucks are frequently used, which are designed to reverse the rotation of the tap when the latter has reached the required depth. One form of tap-holding chuck is so arranged that the tap automatically stops when it strikes the bottom of the hole or when an adjustable depth gage comes against the top of the work. The raising of the spindle then reverses the tap which backs out at an increased speed.

TAPPING MACHINE.—Machines designed especially for tapping or for drilling and tapping holes are made in quite a variety of designs. Some of these machines are intended for one class of work, like the tapping of nuts, whereas others are adapted to tapping operations of a general nature; there are vertical and horizontal, and single- and multiple-spindle types. Tapping machines also vary in regard to the mechanism for obtaining the forward and reverse motions of the tap spindle and the method of controlling three motions.

TEE.—A fitting, either cast or wrought, that has one side outlet at right angles to the run. A single outlet branch pipe.
Service Tee: A tee having inside thread on one end and on branch, but outside thread on other end of run. Also known as street tee.

TEMPERING.—Reheating, after hardening to some temperature below the critical temperature range followed by any rate of cooling. Although the terms "tempering" and "drawing" are practically synonymous as used in commercial practice, the term "tempering" is preferred.

THREAD ANGLE. — The angle of a screw thread is the angle included between the sides of the thread measured in the plane of the axis. The helix angle is the angle made by the helix of the thread (usually at the pitch diameter) and it depends upon the relation between the diameter of the screw thread and the lead.

TIN-PLATE GAGE. — The Tin-plate Gage is a weight gage based upon pounds per "base-box." It is used for tin plate (sheet steel coated with tin) and roofing terne plates.

TOLERANCE.—Tolerance is the amount of variation permitted on dimensions or surfaces of machine parts. The tolerance is equal to the difference between the maximum and minimum limits of any specified dimension. For example, if the maximum limit for the diameter of a shaft is 2.000 inches and its minimum limit 1.990 inches, the tolerance for this diameter is 0.010 inch. As applied to the fitting of machine parts, the word tolerance means the amount that duplicate parts are allowed to vary in size in connection with manufacturing operations, owing to unavoidable imperfections of workmanship. Tolerance may also be defined as the amount that duplicate parts are permitted to vary in size in order to secure sufficient accuracy without unnecessary refinement. The terms "tolerance" and "allowance" are often used interchangeably, but, according to common usage, *allowance* is a difference in dimensions, prescribed in order to secure various classes of fits between different parts.

Unilateral Tolerance: The term "unilateral tolerance" means that the total tolerance, as related to a basic dimension, is in *one* direction only.

Bilateral Tolerance: A bilateral tolerance is divided relative to the basic dimension so that it is partly plus and partly minus.

TOOL STEEL. — Tool steel, as the term is used in the machine-building industry, may be defined in a general way as any steel that is suitable to be used as a cutting tool, or a steel which contains a sufficient amount of carbon so that it will harden if heated above a certain temperature and rapidly cooled. This broad definition includes, under the head of tool steel, both high-speed steels and plain carbon steels.

TRANSLATING GEARS.—When a lathe is to be used for cutting threads in accordance with both the English and metric systems of measurements, what are known as "translating gears" are sometimes used. These gears have 50 and 127 teeth, respectively, these numbers representing the relation between the English and metric systems of measurement.

TREPANNING. — When a comparatively large hole must be cut "from the solid" a trepanning tool is sometimes used. This tool is so designed that it forms a hole by cutting a narrow groove, the central part or core being taken out as a solid piece.

T-SLOT.—T-shaped slots are formed in the tables and bedplates of different types of machine tools to receive the T-bolts used to hold either the work or a fixture in position during the machining operation. The American standard covers T-bolts and slots.

TURRET LATHE SIZE.—There are two general methods of designating the size of a turret lathe. If the machine is intended primarily for operating on bar stock which is fed through the spindle, the size, as listed by manufacturers, indicates approximately, at least, the maximum diameter of stock that will pass through the spindle, and the maximum length that can be turned. The size of a turret lathe intended more especially for chucking operations indicates the maximum diameter that the machine will swing over the ways of the bed.

TWIST DRILL AND STEEL WIRE GAGE.—A gage that is generally used in the United States for drill rod or twist drills made to wire gage sizes.

UNION.—(1) The usual trade term for a device used to connect pipes. It commonly consists of three pieces which are, first, the thread end fitted with exterior and interior threads; second, the bottom end fitted with interior threads and a small exterior shoulder; and third, the ring which has an inside flange at one end while the other end has an inside thread like that on the exterior of the thread end. A gasket is placed between the thread and bottom ends, which are drawn together by the ring. Unions are very extensively used, because they permit of connections with little disturbance of the pipe positions.

Union Ell: An ell with a male or female union at one end.

Union Joint: A pipe coupling, usually threaded, which permits disconnection without disturbing other sections.

Union Tee: A tee with male or female union at connection on one end of run.

UNITED STATES STANDARD SHEET METAL GAGE.—This is a "weight gage," the actual standard being in ounces per square foot. The thicknesses equivalent to the standard weights originally were based upon the weight of wrought iron; now these thicknesses are based upon the weight of steel. Since steel is heavier than wrought iron, the thicknesses have been reduced accordingly. This gage is used for hot- and cold-rolled uncoated sheets and plates of steel or iron, and for Monel metal.

UNIVERSAL CHUCK.—The universal or concentric type of chuck is extensively used on engine lathes because the simultaneous movement of the chuck jaws makes it possible to quickly grip circular parts so that they are located true or concentric with the lathe spindle. The jaws of a universal chuck all move together and keep the same distance from the center, and they can be adjusted by turning any one of the screws, whereas with the independent type the chuck wrench must be applied to each jaw screw. The *combination chuck* may be changed to operate either as an independent or universal type. The advantage of the universal chuck is that round and other parts of a uniform shape are located in a central position for turning without any adjustment. The independent type is, however, preferable in some respects as it is adapted for holding odd-shaped pieces because each jaw can be set to any required position.

VICKERS HARDNESS TEST.—The Vickers test is similar in principle to the Brinell test although the penetrator is of different form and much lighter loads are applied in making a test. The standard Vickers penetrator is a square-based diamond pyramid having an included point angle of 136 degrees. The numerical value of the hardness number equals the applied load in kilograms divided by the area of the pyramidal impression. A smooth, firmly supported, flat surface is required. The load, which usually is applied for 30 seconds, may either be 5, 10, 20, 30, 50 or 120 kilograms. The 50-kilogram load is usually employed. The hardness number is based upon the diagonal length of the square impression.

WASHBURN & MOEN WIRE GAGE.—This gage is the same as the American Steel & Wire Co.'s gage, which, as approved by the Bureau of Standards at Washington, is now known as the "Steel Wire Gage." This gage applies to all steel wire, and is used to a greater extent than any other wire gage in the United States.

WELDING DEFINITIONS.—The following standard definitions of the American Institute of Electrical Engineers have been approved by the American Standards Association and the National Electrical Manufacturers Association.

Pressure Welding: A process of welding metals in the highly plastic or fluid states by the aid of mechanical pressure. This process includes the resistance welding form of electric welding and the pressure type of thermit welding.

Resistance Welding: A pressure welding process wherein the welding heat is obtained by passing an electric current between the contact areas to be welded.

Butt Welding: A resistance welding process wherein a butt joint is employed.

Spot Welding: A resistance welding process wherein the weld is made in one or more spots by the localization of the electric current between the contact points.

Flash Welding: A resistance butt welding process wherein the welding heat is developed by the passage of current in the form of an arc across a short gap between the surfaces to be welded, these surfaces being kept slightly separated until they have flashed off to parallelism and have reached the desired temperature. The electrical circuit is then opened and the upsetting movement takes place. The operation of the machine may be manual, semi-automatic or fully automatic. The name "flash" arises from the fact that during the heating period oxidizing metal is thrown off in a shower of sparks.

Seam Welding: A resistance welding process wherein the weld is made lineally between two contact rollers or a contact roller and a contact bar.

Percussive Welding: A resistance welding process wherein electric energy is suddenly discharged across the contact area or areas to be welded and a hammer blow is applied simultaneously with or immediately following the electrical discharge.

Fusion Welding: A process of welding metals in the molten, or molten and vapor state, without the application of mechanical pressure or blows.

Arc Welding: A fusion welding process wherein the welding heat is obtained from an electric arc formed either between the base metal and an electrode, or between two electrodes with or without the use of gases.

Carbon Arc Welding: An arc welding process wherein a hard carbon or graphite electrode is used and filler metal, if required, is supplied by a welding rod.

Shielded Carbon Arc Welding: A carbon arc welding process wherein the molten filler and weld metals are effectively protected from the air by supplemental means.

Metal Arc Welding: An arc welding process wherein the electrode used is a metal rod or wire which, when melted by the arc, supplies the filler metal in the weld.

Shielded Metal Arc Welding: A metal arc welding process wherein the molten filler and weld metals are effectively protected from the air by supplemental meáns.

Atomic Hydrogen Welding: A fusion welding process wherein the heat of an electric arc between two suitable electrodes is used to dis-associate molecular hydrogen into its atomic form, which on recombination in the molecular form gives up the energy required to dis-associate it, producing a flame of very high temperature and at the same time bathing the molten metal in hydrogen. It may be considered as a combination of the gas- and arc-welding processes.

WHEEL-TURNING LATHE.—A "wheel lathe" is a special design used for turning locomotive driving wheels and car wheels after they have been pressed onto the axle. Lathes of this class are of duplex form, there being two driving heads and two tool-rests, so that both wheels may be turned simultaneously.

WIRE FEED.—The name wire feed is often applied to the mechanism for feeding a rod through the spindle of a screw machine. The term "wire-feed screw machine" is sometimes used to indicate a design having a mechanism for automatically feeding the stock through the spindle, whereas a machine not having this stock-feeding mechanism may be designated as a *plain screw* machine.

WOODRUFF KEYS.—In the Woodruff key system, half-circular disks of steel are used as keys, the half-circular side of the key being inserted into the keyseat. Part of the key projects and enters into a keyway in the part to be keyed to the shaft in the ordinary way. The advantage of this method of keys is that the keyway is easily milled by simply sinking a milling cutter, of the same diameter as the diameter of the stock from which the keys are made, into the shaft. The keys are also very cheaply made, as they are simply cut off from round bar stock and milled apart in the center. Dimensions of Woodruff keys are given in engineering handbooks.

WORKING DEPTH OF GEAR TOOTH.—Depth to which a tooth extends into the tooth space of a mating gear when the center distance is standard—equals twice the addendum.

WORKING GAGE.—"Working gages" are those which are used in testing the work for size during the actual manufacture of the part. This type of gage has a greater amount allowed for wear than any other type, and hence the actual tolerance on the work between the maximum and minimum gage is smaller, by the amount allowed for wear on the gage, than the actual amount specified on the drawing.

WYE (Y).—A fitting either cast or wrought that has one side outlet at any angle other than 90 degrees. The angle is usually 45 degrees, unless another angle is specified. The fitting is usually indicated by the letter Y.

ZINC GAGE.—This gage applies to sheet zinc only. For zinc wire the American or Brown & Sharpe wire gage is employed.

INDEX

Abrasives, grinding wheel, 331
Adapter, definition, 513
 for holding end mills, 112
Allowance, definition, 514
Alloy steel, 470
Aluminum-oxide abrasives, 331
American Standard gear-tooth systems, 286
American Standard section lines for drawings, 548
Angle diameter, definition, 514
Angle, helix, 181
 calculating, 182
Angle, lead, 183
Angular indexing, 158
Angular milling cutters, milling, 195
Annealing tool steels, 480
APT language, 511
APT program, 511
Arbor, definition, 515
Arnold grinding gage, size measuring, 345
Attachment, rack-cutting, for milling machine, 238

Basic dimension, definition, 517
Baths, molten, for heat-treating steels, 482 to 484
Bell-mouthed holes, causes of, in grinding, 363
Bevel gear cutting, angular position for milling of teeth, 242
 correcting milled teeth by filing, 249
 cutting by generating process, 299
 cutting very large sizes, 266
 finishing cuts on planer of templet type, 270
 milling the teeth, 241
 offset adjustment for milling teeth, 246
 planer of two-tool templet type, 276
 roughing cuts on planer of templet type, 268
 selection of formed milling cutter, 243
 templet type of machine for very large sizes, 266
Bevel gearing, pitch, 245
 pitch diameter measurement, 244
Blind hole, definition, 518
Boring, drilling and milling machine, 132
Box-tool, definition, 519
Brinell hardness test, 520
Broaches, burnishing, 438
 general information, 437
 sharpening, 438
Broaching, cylinder blocks, 435
 external surface broaching, 425
 helical groove or spline broaching, 433
 internal surface broaching, 422

slab and progressive broaching, 429
Broaching machines, duplex type, 427
 general types, 424
Brush surface analyzer, 421
Bushings, methods of grinding, 362

Cam milling, plate, 197
Cape chisel, 440
Carbide cutting tools, 475
Carbon steel, 470
Cemented-carbide cutting tools, 475
Centerless grinding, external, 345
 internal, 367
Change gears, definition, 523
 for gear-hobbing machines, 292
 for helical milling, 179
Chip breaker, definition, 523
Chipper, pneumatic, 444
Chipping with air-driven chipper, 444
 with hammer and chisel, 442
 with pneumatic hammer, 443
Chisels, metal cutting, 440
 hardening and tempering, 445
 steel for, 445
Chordal addendum, gear tooth, 228, 229
Chordal measurements over two or more spur gear teeth, 230
Chordal thickness of gear teeth, table, 228
Chucks, magnetic, on grinding machines, 353
Circular pitch, gear teeth, 215, 218
 table giving equivalent dimensions, 219
Circular pitch, normal, helical gears, 251
Clearance angles, rules for obtaining on milling cutters, 377
Clearance drill, meaning of term, 6
 table of diameters for machine screws, 4
Clearance, milling cutter teeth, 372
 reamer teeth, 391
"Climb-cut" milling, 118
"Closed loop" in numerical control, 508
Clutch, saw-tooth, milling, 191
Cobalt high-speed steel, 473
Cold chisels, metal cutting, 440
 hardening and tempering, 445
 steel for, 445
Comparator, definition, 524
Comparoscope, 421
Compound indexing, 143, 146
 when work is indexed around several times, 149
Computer program in numerical control, 508
Computer programming, preparing a contour part program in, 511
 preparing point-to-point manuscript for, 508 to 510

Contour part program for a computer, 511
Cooling fluid, application in cylindrical grinding, 318
cylindrical grinding machine nozzles, 319
internal grinding, 364
Counterboring, definition, 525
Countersinking, definition, 525
Crankpin grinding, 343
Crankshaft grinding, 343
Cutter grinding, 371
Cutters, formed, involute, for spur gears, 224
Cutters, gang, for milling rack teeth, 239
Cutters, milling, 107
Cutters, roughing or stocking, for spur gears, 234
Cylindrical grinding, 314
steadyrests, 325
work speed, 330

Definitions of shop terms, 513
Diameter of spur gear, pitch, rule for finding, 216
Diametral pitch, gear teeth, 215, 218
definition, 526
finding, outside diameter and number of teeth are known, 217
table giving equivalent dimensions, 218
normal, helical gears, 251, 526
Diamond dust, grading for lapping, 399
Diamond tools, grinding wheel truing, 340
Die-holders, thread cutting dies, 29
Die-sinking, 126
Dies and molds, machines for milling, 205
Dies, thread cutting, 15
automatic or self-opening types, 24
classes of work used for, 15
different types, 16
lead-screw control, 23
radial and tangential chasers, 19
tapering threads, 26
throat of chaser, 20
types of die-holders, 29
Differential indexing, 153
ratio of gearing, 155
Disk grinding, 356
Dividing or indexing head, 135
angular position for milling teeth of cutters, 193, 195
position for plate cam milling, 197
use for helical milling, 174
Drill, tap, rules for determining diameter, 2

Emery, 528
End-mills, 111
angular position of index-head for milling teeth, 193

Feedback principle in numerical control, 499
Fellows gear shaper, operating principle, 279
File definitions, 529

Files, 446
cant-files and cant-saw, 453
cleaning teeth, 459
cross-cut, 454
double cut, 446
featheredge, 455
flat type, 447
ginsaw, 454
grades and shapes for general use, 455
gulleting type, 453
half-round, 453
handsaw, 454
hand type, 447
knife, 454
mill type, 452
nominal length, 456
pillar type, 452
pit saw, 453
rotary, 461
round type, 453
single-cut, 446
square type, 452
standard shapes, 448 to 451
three-square, 453
warding type, 452
Filing, general practice, 455 to 461
draw-filing, 458
Filing machines, 462
Finish on metal surface, indicating quality, 419
Fits, classes obtained with ground thread taps, 11
Floating tool-holder, definition, 530
Fly cutter, definition, 531
Fly cutter for cutting worm-gears, 305
Formed milling cutters, 109
application, 119
definition of formed cutter, 531
for milling bevel gears, 243
for milling helical gears, 254
for milling spur gears, 224
grinding so as to retain shape, 384
grinding when teeth have rake, 387

Gage-blocks, lapping, precision, 410
Gage terms, standard, definitions, 532
Gang cutters for milling rack teeth, 239
Gang cutters for spur gear cutting, 236
Gear cutting, 212 to 308
generating methods, 278 to 308
hobbing process, 288
milling with formed cutters, 212
spur gear, checking size, 227 to 232
Gear cutting attachment, milling machine, 222
Gear cutting machine, automatic formed-cutter type, 233
Gearing, pressure angle, as related to gear cutting, 283
Gearing, ratio, 544
Gear shaper, Fellows, operating principle, 279
Gear teeth, American Standard, 286
dimensions for different pitches, 218, 219
full-depth and stub, 286
table of chordal thicknesses, 228

Gear Teeth (Continued)
   teeth cast to approximate shape, 265
   why standards are represented by rack
      teeth, 285
Gear-tooth caliper, vernier type, 227
Gear-tooth grinding, general methods, 309
Generating methods of forming gear
   teeth, 278
Gorton "duplicator" for die and mold
   milling, 205
Graduating on milling machine by indexing, 163, 165
Grinders, application of portable hand
   type, 467
   portable, electric, 348, 366
Grinding allowances, 323
Grinding attachments, radial, for cutters,
   383
Grinding, cylindrical, 314
   external, centerless, 345
   internal, 357
   irregular shapes, 343
   surface, 350
   tapering parts, 321
   with portable grinder on lathe, 348
   with portable hand-manipulated grind-
      er, 467
Grinding crankpins, 343
Grinding gage, Arnold, size-measuring,
   345
Grinding gear teeth, 309
Grinding milling cutters, 371
Grinding reamers, 390
Grinding screw threads, 43
Grinding wheels, abrasives used, 331
   glazed and loaded, 342
   grade and grain, 332
   internal grinding, 359
   selection, 333
   shapes and sizes, 335 to 339
   special shapes for hand grinding, 469
   speeds, 329
   standard shapes, 337
   truing, 340

Hammers, weight of ball peen, 443
Hardness tests, 520, 546, 558
Haynes Stellite, 474
Heat-treatment, cold chisels, 445
Heat-treatment of steels, 476
Helical curve, 174
Helical flutes and teeth on reamers, 394
Helical gear cutting, angular position of
   hob for cutting, 298
   angular position of milling machine
      table, 255
   application of milling process, 250
   checking tooth thickness, 257
   depth of tooth, 257
   hobbing process, 295
   lead of teeth, 255
   number of formed milling cutter, 254
   pitch of formed milling cutter, 251
   rule for finding outside diameter of
      gear, 254
   vertical milling attachment for large
      helix angles, 257, 258

Helical gears, helix angle, 256
   pitch diameter, 253
Helical milling, 174
   calculating change gears, 179
   lead of helix given, 176
   position of cutter, 186
   position of machine table, 188
Helix angle, 181
   calculating, 182
   helical gear, 256
   difference between helix angle and lead
      angle, 535, 537
High-speed steel, 471
   cobalt high-speed steel, 473
   hardening high-speed steel, 486
   tempering high-speed steel, 488
Hobbing process of cutting gears, 288
   angular position of hob, 292
   hobbing helical gears, 295
   hob rotation relative to gear, 292
   use of multiple-threaded hobs, 294
Hobbing teeth of worm-wheel on milling
      machine, 261
Hob grinding attachments, 388, 389
Hobs for gear cutting, 288
   right-hand and left-hand, 298
   multiple-threaded form, 294
   worm-gear, 303, 307
Honing method of finishing surfaces, 415
Horizontal boring, drilling and milling
      machine, 132

In-cut milling, 118
Index-head, 135
   angular position for milling angular
      cutters, 195
   angular position for milling cutters, 193
   helical milling, 174
   position for plate cam milling, 197
   sector for fractional turns of crank, 141
   special, for milling spur gears, 222
Indexing, 135
   angular, 158
   compound method, 143, 146
   compound, when work is indexed
      around several times, 149
   differential, 153
   direct or rapid method, 138
   graduating on milling machine, 163, 165
   in cutting rack teeth on milling ma-
      chine, 238
   in cutting spur gears, 221
   plain or simple method, 139
   simple and compound systems com-
      bined, 152
   tapering parts, 161
   when irregular spacing is required, 164,
      169
Internal gears, cutting on Fellows gear
      shaper, 279
   cutting very large sizes, 264
Internal grinding, 357
   allowances, 363
   centerless, 367
   cooling fluids, 364
   planetary method, 365
   wheels used, 359

Involute cutters, series for cutting spur gears, 224

Keller type of machine for irregular form milling, 209
Knurling, definition, 536

Land of gear tooth, 536
Land of milling cutter tooth, 372
Land of tap or cutter, 536
Lapping, 396
  abrasives used, 398
  charging laps with abrasive, 400
  charging laps for flat surfaces, 407
  diamond dust grading, 399
  flat surfaces, general procedure, 407
  flat surfaces by rotary method, 408
  forms of laps used, 401
  laps used for cylindrical plugs, 402
  laps for holes, 403
  laps for flat surfaces, 406
  lap materials, 397
  precision gage-blocks, 410
  rotating type of lap, 403
  tapering or conical holes, 404
Lathe size, designating, 537
Lead angle, 183
  difference between lead angle and helix angle, 535, 537
Lead of helical gear teeth, rule for finding, 255
Lead of a milling machine, 177
Line of action in gearing, 283

Machine screws, tap drill and clearance drill diameters, 4
Magnaflux, definition, 539
Magnetic chucks on grinding machines, 353
Mandrel, definition, 539
Marking materials used for scraping, 465
Micro-inch value, relation to quality of surface finish, 419, 420
Milling cutters, 107
  angular position of index-head for milling teeth, 193, 195
  die-sinking type, 128
  end-mills, 111
  formed type, 109
  inserted-tooth type, 109
  milling helical teeth, 184
  plain type, 107
  right-hand and left-hand, 110
  side type, 107
Milling cutter grinding, 371
  angular cutters, 380
  corner beveling and rounding, 383
  cylindrical cutters with helical teeth, 378
  face grinding metal-slitting saw, 389
  formed cutters, 384, 387
  rotation of wheel relative to cutter, 373
  rules for obtaining clearance angles, 377
  side milling and face cutters, 379
  tooth-rest position, 376
Milling process, 100
  "climb-cut", 118

cutter rotation relative to feeding movement, 117
cutting bevel gears, 241
cutting helical gears, 250
cutting spur gears, 212
cutting worm gears, 259
gang method, 113
helical milling, 174, 179
irregular contours, 200
planetary method, 130
plate cams, 197
rotary method, 125
saw-tooth clutches, 191
with formed cutters, 119
work-holding devices, 106
Milling machine attachments, 122, 129
Milling machines, chief functions, 101
  die-sinking type, 126
  horizontal type, 115
  lead of, 177
  plain type, 105
  profiling type, 200
  universal type, 105
  vertical spindle type, 121
Milling screw threads, 33
Module, definition, 540
Molds and dies, machines for milling, 205

Nitriding high-speed steel, 492
Normal circular pitch of helical gears, 251
Normal diametral pitch of helical gears, 251
Normalizing, definition, 541
Nozzles, cooling fluid, for cylindrical grinding machines, 319
Numerical control, APT in, 511
  "closed loop" in, 499
  computer program in, 508
  "continuous path," 495
  "contouring," 495
  definition of, 493
  feedback principle in, 499
  part program in, 507
  "point-to-point," 495
  "positioning," 495
  preparing a contour part program, 511
  preparing point-to-point manuscript, 508 to 510
  preparing tapes for, 501 to 505
  programming a point-to-point part manually in, 505 to 507
  punched tapes for, 494
  types of, 495
Numerically controlled drill press, 501
Numerically controlled machine, working principle of, 499
Numerically controlled vertical boring mill, 501
Numerically controlled vertical profiler, 501
Numerically controlled vertical turret lathe, 501

Pack-hardening, definition, 541
Pantograph type of profiling machine, 207

Part program in numerical control, 507
Pilot for cutting tools, definition, 541
Pin method of checking spur gear sizes, 232
Pitch circles of gears, 215
Pitch, bevel gearing, 245
diametral and circular, 215, 218, 219
rack teeth, 237
various definitions, 542
Pitch diameter, bevel gearing, 244
definition, 542
helical gear, 253
spur gear, 213
Planamilling, 130
Planers, 65
size designation, 66
Planer type of gear cutting machine, 262, 276
Planetary method of milling, 130
Planetary method of thread milling, 42
Planetary type of internal grinder, 365
Planing process, 62
angle-plates for holding work, 71
Planing, attachments for circular, 87
Planing, clamping work, 66
cuts for roughing, 63
cuts for finishing, 65
gang or multiple, 82
"gooseneck" type of tool, 77
principles governing tool grinding, 91
speeds, 83, 85
stop-pins and braces, 70
types of tools used, 74
use of shapers, 93
V-blocks for holding cylindrical parts, 72
vertical or angular surfaces, 79
with two or more tools simultaneously, 80
work-holding fixtures, 69
Point-to-point manuscript preparation for computer programming, 508 to 510
Pressure angle of gearing, 283
effect on tooth form, 284
how obtained in cutting gears, 284
Profiling machines, 200
pantograph type, 207
three-dimensional type, 208
Profiling operations, examples, 202
Profilometer for indicating quality of finish, 421
Programming a point-to-point part manually in numerical control, 505 to 507
Programming, APT in, 511
differences in types of, 507
preparing a part program for a computer, 511
preparing point-to-point manuscript for computer, 508 to 510
Punched tapes for numerical control, 494
Pyrometers, 484

Rack teeth, cutting on milling machine, 238
cutting on planer, 86
generating by using gear-shaped cutter, 282

Rack teeth, pitch of, 237
Rack teeth, why the basis for gear-tooth standards, 285
Radial grinding attachments for cutters, 383
Ratio of gearing, 544
Reamer grinding or sharpening, 390
taper reamers, 393
Reamers, indexing for irregular spacing of teeth, 164, 169
Reamers, why some have helical flutes and teeth, 394
Rockwell hardness test, 546
Rolling screw threads, 51
Root mean square value, meaning of term, 419
Rotary files, power-driven, 461

Scraping surfaces of machine parts, 463
marking materials used, 465
ornamental effects, 467
scrapers for finishing surfaces, 463
Screw thread grinding, 43
Screw thread milling, 33
Screw thread rolling, 51
Screw threads, fits obtained with ground thread taps, 11
Section lines, standard, for different materials, 548
Sector of indexing or dividing head, 141
Shapers, application, 93
draw-cut type, 95
universal type, 95
vertical type, 96
Sharpening milling cutters, 371
Shop terms, definitions, 513
Side mills, angle of index-head for milling teeth, 193
Silicon-carbide abrasives, 331
Slotting attachment, milling machine, 129
Slotter, main features and general application, 97
Spiral gears, hobbing process of cutting, 295
Spiral milling with index-head, 174
Spot-facing, definition, 553
Spur gear cutting, formed-cutter method, 212
formed-cutter type, of automatic machine, 233
gang cutters, 236
generating methods of cutting, 278
hobbing process of cutting, 288
indexing gear on milling machine, 221
measuring chordal thickness of teeth, 227, 229
measuring over pins or rolls to check size, 232
measuring over two or more teeth, 230
pitch and number of formed cutter, 223
rack-shaped cutter, 280
roughing and finishing cuts, 226
roughing or stocking cutters, 234
teeth cast to approximate shape required, 265
templet planers for very large sizes, 262

Spur Gear Cutting (Continued)
total depth of gear-tooth spaces, 218, 219, 225, 286
Spur gears, center-to-center distance, 220
how given pitch is obtained, 217
outside diameter, 220
pitch diameter, 213
Standards for gear teeth, why represented by rack teeth, 285
Steadyrests for cylindrical grinding, 325
Steels, carbon and alloy, 470
for metal-cutting tools, 470
high speed, 471
high-speed, cobalt type, 473
Steels, heat-treatment, 476
hardening, high speed, 486
tempering, high speed, 488
Stellite, 474
Stocking or roughing cutters for spur gears, 234
Stub-tooth gear, American Standard, 286
Superfinishing process, 418
Surface broaching, meaning of term, 425
Surface finish, indicating quality, 419
Surface grinding, 350

Tantalum-carbide tools, 475
Tap drills, diameters for machine screws, 4
rules for determining diameter, 2
Taper grinding, 321
Tapers, shanks of end-mills, 112
Taper thread-cutting dies, 26
Tapes for numerical control, 494
Tap-holders, floating type, 8
Tapping attachments, 9, 10
Tapping machines, general features, 13
Taps, collapsing, 12
ground thread taps, 11
methods of holding taps, 8
methods of driving taps, 6
reversal of tap rotation, 9
taper, plug and bottom taps, 7
Tempering tool steels, 479
Templet type of gear planer, 262
Thread angle, definition, 556
Thread cutting with dies, 15

Thread cutting with taps, 1
Thread grinding, 43
accuracy obtained, 49
"from the solid" or without previous cutting, 47
multi-edged wheel method, 47
single-wheel method, 43
taper multi-edged wheel method, 48
Thread milling, 33
conditions governing application, 40
multiple-cutter method, 35
planetary method, 42
single-cutter method, 33
with gear hobbing machine, 41
Thread milling machines, main features, 39
Thread rolling, 51
application to different classes of screw threads, 55
blank diameter, 54
circular-die method, 53, 57
flat-die method, 51
steels used for thread rolling, 56
when automatic screw machine is used, 56
Titanium-carbide cutting tools, 475
Tolerance, definition, 556
Tool steels, 470
heat-treatment, 476
Tungsten-carbide cutting tools, 475

V-blocks for holding cylindrical parts on planer, 72
Vernier caliper for measuring thickness of gear teeth, 227
Vernier caliper, chordal measurement over spur gear teeth, 230
Vertical milling attachment, milling helical gears, 257, 258
Vickers hardness test, 558
Vise, height above floor, 461

Welding, definitions, 559
Worm-gear cutting, general methods, 302
hobbing teeth on milling machine, 259 to 261
Worm thread, lead angle, 259